Plant Extracts

Plant Extracts: Chemical Composition, Bioactivity and Potential Applications

Editors

Francisca Rodrigues
Cristina Delerue-Matos

MDPI • Basel • Beijing • Wuhan • Barcelona • Belgrade • Manchester • Tokyo • Cluj • Tianjin

Editors
Francisca Rodrigues
LAQV - ISEP
REQUIMTE
Porto
Portugal

Cristina Delerue-Matos
LAQV - ISEP
REQUIMTE
Porto
Portugal

Editorial Office
MDPI
St. Alban-Anlage 66
4052 Basel, Switzerland

This is a reprint of articles from the Special Issue published online in the open access journal *Foods* (ISSN 2304-8158) (available at: www.mdpi.com/journal/foods/special_issues/plant_bioactivity_applications).

For citation purposes, cite each article independently as indicated on the article page online and as indicated below:

LastName, A.A.; LastName, B.B.; LastName, C.C. Article Title. *Journal Name* **Year**, *Volume Number*, Page Range.

ISBN 978-3-0365-2877-9 (Hbk)
ISBN 978-3-0365-2876-2 (PDF)

Preface to "Plant Extracts: Chemical Composition, Bioactivity and Potential Applications"

Society has recently demonstrated a high level of awareness and responsibility concerning environmental issues. The interest in bioactive compounds extracted from natural sources has increased due to their potential application as active ingredients in several industries, particularly the cosmetic, food, and pharmaceutical industries. Plants are rich sources of phenolic compounds that have been widely studied due to their health-promoting properties, namely antioxidant, anti-carcinogenic, and anti-inflammatory activities, among others. Extraction is usually the limiting analytical step in the yield of bioactive compounds. From a green point of view, many extraction techniques have been employed as potential candidates to replace conventional methods, such as ultrasound-assisted extraction (UAE), pressurized liquid extraction (PLE), microwave-assisted extraction (MAE), supercritical fluid extraction (SFE), pulsed electric field extraction, and enzyme-assisted extraction. In this Special Issue, we focus our attention on the chemical characterization of plant extracts and their bioactive composition, focusing also on in-vitro cell assays and molecular tools. The issue comprises original research articles, as well as a review, on topics such as phenolic profile, radical scavenging capacity, in vitro cell assays, comet assay, and antimicrobial capacity. We close this Special Issue with a review paper that focuses on the pharmacological activities of quercetin, one of the principal polyphenols. With this, we aim to provide a contemporary overview of the advantages of bioactive compounds extracted from plants.

Francisca Rodrigues, Cristina Delerue-Matos

Editors

 *foods*

Article

Evaluation of the Extraction Temperature Influence on Polyphenolic Profiles of Vine-Canes (*Vitis vinifera*) Subcritical Water Extracts

Olena Dorosh [1,†], Manuela M. Moreira [1,†], Diana Pinto [1], Andreia F. Peixoto [2], Cristina Freire [2], Paulo Costa [3], Francisca Rodrigues [1,*] and Cristina Delerue-Matos [1]

[1] REQUIMTE/LAQV, Instituto Superior de Engenharia do Instituto Politécnico do Porto, Rua Dr. António Bernardino de Almeida, 431, 4249-015 Porto, Portugal; olena.dorosh@graq.isep.ipp.pt (O.D.); manuela.moreira@graq.isep.ipp.pt (M.M.M.); diana.pinto@graq.isep.ipp.pt (D.P.); cmm@isep.ipp.pt (C.D.-M.)

[2] REQUIMTE/LAQV, Departamento. de Química e Bioquímica, Faculdade de Ciências, Universidade do Porto, Rua do Campo Alegre s/n, 4169-007 Porto, Portugal; andreia.peixoto@fc.up.pt (A.F.P.); acfreire@fc.up.pt (C.F.)

[3] REQUIMTE/UCIBIO, MedTech-Laboratory of Pharmaceutical Technology, Department of Drug Sciences, Faculty of Pharmacy, University of Porto, Rua de Jorge Viterbo Ferreira nº. 228, 4050-313 Porto, Portugal; pccosta@ff.up.pt

* Correspondence: francisca.rodrigues@graq.isep.ipp.pt; Tel.: +351-228340500

† The authors contributed equally to the present work.

Received: 28 May 2020; Accepted: 1 July 2020; Published: 3 July 2020

Abstract: This work focused on evaluating the possibility of using vineyard pruning wastes from two Portuguese *Vitis vinifera* varieties; Touriga Nacional (TN) and Tinta Roriz (TR), as new potential ingredients for the nutraceutical industry. An environmentally friendly extraction technique; namely subcritical-water extraction (SWE), was employed. The overall results indicate that phenolic acids were the major class of compounds quantified; being gallic acid the principal one. The highest value for total phenolic content (TPC) was obtained for the TR extract at 250 °C (181 ± 12 mg GAE/g dw). In terms of antioxidant activity; the DPPH values for the extracts obtained at 250 °C were approximately 4-fold higher than the ones obtained at 125 °C; with TR extract presenting the highest value (203 ± 22 mg TE/g dw). Thus, the TR extract obtained through SWE at 250 °C was selected to evaluate the scavenging activity and the in vitro effects on cells due to the best results achieved in the previous assays. This extract presented the ability to scavenge reactive oxygen species ($O_2^{•-}$, HOCl and ROO$^{•}$). No adverse effects were observed in HFF-1 viability after exposure to extract concentrations below 100 µg/mL. This work demonstrated that vine-canes extracts could be a potential ingredient to nutraceutical industry

Keywords: vine-canes; subcritical water extraction; scavenging capacity; in-vitro assays; byproducts

1. Introduction

Contemporary society is tightly bonded to over-consumerism being characterized by a mass-production of goods and consequently their over consumption. This linear economy depends on two basic assumptions: (i) there will always be resources that can be extracted and (ii) there will always be a place where it is possible to get rid of the materials that are not wanted anymore. Nevertheless, currently, due to the development of science and environmental awareness, there is a growing conscience that these two assumptions are not real and that the linear economy is not sustainable [1]. In fact, human population has grown exponentially in the last two centuries increasing the resources consumption [2]. The increased demand over natural resources has been negatively affecting Earth's overshoot day. To reach a more sustainable world status and to preserve what is still left for future generations, it is imperative the transition from an economy based on fossil resources to a concept of circular economy. A growing number of companies, including the food industry, are working to overcome this challenge and transform their process of production in a more environmentally ethical practice. Food and beverage industries are the principal manufacturing sector in the European Union (EU) generating high amounts of byproducts for which profitable solutions need to be found [3,4]. Until now, the general application for the agro-food industry waste has been animal feed (that may not adjust to the nutritional requirements), combustion feedstock or fertilizers, causing major environmental issues [5,6]. Nevertheless, these byproducts can be used as renewable natural resources for many applications, such as low-cost adsorbents, nutraceuticals, supplement food products and ready meals, leading these industries to a concept more related to a circular economy [6,7].

Grapes (*Vitis vinifera*) are one of the principal fruits produced around the world [8]. According to OIV (International Organization of Vine and Wine), in the last years, Portugal is one of the principal world wine producers, representing this sector as a huge impact on the economy [9]. Consequently, large amounts of wine wastes, namely skins, seeds and stems are produced every year, representing approximately 20% of the total weight of processed grapes [10]. These undervalued byproducts are rich in bioactive compounds, particularly polyphenols [11–15], that could be used in several applications such as antioxidants in food, cosmetic or even pharmaceutical industries [16]. Depending on the vine varieties, around 1.75 tons of vine-cane wastes are produced for each hectare of vineyard. After harvesting season, many agricultural byproducts, including vine-canes, are usually incorporated in the soil, enhancing the soil health due to the degradation of organic matter and reducing the necessities of organic fertilizers and/or correctives [17]. Following the need to find a better end for vine-canes, different approaches were already explored, such as the biochar production, biofuels, pulp for paper sheets and particle board [18–20].

Vine-canes are constituted by two main fractions, holocellulose (cellulose and hemicellulose) and lignin, which correspond to approximately 68% and 20% of the total vine-canes weight, respectively [21]. The lignin degradation process releases phenolic compounds of low molecular weight, alcohols, aldehydes, ketones or acids [16]. By employing the appropriate extraction technique, considerable amounts of these valuable compounds can be recovered from the lignin fraction of vine-canes and afterwards used in added valued products. Subcritical water extraction (SWE) can be exploited as a sustainable and clean technique to achieve this goal. SWE is a pressurized liquid extraction technique that uses water at high temperatures (100–374 °C) and pressures (1–22.1 MPa), but below its critical point (374 °C and 22.1 MPa). The use of water as extracting solvent makes this technique safe, cost-effective and environmentally interesting, particularly for the extraction of phenolic compounds to be used in products for human consumption. For example, Gabaston et al. investigated the effect of different times (5, 15 and 30 min) and temperatures (100, 130, 160 and 190 °C) in SWE [22]. All extractions were conducted with 5 g of vine-canes powder in a 34 mL cartridge. According to the authors, the best results were obtained at 160 °C for 5 min (3.62 g of stilbenes/kg dry weight (dw)) [22]. Indeed, in our previous study, we compared three extraction techniques, namely microwave-assisted extraction (MAE), SWE and conventional extraction (CE), in what concerns to phenolic compounds extracted from vine-canes [12]. The obtained results revealed the advantages and potentialities of SWE, when compared to the other techniques. In this study, we focused on the extraction of vine-canes from two Portuguese varieties (Touriga Nacional, TN and Tinta Roriz, TR) from the Dão region using SWE performed at two different temperatures (125 and 250 °C) in order to maximize the bioactive compounds extraction. Opposite to our previous study, where we establish a specific temperature for the SWE (150 °C) in this work we aim to compare different extraction temperatures taking in consideration the results reported by other authors highlight temperature as the most influencing parameter on SWE [23–25]. Thus, based on the previous studies, we decided to explore the thermo-chemical conversion reactions effects on the recovery of antioxidant rich products using subcritical water conditions. Based on this, the antioxidant activities and the phenolic profile of the two vine-canes varieties were investigated. The best extract was selected for further assays, being screened the scavenging activity against oxygen radical species as well as the cell viability effects. Overall, this work follows a sustainable approach for the valorization of vine-canes from the grape industry in order to promote their added value and circular economy.

2. Materials and Methods

2.1. Samples Collection and Preparation

Vine-canes samples were kindly provided by Sogrape Vinhos, S.A. (Portugal). Both *V. vinifera* vine-canes varieties, namely TN and TR, were obtained in Quinta dos Carvalhais, located in Mangualde (North of Portugal), in November 2015 by randomized selection. Samples were oven-dried (Model no. 2000208, J.P. Selecta, Barcelona, Spain) at 50 °C for 24 h and milled (Retsch ZM200) to a particle size smaller than 1 mm. The fine particles obtained after milling were stored in sealed bags at room temperature until use.

2.2. Subcritical Water Extraction

SWE was conducted in a Parr Series 4560 Reactor connected to the Parr 4848 Reactor Controller (Figure 1).

Figure 1. Subcritical water extractor used.

The extractions were performed using 40 g of milled vine-canes and 400 mL of water. Two different extraction temperatures were tested, 125 and 250 °C, for 50 min after the sample reached the desired temperature at 250 rpm. After extraction, the system was cooled down and the extract was filtered and centrifuged at 15,763× *g* (Heraeus Megafuge 16 Centrifuge Series, Thermo Scientific, Waltham, MA, USA) for 15 min at 4 °C. Then, the extract was stored at −80 °C and lyophilized (Edwards lyophilizer) for 48 h, after being stored at 4 °C until further use.

2.3. Determination of Total Phenolic and Flavonoid Contents

The total phenolic content (TPC) was determined according to Singleton and Rossi [26], with minor modifications described by Paz et al. [27]. The reaction solution consisted of 25 µL of deionized water (blank), standard or sample, 75 µL of deionized water and 25 µL of diluted Folin–Ciocalteu reagent (1:1). After 6 minutes in the dark, 100 µL of a sodium carbonate solution (0.708 M) were added to each well. The microplate (BioTek Instruments, Inc., Winooski, VT, USA) was kept away from the light for 90 min and then the absorbance was measured at 760 nm. Calibration curves were done using gallic acid (GA) as standard. Results were expressed as milligrams of gallic acid equivalents (GAE) per gram of dw (mg GAE/g dw).

The total flavonoid content (TFC) was performed according to Paz et al. [27]. The procedure consisted in adding to each well 100 µL of deionized water, 10 µL of sodium nitrite solution (0.725 M) and 25 µL of deionized water (blank) or standard or sample. After 5 min in the dark, 15 µL of aluminum chloride (0.75 M) were added to each well, and after 1 minute of reaction, also in the dark, 50 µL of sodium hydroxide (1.0 M) were added. The absorbance was measured at 510 nm. Epicatechin was used as standard. Results were expressed as mg of epicatechin equivalents (EE) per gram of dw (mg EE/g dw).

2.4. Determination of Antioxidant Activity and DPPH Free Radical Scavenging Assay

The FRAP assay was based on the protocol reported by Benzie and Strain [28], with minor modifications described by Paz et al. [27]. FRAP reagent was prepared using a mixture of acetate buffer (pH 3.6; 0.3 M), 2,4,6-Tri(2-pyridyl)-s-triazine (TPTZ; 0.01 M) in HCl solution (0.04 M) and $FeCl_3.6H_2O$ (0.27 M) in a 10:1:1 ratio. One hundred and eighty microliters of FRAP reagent were added to each well of the microplate along with 20 µL of deionized water (blank) or standard or sample. Absorbance was measured at 593 nm, after incubating in the dark at 37 °C for 10 min. A calibration curve was prepared with ascorbic acid (AA). Results were expressed as milligrams of AA equivalents (AAE) per gram of dw (mg AAE/g dw).

DPPH-RSA was performed following the protocol described by Paz et al. [27]. For that, 200 µL of an ethanolic solution of DPPH (0.1 M) were added to 25 µL of the standard or sample. The blank contained 225 µL of ethanol and the control 225 µL of the DPPH reagent. The reaction solution was incubated for 30 min in the dark. DPPH-RSA was determined spectrophotometrically at 517 nm. Calibration curve was made with Trolox. Results were expressed in milligrams of Trolox equivalents (TE) per gram of dw (mg TE/g dw).

2.5. Qualitative and Quantitative Polyphenol Characterization

The phenolic profile of subcritical water extracts was obtained by high performance liquid chromatography (HPLC) with photodiode array (PDA) detection employing the method described by Moreira et al. [12]. A Shimadzu HPLC system equipped with a Phenomenex Gemini C_{18} column (250 mm × 4.6 mm, 5 µm) and a guard column with the same characteristics, that were kept at 25 °C, were used. Individual phenolic compounds were prepared in methanol and their mixtures for calibration curves construction were obtained by dilution of appropriate amounts of the stock solutions in methanol:water 50:50 (*v/v*). The mobile phase was composed by methanol (A) and water (B) both with 0.1% formic acid, which were previously filtered (0.20 µm nylon filter, Supelco, Bellefonte, PA, USA) and degassed for 15 min in an ultrasonic bath (Raypa® trade, Terrassa, Spain). A gradient program, at a flow rate of 1.0 mL/min and 20 µL of injection volume were used and the identification of detected peaks in subcritical water extracts was performed by comparing their retention time and UV-vis spectra with the ones of pure standards. GA, protocatechuic acid, (+)-catechin, 4-hydroxyphenilacetic acid, 4-hydroxybenzoic acid, 4-hydroxybenzaldehyde, vanillic acid, syringic acid, (-)-epicatechin, naringin, phloridzin, cinnamic acid, naringenin, phloretin and pinocembrin were quantified at 280 nm; chlorogenic acid, caffeic acid, *p*-coumaric acid, ferulic acid, sinapic acid, resveratrol and tiliroside at 320 nm and quercetin-3-*O*-glucopyranoside, rutin, ellagic acid, myricetin, kaempferol-3-*O*-glucoside, kaempferol-3-*O*-rutinoside, quercetin and kaempferol at 360 nm and their amount was expressed as mg/100 g dw.

2.6. Reactive Oxygen Species Scavenging Capacity

2.6.1. Superoxide Radical Scavenging Assay

The superoxide radical ($O_2^{\bullet-}$) scavenging assay was performed according to Pistón et al. [29]. The reaction mixture was prepared by adding to each well the following reagents dissolved in phosphate buffer (19×10^{-3} M, pH 7.4): 50 µL of NADH (166×10^{-6} M); 150 µL of nitroblue tetrazolium (NBT; 43×10^{-6} M); 50 µL of tested extract at different concentrations or phosphate buffer for the blank or positive controls and finally 50 µL of PMS (2.7×10^{-6} M). Absorbance was read at 560 nm for 6 min at 37 °C in the microplate reader. GA and catechin were used as positive controls. The observed effects were expressed as inhibition percentages of the NBT reduction to diformazan. Results were expressed as the necessary concentration of controls and subcritical water extract of TR variety obtained at 250 °C to inhibit 50%, IC_{50}, of the NBT reduction to diformazan.

2.6.2. Hypochlorous Acid Scavenging Activity

Hypochlorous acid (HOCl) scavenging activity was measured using a fluorescent methodology previously described by Pistón et al., based on the HOCl-induced oxidation of dihydrorhodamine (DHR) to rhodamine [29]. GA and catechin were used as positive controls. A 1% (*m/v*) NaOCl solution was used to prepare the HOCl solution by adjusting the pH to 6.2 with addition of H_2SO_4. The assay was directly performed in a 96-well microplate and the reagents previously dissolved in phosphate buffer (100×10^{-3} M, pH 7.4). In each well, the reaction mixture consisted of: 150 µL of phosphate buffer (100×10^{-3} M); 50 µL of tested extract at different concentrations or phosphate buffer for the blank or positive controls; 50 µL of DHR (5×10^{-6} M) and finally 50 µL of HOCl (5×10^{-6} M). Absorbance was read at 560 nm for 6 min at 37 °C in the microplate reader. Results were expressed as the inhibition, in IC_{50}, of HOCl-induced oxidation of DHR.

2.6.3. Peroxyl Radical Scavenging Activity

Peroxyl radical ($ROO^{\bullet}$) was generated by thermo-decomposition of AAPH at 37 °C. The $ROO^{\bullet}$ scavenging activity, also known as the oxygen radical absorbance capacity (ORAC) assay, was measured by the fluorescence decay of fluorescein as previously described by Ou et al. [30]. Reaction mixtures contained the following reagents dissolved in potassium phosphate buffer (75×10^{-3} M, pH 7.4): 150 µL of fluorescein (61.2×10^{-9} M); 25 µL of the tested extract at different concentrations or phosphate buffer for the blank or positive controls and 25 µL of 2,2′-Azobis(2-amidinopropane) dihydrochloride (AAPH; 19.1×10^{-3} M). After preparing the reaction mixture, the 96-well plate was incubated in the microplate reader during 2 h at 37 °C, where the decrease in fluorescence was measured every minute. The excitation and emission wavelengths used were 528 ± 20 nm and 485 ± 20 nm, respectively. GA and catechin were used as positive controls, while Trolox was employed as standard. Results were expressed in ratio values: slope of the sample/slope obtained for Trolox.

2.7. Cell Viability Assay

The 3-(4,5-dimethylthiazol-2-yl)-5-(3-carboxymethoxyphenyl)-2-(4-sulfophenyl)-2H-tetrazolium (MTT) assay was employed to access the cell viability after exposure to the extract. HFF-1 was purchased from ATCC (ATCC Number: SCRC-1041; ATCC, Manassas, VA, USA). Passages 24–26 were used. The assay was performed according to Rodrigues et al. [31]. HFF-1 was grown in DMEM medium, previously fortified with 10% of heat inactivated fetal calf serum, 1% of non-essential amino acids and 1% of antibiotic. Cells were maintained in an incubator (CO_2 Incubator MCO-18AC, Panasonic, Osaka, Japan) at 5% CO_2 and 37 °C and the culture medium was changed every 2 days until cells reached a good confluence. Afterwards, cells were cultured in 96-well microplates at a density of 2.5×10^4 cells/mL for 24 h at 37 °C with 5% of CO_2, in order to provide conditions for an exponential cell growth. After the period of multiplication and adherence, the medium was removed, and cells were washed with phosphate-buffer saline (PBS) solution. Following, cells were incubated with different concentrations of extract (0.1–1000 µg/mL) dissolved in the DMEM medium for 24 h at 37 °C. After the incubation period, extracts were removed and cells were washed again with PBS and subsequently MTT was added to each well. The microplate was incubated for 4 h at 37 °C with 5% of CO_2 in the dark. After that period of incubation, DMSO was added and the microplate was put into agitation for 10 min to solubilize the MTT crystals. Positive control was made by incubating the cells only in culture medium, while negative control was made by incubating the cells only in Triton X-100. The absorbance was measured at 490 nm and at 630 nm to subtract the background. Results were expressed as percentages of cell viability.

2.8. Statistical Analysis

The statistical analysis was performed using IBM SPSS Statistics for Windows software (Version 24.0, IBM Corp., Armonk, NY, USA). Data was reported as mean ± standard deviation (SD) of three replications. The normal distribution and the homogeneity of variances were assessed by Shapiro–Wilk's and Levene's tests, respectively. For all assays, the data were normal and the homogeneity of variances confirmed. To evaluate the differences between samples, the one-way ANOVA was used. Tukey's HSD test was employed for the post hoc comparisons of the means, being $p < 0.05$ accepted as denoting significance. To compare the same sample at different temperatures, a t-test was employed, being $p < 0.05$ accepted as denoting significance. GraphPad Prism 5 software (GraphPad, La Jolla, CA, USA) was employed to calculate the IC_{50} values of ROS scavenging activity.

3. Results and Discussion

3.1. Total Phenolic and Flavonoid Contents, Antioxidant and Antiradical Activities

In the present study, the influence of the extraction temperature parameter was evaluated, since previous studies have demonstrated that the temperature increase in SWE has more impact in the extraction efficiency than the solid–liquid ratio and the extraction time employed [23–25]. For example, Jokić et al. reported that the best result for the TPC assay was obtained when the SWE temperature was 250 °C [23]. Taking into consideration the published studies, two extreme values of temperature were employed in the present work aiming to understand their influence in the phenolic profile characterization and antioxidant activity. Table 1 summarizes the values obtained for TPC, TFC, DPPH-RSA and FRAP assays for extracts obtained by SWE performed in TN and TR vine-canes varieties, with two different extraction temperatures (125 and 250 °C).

Table 1. Total phenolic content (TPC, results expressed in mg gallic acid equivalents (GAE)/g dw), total flavonoid content (TFC, results expressed in mg epicatechin equivalents (EE)/g dw), 2,2-diphenyl-1-picrylhydrazyl radical scavenging activity (DPPH-RSA, results expressed in milligram Trolox equivalents (TE)/g dw) and ferric reduction antioxidant power (FRAP, results expressed in milligram ascorbic acid equivalents (AAE)/g dw) of subcritical water extracts from Touriga Nacional (TN) and Tinta Roriz (TR) vine-canes varieties. Results were expressed as mean ± standard deviation.

Temp. (°C)	TPC (mg GAE/g dw)		TFC (mg EE/g dw)		DPPH-RSA (mg TE/g dw)		FRAP (mg AAE/g dw)	
	TN	TR	TN	TR	TN	TR	TN	TR
125	47 ± 7	55 ± 2 [#]	17 ± 2	20 ± 1 [#]	51 ± 9	56 ± 2 [#]	46 ± 7	53 ± 2 [#]
250	165 ± 8 [*]	181 ± 12 [*,#]	46 ± 3 [*]	51 ± 6 [*,#]	202 ± 22 [*]	203 ± 22 [*]	186 ± 21 [*]	202 ± 14 [*]

For each assay (TPC, TFC, DPPH-RSA and FRAP), [#] for the same extraction temperature means that the vine-cane variety produces statistically significant differences ($p < 0.05$); [*] for the same vine-cane variety means that different extraction temperatures produce statistically significant differences ($p < 0.05$).

According to Table 1, the increase of temperature from 125 to 250 °C resulted at least in 3.3-fold higher TPC values for TN and TR vine-canes varieties. Significant differences ($p < 0.05$) between the quantity of phenolic compounds extracted at 125 and 250 °C were observed. Considering the different vine-canes varieties studied, TR presented the highest TPC values ($p < 0.05$) at both tested temperatures. The highest value obtained was 181 ± 12 mg GAE/g dw for the TR extract obtained at 250 °C. These results are in line with Dorosh et al. as well as Moreira et al. that observed the same tendency for vine-canes extraction using SWE and other advanced extraction technologies: TR extracts present higher TPC values than the TN variety [12,32].

Considering the two variables studied, the values obtained for TFC were in accordance with the ones obtained for the TPC assay. As shown in Table 1, significant differences ($p < 0.05$) were observed between the TFC values obtained at 125 and 250 °C for the same vine-cane variety, as well as for the two different vine-canes varieties using the same extraction temperatures. The highest TFC was reported for TR extract obtained at 250 °C (51 ± 6 mg EE/g dw). However, the TFC values of the present study are not in agreement with the ones reported by Moreira et al. [12] since a higher TFC was reported by the authors for TN extracts obtained by SWE at 150 °C. This difference could be due to the environmental conditions observed in the harvest year, for example, as reported by other authors for other seasonal matrices also produced in the same region [33].

In what concerns to DPPH-radical scavenging assay, the capacity to reduce this radical was similar ($p > 0.05$) for both vine-canes extracts obtained at the highest temperature studied (250 °C). Nevertheless, at 125 °C, TN and TR extracts presented significant differences ($p < 0.05$). Regarding the comparison of different extraction temperatures for the same vine-cane variety, the DPPH values of the extracts obtained at 250 °C were approximately 4-fold higher than the ones obtained at 125 °C, with TR extract presenting the highest value (203 ± 22 mg TE/g dw). In fact, the increase of SWE temperature seems to be the major factor influencing the antioxidant capacity of the extracts, as previously reported by other authors [23–25]. The employment of high temperatures probably resulted in the formation of new compounds, which could justify the higher antioxidant activity observed [16,34]. Additionally, it was observed that the extracts with the highest TPC and TFC showed also higher DPPH$^\bullet$ scavenging capacity. The increase in antioxidant capacity caused by a higher polyphenols content was also previously reported by several other authors [11,12,35].

According to the results presented in Table 1, the obtained values for FRAP assay were in accordance to the ones obtained for DPPH-RSA assay. For instance, considering the same vine-cane variety, the results obtained for the extracts at 250 °C were almost 4-fold higher than the ones obtained at 125 °C ($p < 0.05$). Additionally, no significant differences ($p > 0.05$) were observed between the values obtained for TN and TR for the highest extraction temperature tested (250 °C). Nevertheless, significant differences ($p < 0.05$) between the two varieties were reported for extracts performed at 125 °C. The highest FRAP value was obtained for TR extract at 250 °C (202 ± 14 mg AAE/g dw), which is the same subcritical water extract that presented the highest results for the three assays previously discussed. In fact, TR variety had higher antioxidant activity than TN, which was also previously reported by Moreira et al. [12].

Based on the data previously discussed, it is possible to conclude that higher extraction temperatures resulted in higher amounts of polyphenols and flavonoids as well as higher antioxidant properties. Further, it was also demonstrated that the vine-cane variety used exerts a significant influence in the obtained results, with TR variety being a better matrix to recover bioactive compounds.

3.2. Identification of Phenolic Compounds by HPLC-PDA

Due to the significant differences in the TPC, TFC, DPPH-RSA and FRAP results of the extracts obtained at 125 and 250 °C, HPLC-PDA analyses were performed to understand which phenolic compounds were the main contributors to the antioxidant properties. Figure 2 presents the HPLC chromatograms obtained for the polyphenol's standard mixture (Figure 2A) and for TR extract obtained at 250 °C (Figure 2B). Table 2 summarizes the obtained results.

According to Table 2, phenolic acids were the major class of compounds identified and quantified, corresponding to 43% and 78% and to 38% and 80% of the total amount of phenolic compounds for TN and TR extracts obtained at 125 and 250 °C, respectively. GA was the main phenolic acid quantified, especially in the extracts obtained at 250 °C (891 ± 45 and 1066 ± 53 mg of GA/100 g dw for TN and TR respectively). In fact, GA is a compound commonly found in wines, being responsible for its characteristic astringency [14,16]. In a recent study, we reported a higher amount of phenolic acids for TN subcritical water extract obtained at 150 °C (790 ± 40 mg of phenolic acids/100 g dw) than for the CE and MAE extracts (77.3 ± 3.9 and 265 ± 13 mg of phenolic acids/100 g dw, respectively) [12]. In the same study we also observed higher amounts of phenolic compounds for TN than for TR variety contrary to the results presented in Table 2 [12].

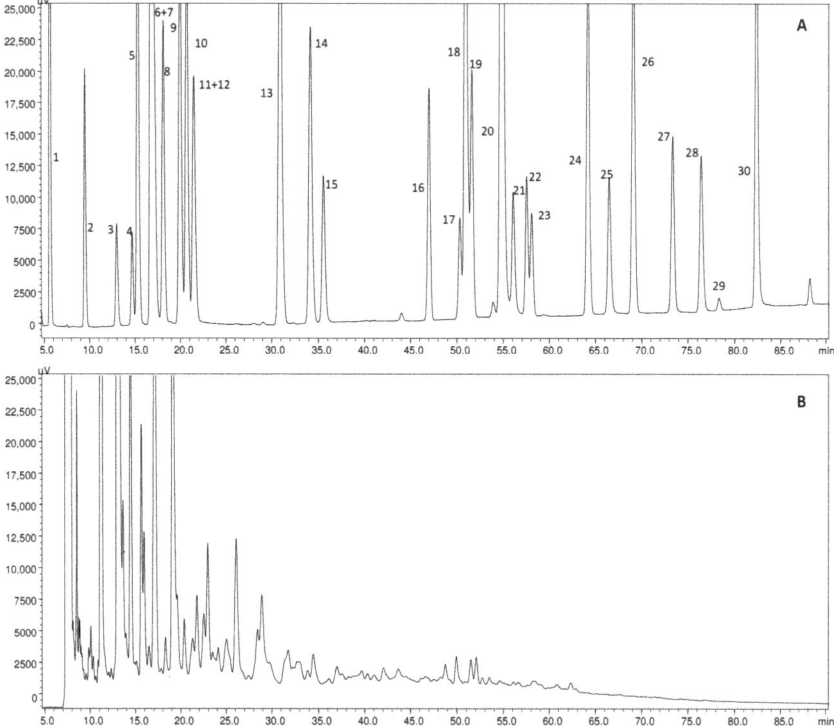

Figure 2. HPLC chromatograms at 280 nm for (**A**) polyphenols standard mixture and (**B**) Tinta Roriz subcritical water extract at 250 °C; peak identification: (1) gallic acid, (2) protocatechuic acid, (3) (+)-catechin, (4) 4-hydroxyphenilacetic acid, (5) 4-hydroxybenzoic acid, (6) 4-hydroxybenzaldehyde, (7) chlorogenic acid, (8) vanillic acid, (9) caffeic acid, (10) syringic acid, (11) (−)-epicatechin, (12) *p*-coumaric acid, (13) ferulic acid, (14) sinapic acid, (15) naringin, (16) rutin, (17) resveratrol, (18) quercetin-3-*O*-glucopyranoside, (19) phloridzin, (20) cinnamic acid, (21) ellagic acid, (22) myricetin, (23) kaempferol-3-*O*-glucoside, (24) kaempferol-3-*O*-rutinoside, (25) naringenin, (26) quercetin, (27) phloretin, (28) tiliroside, (29) kaempferol and (30) pinocembrin.

The results discussed above suggest that the applied extraction technique significantly influences the amount of recovered bioactive compounds. Besides phenolic acids, flavanols and flavonols were the two main subclasses of flavonoid compounds present in higher amounts in the analyzed extracts. Moreover, comparing the amount of flavonols extracted by SWE at 125 °C in the present study with the amount reported in our last study at 150 °C, a higher content of flavonols was recovered when lower extraction temperature was employed, demonstrating that temperatures higher than 125 °C may cause degradation of these compounds [12]. Gabaston et al. focused on the extraction of stilbenes using SWE and obtained an amount of 362 mg of stilbenes/100 g dw, when performing the extraction at 160 °C for 5 min [22]. For the same conditions, the amount of resveratrol recovered was 130 mg/100 g dw, which is approximately 8 and 10 times more than the obtained for the extraction at 250 °C with TN and TR, respectively. Additionally, the authors pointed out that high temperatures and long periods of extraction may lead to the degradation of these compounds, which can justify the results obtained for TN and TR and presented in this paper.

Table 2. Content of the identified phenolic compounds in Touriga Nacional (TN) and Tinta Roriz (TR) extracts obtained through subcritical water extraction (SWE) at 125 and 250 °C. Results were expressed as mean ± standard deviations (milligrams of compound/100g dw, *n*=3).

Compound	TN		TR	
	125 °C	250 °C	125 °C	250 °C
Phenolic acids				
Gallic acid	60.1 ± 3.0	891 ± 45	74.2 ± 3.7	1066 ± 53
Protocatechuic acid	33.8 ± 1.7	14.5 ± 0.7	33.4 ± 1.7	21.2 ± 1.1
4-hydroxyphenilacetic acid	16.8 ± 0.8	62.6 ± 3.1	49.2 ± 2.5	134 ± 7
4-hydroxybenzoic acid	9.2 ± 0.5	22.6 ± 1.1	8.4 ± 0.4	44.9 ± 2.2
4-hydroxybenzaldehyde	4.9 ± 0.2	7.5 ± 0.4	4.7 ± 0.2	9.8 ± 0.5
Chlorogenic acid	6.2 ± 0.3	23.2 ± 1.2	7.4 ± 0.4	44.3 ± 2.2
Vanillic acid	15.0 ± 0.7	15.6 ± 0.8	13.5 ± 0.7	31.6 ± 1.6
Caffeic acid	14.9 ± 0.7	13.6 ± 0.7	14.4 ± 0.7	19.6 ± 1.0
Syringic acid	ND [a]	37.9 ± 1.9	<LOD [b]	65.6 ± 3.3
p-coumaric acid	17.2 ± 0.9	16.0 ± 0.8	22.9 ± 1.1	21.7 ± 1.1
Ferulic acid	21.9 ± 1.1	18.9 ± 0.9	24.5 ± 1.2	19.6 ± 1.0
Sinapic acid	17.1 ± 0.9	14.4 ± 0.7	22.0 ± 1.0	12.2 ± 0.6
Cinnamic acid	11.1 ± 0.6	8.1 ± 0.4	12.2 ± 0.6	11.1 ± 0.5
∑Phenolic acids	228 ± 11	1145 ± 57	286 ± 14	1502 ± 75
Flavanols				
(+)-Catechin	102 ± 5	181 ± 9	245 ± 12	216 ± 11
(-)-Epicatechin	3.9 ± 0.2	3.1 ± 0.2	17.2 ± 0.9	14.6 ± 0.7
∑Flavanols	106 ± 5	184 ± 9	262 ± 13	231 ± 12
Flavanones				
Naringin	<LOD	8.6 ± 0.4	<LOD	14.7 ± 0.7
Naringenin	4.5 ± 0.2	2.4 ± 0.1	6.1 ± 0.3	3.3 ± 0.2
∑Flavanones	4.5 ± 0.2	11.0 ± 0.6	6.1 ± 0.3	18.0 ± 0.9
Flavonols				
Rutin	3.1 ± 0.2	9.2 ± 0.5	1.4 ± 0.1	15.4 ± 0.8
Quercetin-3-*O*-glucopyranoside	4.4 ± 0.2	5.2 ± 0.3	4.1 ± 0.2	10.7 ± 0.5
Myricetin	84.3 ± 4.2	14.6 ± 0.7	86.4 ± 4.3	16.2 ± 0.8
Kaempferol-3-*O*-glucoside	9.4 ± 0.5	ND	10.1 ± 0.5	8.6 ± 0.4
Kaempferol-3-*O*-rutinoside	4.3 ± 0.2	2.3 ± 0.1	4.7 ± 0.2	7.0 ± 0.4
Quercetin	40.9 ± 2.0	24.9 ± 1.2	40.7 ± 2.0	30.8 ± 1.5
Kaempferol	34.6 ± 1.7	23.2 ± 1.2	41.6 ± 2.1	24.2 ± 1.2
∑Flavonols	181 ± 9	79.4 ± 4.0	189 ± 9	113 ± 6
Stilbenes				
Resveratrol	8.0 ± 0.4	15.8 ± 0.8	10.5 ± 0.5	13.1 ± 0.7
∑Stilbenes	8.0 ± 0.4	15.8 ± 0.8	10.5 ± 0.5	13.1 ± 0.7
Others				
Phloridzin	<LOD	3.7 ± 0.2	<LOD	8.0 ± 0.4
Phloretin	ND	<LOD	<LOD	<LOD
∑Others	0.0 ± 0.0	3.7 ± 0.2	0.0 ± 0.0	8.0 ± 0.4
∑All phenolic compounds	527	1440	755	1884

[a] ND: not detected; [b] LOD: limit of detection.

Thus, the HPLC results showed that the amount of phenolic compounds quantified for the TR vine-cane variety was at least 1.31 times higher than the ones reported for TN. Regarding the two extraction temperatures tested (125 and 250 °C), the extracts obtained with the highest temperature presented higher amounts of phenolic compounds, which was already expected as the values obtained for the TPC, TFC, DPPH-RSA and FRAP assays for these extracts were also higher. In this way, the TR extract obtained at 250 °C was selected for the further assays.

3.3. Capacity of Scavenging Reactive Oxygen Species

Reactive oxygen species (ROS) are produced as side products of metabolic reactions. Nevertheless, in some cases these compounds are not naturally neutralized by cells, interacting with other molecules and interfering in metabolic pathways, which results in oxidative damage to cellular biomolecules [36]. Therefore, it is critical to perform assays that evaluate the scavenging capacity of ROS by extracts. Table 3 summarizes the results obtained for ROS scavenging assays.

Table 3. Superoxide ($O_2{}^{\bullet-}$), hypochlorous acid (HOCl) and peroxyl radical ($ROO^\bullet$) scavenging activities of Tinta Roriz (TR) subcritical water extract obtained at 250 °C.

Reactive Species	$O_2{}^{\bullet-}$	HOCl	$ROO^\bullet$
	IC_{50} (µg/mL)		S_{sample}/S_{Trolox} [a]
TR sample	83.67 ± 5.84	33.94 ± 2.95	0.024 ± 0.001
Gallic acid	5.18 ± 0.19	1.25 ± 0.05	1.119 ± 0.005
Catechin	48.99 ± 0.75	0.18 ± 0.01	7.592 ± 0.074

IC_{50} = in-vitro inhibitory concentration, expressed in µg/mL, required to scavenge 50% of the generated reactive oxygen species (mean ± SD, *n* = 3). [a] Results for $ROO^\bullet$ scavenging activity are expressed as slop ratio between samples or positive controls and Trolox. S_{sample} = slope of extract/positive controls curves and S_{Trolox} = slope of Trolox curve.

3.3.1. Superoxide Radical Scavenging Assay

As can be observed in Table 3, the TR extract presented a lower scavenging capacity of $O_2{}^{\bullet-}$ (IC_{50} = 83.67 ± 5.84 µg/mL) than the positive controls employed (GA and catechin), which means that higher concentrations of extract would be needed to obtain the same results of the positive controls. Farhadi et al. assessed the percentage of $O_2{}^{\bullet-}$ scavenging activity of skins, pulps, seeds, canes and leaves of five native grape cultivars in west Azerbaijan province (Iran) [37]. Even though the extracts obtained from grape skins exhibited the highest $O_2{}^{\bullet-}$ scavenging activity (with values ranging from 86.15% to 89.92% of inhibition), vine-canes also proved to be a promising matrix (with inhibition percentages ranging from 81.46% to 86.34%). In another study performed by Barros et al., grape stem extracts were also proposed as good $O_2{}^{\bullet-}$ scavengers [38]. The authors analyzed grape stems from seven Portuguese *V. vinifera* L. varieties, four red and three white. The IC_{50} values ranged from 970 to 2010 µg/mL for the Rabigato and Tinta Barroca grape stems varieties, respectively, being 11-fold higher than the result obtained in the present work, which demonstrates that extracts obtained from vine-canes through SWE were more efficient in scavenging $O_2{}^{\bullet-}$.

3.3.2. Hypochlorous Acid Scavenging Assay

Regarding the HOCl scavenging assay, catechin was the best scavenger presenting an IC_{50} value of 0.18 ± 0.01 µg/mL. Similarly to the $O_2{}^{\bullet-}$ assay, the TR extract showed a lower HOCl quenching power (IC_{50} = 33.94 ± 2.95 µg/mL) than catechin and GA (IC_{50} = 1.25 ± 0.05 µg/mL). Wada et al. evaluated the scavenging capacity of hypochlorite ion (ClO^-) by grape seed extracts using a concentration range of 0.02–2 mg/mL [39]. Two commercial grape seed extracts purchased from Mitsubishi Chemical Corporation (Kanagawa) were used in this study: extract A (proanthocyanin 99%) and B (proanthocyanin >80%). The ClO^- quenching capacity of grape seed extracts A and B at 1.0 mg/mL were 27.7% ± 4.2% and 22.0% ± 3.7%, respectively. The authors used five reference compounds to compare the results obtained for the extracts, namely trans-resveratrol, chalcone, cyanidin, delphinidin and pelargonidin. These grape seeds extracts exhibited lower ClO^- quenching effects than trans-resveratrol, cyanidin and delphinidin, but higher scavenging efficiency compared to chalcone [39].

3.3.3. Peroxyl Radical Scavenging Assay

According to the obtained results (Table 3), the TR extract showed a low quenching capacity against $ROO^\bullet$ with S_{sample}/S_{Trolox} lower than 1 (S_{sample}/S_{Trolox} = 0.024 ± 0.001). Noteworthy, catechin presented the highest scavenging ability for $ROO^\bullet$ (S_{sample}/S_{Trolox} = 7.592 ± 0.074), followed by GA (S_{sample}/S_{Trolox} = 1.119 ± 0.005). In fact, the obtained subcritical water extracts were mainly composed by gallic acid and catechin, which were the main contributors to the phenolic profile of TR. However, despite their higher amount in comparison to the other phenolic compounds identified and quantified in the extracts, the levels found and recovered were lower to produce the same effect as the positive standards alone. Additionally, a synergistic effect of all the compounds present in TR extract could also negatively affect the capacity to scavenge the peroxyl radical ($ROO^{\bullet}$).

Although the TR extract exhibited a low capacity to quench $ROO^{\bullet}$, previous studies described no scavenging activity of different byproducts for this radical [40,41]. Tournour et al. also determined the ORAC values of extracts from grape pomace of four Douro's *V. vinifera* L. varieties, including TN and TR [42]. The obtained results corresponded to 2337 ± 368 and 1054 ± 199 μmol TE/g of dw, for the TN and TR varieties, respectively. According to the authors, TN presented the best ORAC values, being therefore also better than the ones obtained for TR. Nevertheless, the highest TPC and the best result for the DPPH-RSA assay, obtained in the study aforementioned, also corresponded to the TN variety, which suggest that the radical scavenging capacity is directly related to the results obtained by both assays. In the present study, the TR variety was the one that achieved the highest values for the TPC and DPPH assays, which supports the choice made to use this extract for further studies. Barros et al. also assessed the ORAC values of seven Portuguese grape stems varieties extracts, obtained through sonication baths [38]. The results obtained ranged from 40.26 to 150.79 mM Trolox/100 g of dried grape stems. A direct comparison with the values from the present work cannot be made, but the results reported by these authors supported the ones obtained in $O_2^{\bullet-}$ scavenging assay, where higher values were reported for the red vine varieties.

Among the ROS studied, the highest scavenging efficiencies of the TR extract were observed for HOCl (IC_{50} = 33.94 ± 2.95 μg/mL) and $O_2^{\bullet-}$ (IC_{50} = 83.67 ± 5.84 μg/mL). According to the results observed for these radical scavenging assays, TR vine-canes extracts obtained by SWE at 250 °C showed promising results for applications in health-related products.

3.4. Cell Viability Studies

The cell viability results are represented in Figure 3 and were obtained after exposure of HFF-1 to different concentrations of TR extract (0.1; 1.0; 10; 100 and 1000 μg/mL).

As it is possible to observe, extract concentrations under 100 μg/mL did not result in a reduction of HFF-1 cellular viability. For these concentrations, the cell viability was higher than 100%. However, at a concentration of 1000 μg/mL there was a considerable reduction of cell viability to 52.15% ($p < 0.05$). Manca et al. incorporated grape seed and vine-canes extracts in vesicular systems designed for topical applications and analyzed their cell viability in HFF-1 [43]. In the MTT test, the authors incubated the cells solely with the extracts to serve as a comparison value and with the vesicles in four different concentrations, namely 0.2, 2, 20 and 40 μg/mL, for 48 h. After the incubation time, the best values were obtained for the cells incubated only with the grape seed and vine-canes extracts (>100%). These results are in accordance with the ones obtained for the TR extract, since until the concentration of 100 μg/mL the cell viability was also above 100%.

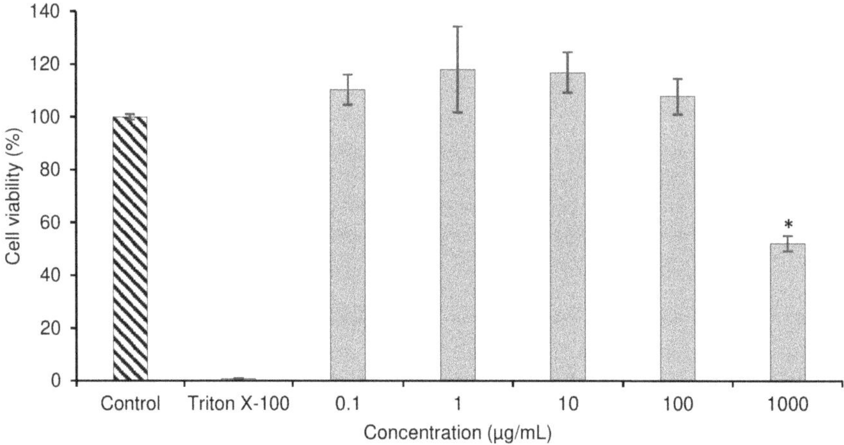

Figure 3. Effect of different concentrations of TR extract on the metabolic activity of HFF-1 cells, measured by the MTT assay. Values were expressed as mean ± SD (*n*=4). * means significant differences (*p* < 0.05).

4. Conclusions

The present work proposes an alternative application for the reuse of vine-canes, an agro-industrial byproduct derived from grape production. Grapes are one of the major fruit crops produced throughout the world and after harvesting season the vines need to be pruned, which generates every year large amounts of wastes. The challenge is to find alternative solutions to the presently used applications (incorporation in the soil), considering not only economic profits but, mostly, environmental impacts. The results obtained proved that SWE is a suitable environmentally friendly technique for vine-canes extraction of bioactive compounds with antioxidant properties. The TR extract obtained the highest results with the highest temperature tested (250 °C), achieving a TPC of 181 ± 12 mg GAE/g dw and an antioxidant activity through the FRAP assay of 202 ± 14 mg AAE/g dw. In general, a tendency was observed: extracts with higher phenolic content also presented higher flavonoid content, higher DPPH-RSA and FRAP values. Indeed, extracts obtained at 250 °C presented higher results than the ones obtained at 125 °C.

Regarding the HPLC-PDA analysis, the TR variety presented the higher amount of individual phenolic compounds for the highest extraction temperature tested. The phenolic compound present in higher amounts was GA (1145 ± 57 and 1502 ± 75 mg of GA/100 g of dw for the TN and TR varieties extracts obtained at 250 °C, respectively), which seems to be the main compound contributing for the high antioxidant activity of these extracts. Taking into consideration the results obtained, TR extract prepared at 250 °C was selected for the radicals scavenging and cell assays, revealing a good scavenging capacity of oxygen species as well as no adverse effects on HFF-1 were detected until a concentration of 100 μg/mL.

Considering the portion that grape production occupies in cultivation of food crops worldwide, the use of vine-canes in added value products could have a significant impact in the sustainability of food and beverage industries. Indeed, final applications in functional foods or cosmetics would be of extreme interest. As future perspectives, it would be interesting to test a scale up for SWE, since at the industrial level it might be more profitable.

Author Contributions: Conceptualization, M.M.M., F.R. and C.D.-M.; Funding acquisition, A.F.P., C.F., F.R., P.C. and C.D.-M.; Investigation, O.D., D.P., M.M.M., F.R., P.C. and A.F.P.; Methodology, O.D. and D.P.; Project administration, A.F.P., C.F., F.R. and C.D.-M.; Resources, A.F.P., C.F., F.R. and C.D.-M.; Supervision, M.M.M., F.R. and C.D.-M.; Writing—original draft preparation, O.D., D.P., M.M.M., F.R.; Writing—review and editing, M.M.M, F.R., A.F.P., C.F., P.C. and C.D.-M. All authors have read and agreed to the published version of the manuscript.

Funding: This research was funded by FCT/MCTES through national funds (UIDB/50006/2020). This work was also financed by the FEDER Funds through the Operational Competitiveness Factors Program—COMPETE and by National Funds through FCT within the scope of the project "PTDC/BII-BIO/30884/2017—POCI-01-0145-FEDER-030884" and project "PTDC/ASP-AGR/29277/2017-POCI-01-0145-FEDER-029277".

Acknowledgments: O.D. is thankful for the research grant from project PTDC/BII-BIO/30884/2017—POCI-01-0145-FEDER-030884. D.P. is thankful for the PhD grant (SFRH/BD/144534/2019) financed by POPH-QREN and subsidized by the European Science Foundation and Ministério da Ciência, Tecnologia e Ensino Superior. M.M.M. (project CEECIND/02702/2017) and A.F.P. (DL57/2016–Norma transitória) are grateful for the financial support financed by national funds through FCT—Fundação para a Ciência e a Tecnologia, I.P. and to REQUIMTE/LAQV. The supply of the vineyard pruning is acknowledged to Sogrape, S.A. The authors are grateful to João Vasconcellos Porto for the provided identification of the studied plant material.

Conflicts of Interest: The authors declare no conflict of interest. The funders had no role in the design of the study; in the collection, analyses or interpretation of data; in the writing of the manuscript or in the decision to publish the results.

References

1. Kuzmina, K.; Prendeville, S.; Walker, D.; Charnley, F. Future scenarios for fast-moving consumer goods in a circular economy. *Futures* **2019**, *107*, 74–88. [CrossRef]

2. Roser, M.; Ritchie, H.; Ortiz-Ospina, E. World Population Growth. Available online: https://ourworldindata.org/world-population-growth (accessed on 28 May 2019).

3. Siracusa, L.; Ruberto, G. Not only what is food is good-polyphenols from edible and nonedible vegetables waste. In *Polyphenols in Plants*, 2nd ed.; Watson, R., Ed.; Elsevier: Amsterdan, The Netherlands, 2018; pp. 3–21.

4. Clark, G. The industrial revolution. In *Handbook of Economic Growth*, 1st ed.; Aghion, P., Steven, D., Eds.; Elsevier: Amsterdan, The Netherlands, 2005; pp. 217–262.

5. Russ, W.; Schnappinger, M. Waste related to the food industry: A challenge in material loops. In *Utilization of By-Products and Treatment of Waste in the Food Industry*, 1st ed.; Oreopoulou, V., Russ, W., Eds.; Springer: Berlin, Germany, 2007; pp. 1–13.

6. Oliveira, L.S.; Franca, A.S. Low-cost adsorbents from agri-food wastes. In *Food Sciences Technology: New Research*, 1st ed.; Greco, L., Bruno, M., Eds.; Nova Science Publishers: Hauppauge, NY, USA, 2008; pp. 171–209.

7. Rodríguez, R.; Jiménez, A.; Fernández-Bolaños, J.; Guillén, R.; Heredia, A. Dietary fibre from vegetable products as source of functional ingredients. *Trends Food Sci. Technol.* **2006**, *17*, 3–15. [CrossRef]

8. Devesa-Rey, R.; Vecino, X.; Varela-Alende, J.L.; Barral, M.T.; Cruz, J.M.; Moldes, A.B. Valorization of winery waste vs. The costs of not recycling. *Waste Managem.* **2011**, *31*, 2327–2335. [CrossRef] [PubMed]

9. OIV—International Organisation of Vine and Wine (OIV). *OIV Statistical Report on World Vitiviniculture*; OIV—International Organisation of Vine and Wine (OIV): Paris, France, 2016.

10. Rajha, H.; Darra, N.; Hobaika, Z.; Boussetta, N.; Vorobiev, E.; Maroun, R.; Louka, N. Extraction of total phenolic compounds, flavonoids, anthocyanins and tannins from grape byproducts by response surface methodology. Influence of solid-liquid ratio, particle size, time, temperature and solvent mixtures on the optimization process. *Food Nutr. Sci.* **2014**, *5*, 397–409. [CrossRef]

11. Gullón, B.; Eibes, G.; Moreira, M.T.; Dávila, I.; Labidi, J.; Gullón, P. Antioxidant and antimicrobial activities of extracts obtained from the refining of autohydrolysis liquors of vine shoots. *Ind. Crops Prod.* **2017**, *107*, 105–113. [CrossRef]

12. Moreira, M.M.; Barroso, M.F.; Porto, J.V.; Ramalhosa, M.J.; Švarc-Gajić, J.; Estevinho, L.; Morais, S.; Delerue-Matos, C. Potential of portuguese vine shoot wastes as natural resources of bioactive compounds. *Sci. Total Environ.* **2018**, *634*, 831–842. [CrossRef]

13. Rajha, H.N.; Boussetta, N.; Louka, N.; Maroun, R.G.; Vorobiev, E. A comparative study of physical pretreatments for the extraction of polyphenols and proteins from vine shoots. *Food Res. Int.* **2014**, *65*, 462–468. [CrossRef]

14. Sanchez-Gomez, R.; Zalacain, A.; Alonso, G.L.; Salinas, M.R. Vine-shoot waste aqueous extracts for re-use in agriculture obtained by different extraction techniques: Phenolic, volatile, and mineral compounds. *J. Agr. Food Chem.* **2014**, *62*, 10861–10872. [CrossRef]

15. Karacabey, E.; Mazza, G.; Bayındırlı, L.; Artık, N. Extraction of bioactive compounds from milled grape canes (*Vitis vinifera*) using a pressurized low-polarity water extractor. *Food Bioprocess Tech.* **2012**, *5*, 359–371. [CrossRef]

16. Delgado-Torre, M.P.; Ferreiro-Vera, C.; Priego-Capote, F.; Perez-Juan, P.M.; Luque de Castro, M.D. Comparison of accelerated methods for the extraction of phenolic compounds from different vine-shoot cultivars. *J. Agr. Food Chem.* **2012**, *60*, 3051–3060. [CrossRef]

17. Jiménez Gómez, S.; Cartagena Causapé, M.C.; Arce Martínez, A. Distribution of nutrients in anaerobic digestion of vine shoots. *Bioresour. Technol.* **1993**, *45*, 93–97. [CrossRef]

18. Tag, A.T.; Duman, G.; Ucar, S.; Yanik, J. Effects of feedstock type and pyrolysis temperature on potential applications of biochar. *J. Anal. Appl. Pyrol.* **2016**, *120*, 200–206. [CrossRef]

19. Dávila, I.; Gullón, B.; Labidi, J.; Gullón, P. Multiproduct biorefinery from vine shoots: Bio-ethanol and lignin production. *Renew. Energy* **2019**, *142*, 612–623. [CrossRef]

20. Fernandes, M.J.; Moreira, M.M.; Paíga, P.; Dias, D.; Bernardo, M.; Carvalho, M.; Lapa, N.; Fonseca, I.; Morais, S.; Figueiredo, S.; et al. Evaluation of the adsorption potential of biochars prepared from forest and agri-food wastes for the removal of fluoxetine. *Bioresour. Technol.* **2019**, *292*, 121973. [CrossRef] [PubMed]

21. Jiménez, L.; Angulo, V.; Ramos, E.; De la Torre, M.J.; Ferrer, J.L. Comparison of various pulping processes for producing pulp from vine shoots. *Ind. Crops Prod.* **2006**, *23*, 122–130. [CrossRef]

22. Gabaston, J.; Leborgne, C.; Valls, J.; Renouf, E.; Richard, T.; Waffo-Teguo, P.; Mérillon, J.-M. Subcritical water extraction of stilbenes from grapevine by-products: A new green chemistry approach. *Ind. Crops. Prod.* **2018**, *126*, 272–279. [CrossRef]

23. Jokić, S.; Gagić, T.; Knez, Ž.; Banožić, M.; Škerget, M. Separation of active compounds from tobacco waste using subcritical water extraction. *J. Supercrit. Fluids* **2019**, *153*, 104593. [CrossRef]

24. Nile, S.H.; Nile, A.; Gansukh, E.; Baskar, V.; Kai, G. Subcritical water extraction of withanosides and withanolides from ashwagandha (*Withania somnifera* l) and their biological activities. *Food Chem. Toxicol.* **2019**, *132*, 110659. [CrossRef]

25. Lee, J.-H.; Ko, M.-J.; Chung, M.-S. Subcritical water extraction of bioactive components from red ginseng (*Panax ginseng* c.A. Meyer). *J. Supercrit. Fluids* **2018**, *133*, 177–183. [CrossRef]

26. Singleton, V.L.; Rossi, J.A.J. Colorimetry of total phenolics with phosphomolybdic–phosphotungstic acid reagents. *Am. J. Enol. Viticult.* **1965**, *16*, 144–158.

27. Paz, M.; Gúllon, P.; Barroso, M.F.; Carvalho, A.P.; Domingues, V.F.; Gomes, A.M.; Becker, H.; Longhinotti, E.; Delerue-Matos, C. Brazilian fruit pulps as functional foods and additives: Evaluation of bioactive compounds. *Food Chem.* **2015**, *172*, 462–468. [CrossRef]

28. Benzie, I.F.F.; Strain, J.J. The ferric reducing ability of plasma (frap) as a measure of "antioxidant power": The FRAP assay. *Anal. Biochem.* **1996**, *239*, 70–76. [CrossRef] [PubMed]

29. Pistón, M.; Machado, I.; Branco, C.S.; Cesio, V.; Heinzen, H.; Ribeiro, D.; Fernandes, E.; Chisté, R.C.; Freitas, M. Infusion, decoction and hydroalcoholic extracts of leaves from artichoke (*Cynara cardunculus* l. Subsp. Cardunculus) are effective scavengers of physiologically relevant ROS and RNS. *Food Res. Int.* **2014**, *64*, 150–156. [CrossRef] [PubMed]

30. Ou, B.; Hampsch-Woodill, M.; Prior, R.L. Development and validation of an improved oxygen radical absorbance capacity assay using fluorescein as the fluorescent probe. *J. Agr. Food Chem.* **2001**, *49*, 4619–4626. [CrossRef]

31. Rodrigues, F.; Palmeira-de-Oliveira, A.; das Neves, J.; Sarmento, B.; Amaral, M.H.; Oliveira, M.B. *Medicago* spp. extracts as promising ingredients for skin care products. *Ind. Crops. Prod.* **2013**, *49*, 634–644. [CrossRef]

32. Dorosh, O.; Moreira, M.M.; Rodrigues, F.; Peixoto, A.F.; Freire, C.; Morais, S.; Delerue-Matos, C. Vine-canes valorisation: Ultrasound-assisted extraction from lab to pilot scale. *Molecules* **2020**, *25*, 1739. [CrossRef] [PubMed]

33. Rodrigues, F.; Santos, J.; Pimentel, F.B.; Braga, N.; Palmeira-de-Oliveira, A.; Oliveira, M.B.P.P. Promising new applications of *Castanea sativa* shell: Nutritional composition, antioxidant activity, amino acids and vitamin e profile. *Food Funct.* **2015**, *6*, 2854–2860. [CrossRef] [PubMed]

34. Plaza, M.; Amigo-Benavent, M.; del Castillo, M.D.; Ibáñez, E.; Herrero, M. Neoformation of antioxidants in glycation model systems treated under subcritical water extraction conditions. *Food Res. Int.* **2010**, *43*, 1123–1129. [CrossRef]

35. Anastasiadi, M.; Pratsinis, H.; Kletsas, D.; Skaltsounis, A.-L.; Haroutounian, S.A. Grape stem extracts: Polyphenolic content and assessment of their *in vitro* antioxidant properties. *LWT Food Sci. Technol.* **2012**, *48*, 316–322. [CrossRef]

36. Nunes, M.A.; Rodrigues, F.; Oliveira, M.B.P.P. Chapter 11—Grape processing by-products as active ingredients for cosmetic proposes. In *Handbook of Grape Processing by-Products*, 1st ed.; Galanakis, C., Ed.; Academic Press: Cambridge, MA, USA, 2017; pp. 267–292.

37. Farhadi, K.; Esmaeilzadeh, F.; Hatami, M.; Forough, M.; Molaie, R. Determination of phenolic compounds content and antioxidant activity in skin, pulp, seed, cane and leaf of five native grape cultivars in west azerbaijan province, iran. *Food Chem.* **2016**, *199*, 847–855. [CrossRef]

38. Barros, A.; Gironés-Vilaplana, A.; Teixeira, A.; Collado-González, J.; Moreno, D.A.; Gil-Izquierdo, A.; Rosa, E.; Domínguez-Perles, R. Evaluation of grape (*Vitis vinifera* L.) stems from portuguese varieties as a resource of (poly)phenolic compounds: A comparative study. *Food Res. Int.* **2014**, *65*, 375–384. [CrossRef]

39. Wada, M.; Kido, H.; Ohyama, K.; Ichibangase, T.; Kishikawa, N.; Ohba, Y.; Nakashima, M.N.; Kuroda, N.; Nakashima, K. Chemiluminescent screening of quenching effects of natural colorants against reactive oxygen species: Evaluation of grape seed, monascus, gardenia and red radish extracts as multi-functional food additives. *Food Chem.* **2007**, *101*, 980–986. [CrossRef]

40. Almeida, D.; Pinto, D.; Santos, J.; Vinha, A.F.; Palmeira, J.; Ferreira, H.N.; Rodrigues, F.; Oliveira, M.B.P.P. Hardy kiwifruit leaves (*Actinidia arguta*): An extraordinary source of value-added compounds for food industry. *Food Chem.* **2018**, *259*, 113–121. [CrossRef] [PubMed]

41. Pinto, D.; Diaz Franco, S.; Silva, A.M.; Cupara, S.; Koskovac, M.; Kojicic, K.; Soares, S.; Rodrigues, F.; Sut, S.; Dall'Acqua, S.; et al. Chemical characterization and bioactive properties of a coffee-like beverage prepared from *Quercus cerris* kernels. *Food Funct.* **2019**, *10*, 2050–2060. [CrossRef] [PubMed]

42. Tournour, H.H.; Segundo, M.A.; Magalhães, L.M.; Barreiros, L.; Queiroz, J.; Cunha, L.M. Valorization of grape pomace: Extraction of bioactive phenolics with antioxidant properties. *Ind. Crops Prod.* **2015**, *74*, 397–406. [CrossRef]

43. Manca, M.L.; Firoznezhad, M.; Caddeo, C.; Marongiu, F.; Escribano-Ferrer, E.; Sarais, G.; Peris, J.E.; Usach, I.; Zaru, M.; Manconi, M.; et al. Phytocomplexes extracted from grape seeds and stalks delivered in phospholipid vesicles tailored for the treatment of skin damages. *Ind. Crops. Prod.* **2019**, *128*, 471–478. [CrossRef]

Article

Antioxidant, Antidiabetic, and Antiobesity Properties, TC7-Cell Cytotoxicity and Uptake of *Achyrocline satureioides* (Marcela) Conventional and High Pressure-Assisted Extracts

Adriana Maite Fernández-Fernández [1], Eliane Dumay [2], Françoise Lazennec [2], Ignacio Migues [3], Horacio Heinzen [3], Patricia Lema [4], Tomás López-Pedemonte [1] and Alejandra Medrano-Fernandez [1,*]

[1] Departamento de Ciencia y Tecnología de Alimentos, Facultad de Química, Universidad de la República, General Flores 2124, Montevideo 11800, Uruguay; afernandez@fq.edu.uy (A.M.F.-F.); tlopez@fq.edu.uy (T.L.-P.)

[2] Ingénierie des Agropolymères et Technologies Emergentes, Équipe de Biochimie et Technologie Alimentaires, Université de Montpellier, 2 Place Eugène Bataillon, 34095 Montpellier, France; eliane.dumay@orange.fr (E.D.); francoise.lazennec@umontpellier.fr (F.L.)

[3] Departamento de Química Orgánica, Facultad de Química, Universidad de la República, General Flores 2124, Montevideo 11800, Uruguay; imigues@fq.edu.uy (I.M.); heinzen@fq.edu.uy (H.H.)

[4] Grupo Tecnologías Aplicadas a la Ingeniería de Alimentos, Facultad de Ingeniería, Universidad de la República, Av Julio Herrera y Reissig 565, Montevideo 11300, Uruguay; plema@fing.edu.uy

* Correspondence: amedrano@fq.edu.uy; Tel.: +598-2924-26-75

check for **updates**

Citation: Fernández-Fernández, A.M.; Dumay, E.; Lazennec, F.; Migues, I.; Heinzen, H.; Lema, P.; López-Pedemonte, T.; Medrano-Fernandez, A. Antioxidant, Antidiabetic, and Antiobesity Properties, TC7-Cell Cytotoxicity and Uptake of *Achyrocline satureioides* (Marcela) Conventional and High Pressure-Assisted Extracts. *Foods* **2021**, *10*, 893. https://doi.org/10.3390/foods10040893

Academic Editors: Francisca Rodrigues and Cristina Delerue-Matos

Received: 3 March 2021
Accepted: 12 April 2021
Published: 19 April 2021

Publisher's Note: MDPI stays neutral with regard to jurisdictional claims in published maps and institutional affiliations.

Abstract: The growing incidence of non-communicable diseases makes the search for natural sources of bioactive compounds a priority for such disease prevention/control. *Achyrocline satureioides* ('marcela'), a plant rich in polyphenols and native to Brazil, Uruguay, Paraguay, and Argentina, could be used for this purpose. Data on its antidiabetic/antiobesity properties and cellular uptake of bioactive compounds are lacking. The potentiality of non-thermal technologies such as high-hydrostatic pressure (HP) to enhance polyphenol extraction retains attention. Thus, in the present study aqueous and ethanolic marcela extracts with/without assisted-HP processing were chemically characterized and assessed for their in vitro antioxidant capacity, antidiabetic and antiobesity activities, as well as cellular cytotoxicity and uptake on intestinal cell monolayers (TC7-cells, a clone of Caco-2 cells). Aqueous and ethanolic conventional extracts presented different polyphenolic profiles characterized mainly by phenolic acids or flavonoids, respectively, as stated by reverse phase-high-performance liquid chromatography (RP-HPLC) analyses. In general, ethanolic extracts presented the strongest bioactive properties and HP had none or a negative effect on in vitro bioactivities comparing to conventional extracts. TC7-cell viability and cellular uptake demonstrated in conventional and HP-assisted extracts, highlighted the biological effects of marcela bioactive compounds on TC7-cell monolayers. TC7-cell studies showed no HP-induced cytotoxicity. In sum, marcela extracts have great potential as functional ingredients for the prevention/treatment of chronic diseases such as diabetes.

Keywords: bioactive compounds; cell metabolism; flavonoids; high-hydrostatic pressure; marcela; phenolic compounds; TC7-cellular uptake

1. Introduction

Achyrocline satureioides (known by the popular name of 'marcela') could be used for the prevention/treatment of non-communicable chronic diseases including cardiovascular diseases, cancers, respiratory diseases, and diabetes [1], which are the main cause of deaths in the current times. Thus, the search for antioxidant, antidiabetic, and antiobesity natural sources is of great importance for their prevention/treatment. Marcela has been studied for its antioxidant, cell cytoprotective effect against oxidants [2], anti-inflammatory, analgesic, antispasmodic, constipating, sedative, immunomodulatory, antiviral, antiherpetic, choleretic and hepatoprotective actions, whereas partial cytotoxicity in mice and rats has been found for aqueous and ethanolic extracts [3]. It is a plant native to Brazil, Uruguay, Paraguay, and Argentina, commonly used as herbal tea [2–4]. Recently, marcela proved to present anti-cancer activity against glioma cell lines (U87, U251 and C6) and to be less cytotoxic to brain cell than gliomas [5]. However, no scientific studies regarding its antidiabetic and antiobesity underlying mechanisms have been reported. Marcela extracts are composed of flavonoids such as quercetin, luteolin and 3-*O*-methylquercetin in their glycosylated and aglycone forms [2], found in both ethanolic [6] and aqueous [2] extracts. These compounds possess several bioactive properties such as cytoprotective activity against oxidant agents [2], but there are no reports on the bioavailability and/or absorption experiments, neither about its cytotoxicity on intestinal cells (as a means to elucidate the effect after their ingestion), which are necessary to assess the potential effectiveness of the marcela bioactive compounds. Once absorbed, these compounds may exert the above-mentioned bioactivities.

Aqueous and ethanolic extracts have shown different polyphenolic profiles as a consequence of different polyphenols solubility correspondent to solvent polarity, with subsequently different bioactive properties and/or biological effectiveness. High hydrostatic pressure (HP) technology proved to increase polyphenolic extraction yields [7] and plant cell membrane damage [8]. HP can also disrupt weak bonds such as hydrophobic bonds subsequently generating conformational changes as well as denaturating cell proteins, which could lead to enhance compounds accessibility during extraction [8]. HP technology could be a resourceful procedure for *Achyrocline satureioides* polyphenols extraction by the use of moderate or no heat treatment [7,9], being especially useful for thermolabile compounds extraction [10]. However, these compounds could suffer modifications during the process. Thus, studies regarding bioactivity, absorption and cytotoxicity are needed in order to state if this innovative technology presents advantages related to conventional extraction procedures as well as to ensure this novel extracts are safe for consumption.

The aim of the present work is to evaluate *Achyrocline satureioides* antioxidant, antidiabetic and antiobesity properties of aqueous and ethanolic extracts compared to HP-assisted extracts, along with the exposure to cultures of intestinal cells in order to elucidate the degree of cytotoxicity (as assessed on cellular metabolic activity and cell membrane integrity) and uptake/absorption of extracted bioactive compounds.

2. Materials and Methods

2.1. Raw Material and Chemicals

Achyrocline satureioides (marcela) commercial samples were purchased in a pharmacy store (La Botica del Señor, Montevideo, Uruguay), and milled with a domestic coffee mill. All the reagents used in physicochemical characterization analyses were of reagent grade. Phenolic acids (gallic, chlorogenic and caffeic acids) and quercetin standards were purchased from Sigma-Aldrich (St. Louis, MO, USA) and used for marcela extract composition by reverse phase-high-performance liquid chromatography (RP-HPLC) and reverse phase-ultra-high-performance liquid chromatography (RP-UHPLC) analyses. Antioxidant assays reagents were purchased from Sigma-Aldrich (St. Louis, MO, USA): Folin reagent, 2,20-azinobis-(3-ethylbenzothiazoline-6-sulfonic acid) diammonium salt (ABTS), 6-hydroxy-2,5,7,8-tetramethylch-roman-2-acid (Trolox), fluorescein (FL) disodium salt, 2,20-azobis (2-methylpropionamidine) dihydrochloride (AAPH). Antidiabetic assays reagents were also purchased from Sigma-Aldrich (St. Louis, MO, USA): bovine serum albumin (BSA), methylglyoxal (MGO), aminoguanidine (AG), α-glucosidase (rat intestine acetone powder), acarbose, 4-methylumbelliferyl-α-D-glucopyranoside, human saliva α-amylase (Type IX-A), starch, maltose standard, 3,5-dinitrosalicylic acid. Antiobesity assay reagents were the following: lipase from porcine pancreas, 4-methylumbelliferyl oleate (4-MUO), and dimethyl sulfoxide (DMSO), which were purchased from Sigma-

Aldrich (St. Louis, MO, USA), and orlistat standard was purchased from Alfa Aesar (Haverhill, MA, USA).

2.2. TC7-Cells and Reagents for Cell Culture

TC7-cells (a clone of Caco-2 cells) was kindly provided by Dr. Rousset (Centre de Recherche des Cordeliers, UMR S872, Paris, France). For cell culture, the following reagents were used.

High-glucose Dulbecco's modified Eagle medium (DMEM) with L-glutamine and pyruvate (Phenol red-DMEM), high-glucose Dulbecco's modified Eagle medium without L-glutamine and pyruvate (Phenol red-free DMEM), Dulbecco's phosphate-buffered saline (DPBS) + Ca^{2+} and Mg^{2+}, Hank's Balanced Salt Solution (HBSS), penicillin-streptomycin mixture, MEM non-essential amino acid and foetal bovine serum (FBS) from Gibco™ were purchased from Life Technologies (Villebon-sur-Yvette, France). For MTT-assay, 3-(4,5-dimethyl-2-thiazolyl)-2,5-diphenyl tetrazolium bromide (MTT) and dimethyl sulfoxide (DMSO) were purchased from Sigma-Aldrich (St-Quentin Fallavier, France). The β-nicotinamide adenine dinucleotide hydrate (NAD), Trizma® base (Tris), L-(+)-lactic acid needed for LDH-assay, and Quercetin (96% dry matter, 95% purity) came from Sigma-Aldrich (St-Quentin Fallavier, France). Triton® X-100 came from Merck (Darmstadt, Germany).

2.3. Methods

2.3.1. Preparation of Marcela Extracts

Marcela aqueous extract (Mac) was obtained by adding 100 g of milled marcela powder to 1000 mL of distilled water. The mix was boiled for 1 h, filtered with paper (Whatman n°1) and the liquid was freeze-dried until constant weight (96 h).

Marcela ethanolic extract (Me) was obtained by adding 100 g of milled marcela powder to 1000 mL of ethanol (95%) followed by maceration at 20 °C for 24 h. The mixture was filtered with paper (Whatman n°1) then rotavaporated (60 °C, 120 rpm, under reduced pressure, approximately 13 kPa) to dryness and 30 mL of distilled water was added to recuperate polyphenols. Afterwards, the liquid was freeze-dried until constant weight (96 h).

High pressure (HP) extracts were prepared using a high-pressure unit Model S-IL-100-250-09W (HP Food Processor, Stansted Fluid Power, Ltd., Harlow, UK) located in Laboratorio Tecnológico del Uruguay pilot food plant (Montevideo, Uruguay). The pressure chamber (2 L volume, 100 mm bore internal diameter, 250 mm long) has inside the canister to hold samples. The vessel body and the pressure-transmitting fluid (water) were kept at treatment temperature (25 °C) by circulating water through an internal heat transfer jacket fitted to the outside of the high-pressure barrel assembly. The temperature of the pressure-transmitting fluid was monitored with a thermocouple positioned at the chamber bottom. Before treatment, samples were individually packed in Cryovac® pouches (Sealed Air®, Charlotte, North Carolina, USA) by adding 10 g of milled marcela powder to 100 mL of phosphate buffer pH 7.9 for aqueous HP extract, or to 100 mL of ethanol (95%) for ethanolic HP extract, then vacuum sealed. Samples were introduced in the pressurization chamber previously thermostated at 25 °C then submitted to 400 MPa and 25 °C for 1 min, in the case of ethanolic HP extract (Me HP), or 200 MPa and 25 °C, at pH 7.9 for 1 min for aqueous HP extract (Mac HP). These conditions were selected in previous trials studying optimum conditions for marcela antioxidant compounds extraction through HP procedure. Pressure was raised from 0.1 MPa at a rate of 100 MPa per 30 s, maintained at the desired pressure level for 1 min then reduced down to 0.1 MPa in less than 30 s. Sample blanks were also prepared in the same way at 0.1 MPa and 25 °C for 1 min but without the pressure-processing step, by adding 10 g of milled marcela powder to 100 mL of phosphate buffer pH 7.9 for aqueous HP blank extract (Mac HP BL), or to 100 mL of ethanol (95%) for ethanolic HP blank extract (Me HP BL). Afterwards, liquid samples were freeze-dried until constant weight (96 h). All the extracts were stored at −20 °C for further analyses.

2.3.2. Proximate Analysis

To characterize the initial sample or raw material (marcela powder), different parameters were determined: fat, protein, ashes, dietary fiber, moisture, and total carbohydrates (by difference using protein, moisture, ashes, and fat content). All determinations were performed at least in triplicate as in Association of Official Analytical Chemists (AOAC) [11] methods. Briefly, protein content was determined by Kjeldhal method using the conversion factor 6.25, moisture was determined using a conventional oven at 105 °C till constant weight, ashes was determined by using a furnace at 525 °C for 8 h, and fat content was obtained by Soxhlet method for 6 h using petroleum ether.

2.3.3. RP-HPLC and RP-UHPLC Analyses

To obtain the polyphenolic profile of extracts, each of the extract samples were eluted by RP-HPLC (Shimadzu, SPD-20A detector and LC-10AT pump) according to De Souza et al. [6] with detection at 370 nm in a Jupiter C18 reverse phase column and an isocratic flow program of 1 mL/min. The mobile phase was composed by methanol:0.16 M phosphoric acid, in a ratio of 53:47 (v/v). The injection volume was 20 µL.

Each of the samples were also eluted by RP-UHPLC according to Reza et al. [12] with some modifications. RP-UHPLC UltiMate 3000 (Thermo Fisher Scientific, Massachusetts, USA) was used with a diode array detection (DAD detector). The reverse phase column was a Thermo Scientific BDS Hypersil C18 (150 × 3 mm, 3 µm particle size) used at 1 mL/min flow rate. Mobile phase was composed by methanol, phosphoric acid (pH 2.81) and acetonitrile in gradient: time 0 min, 5% acetonitrile and 95% phosphoric acid (initial condition); time 10 min, 10% acetonitrile, 10% methanol and 80% phosphoric acid; time 20 min, 20% acetonitrile, 20% methanol and 60% phosphoric acid; time 40 min, 20% acetonitrile, 20% methanol and 60% phosphoric acid; time 45 min, 100% acetonitrile; time 50 min, 100% acetonitrile; time 55 min, 5% acetonitrile and 95% phosphoric acid (to return to the initial conditions). The duration of each run was 55 min. The injection volume was 20 µL. The software used was Dionex Chromeleon 7.1 SR2. Phenolic acids were quantified by detection at 290 nm and quercetin was quantified at 370 nm. Phenolic acids and quercetin were identified and quantified by the use of pure standards and the construction of calibration curves through the detection at 290 and 370 nm for phenolic acids and quercetin, respectively.

2.3.4. Antioxidant Capacity

Total polyphenol content was performed by Folin–Ciocalteau method [13] as described by Fernández-Fernández, Iriondo-DeHond, Dellacassa, Medrano-Fernandez, and del Castillo [14], preparing sample solutions in distilled water and using a gallic acid standard curve (0.05–1.0 mg/mL). Results were expressed as mg GAE/g extract.

The ABTS method [15] was performed as described by Fernández-Fernández et al. [14], preparing samples in phosphate buffer (pH 7.4) and using a Trolox calibration curve (0.25–1.5 mM). Results were expressed as µmol TE/mg extract.

Oxygen radical antioxidant capacity-fluorescein (ORAC-FL) assay was performed by the method of Ou, Hampsch-Woodill, and Prior [16] modified by Dávalos, Bartolomé, and Gómez-Cordovés [17] as described in Fernández-Fernández et al. [14]. The area under the curve (AUC) of fluorescence vs. time were calculated and normalized to the AUC of the blank as follows: $AUC_{antioxidant\ (trolox\ or\ sample)} - AUC_{blank}$. Trolox calibration curve (AUC vs. [Trolox]) was constructed and results were expressed as µmol TE/mg extract.

All samples were prepared in triplicate and each one of the preparations was measured in triplicate.

2.3.5. Antidiabetic and Antiobesity Activities

α-glucosidase inhibition capacity was evaluated as described by Fernández-Fernández et al. [14] as an antidiabetic strategy. Briefly, fluorescence measurements were displayed at

37 °C for 30 min (each minute) at 360 and 460 nm of excitation and emission wave lengths, respectively. Acarbose was used as reference with probed inhibition capacity.

α-amylase inhibition assay was evaluated as another antidiabetic strategy and performed as reported by Li, Yao, Du, Deng, and Li [18] with some modifications described by Fernández-Fernández et al. [19]. The inhibition capacity was calculated by taking positive control as 100% of enzyme activity.

Fluorescent advanced glycation end products (AGEs) formation was evaluated by determining BSA-MGO formation inhibition (antiglycant capacity) as another antidiabetic strategy, obtaining the IC_{50} value [20]. Briefly, sample mixtures consisted of 500 µL BSA stock solution (2 mg/mL in PBS, 1 mg/mL final concentration), 25/50 µL 5/10 mM MGO stock solution (200 mM in PBS, 5 mM final concentration), different volumes of extracts from a sample stock solution of 50 mg/mL (concentrations 0.1-5 mg/mL) plus sufficient volume of PBS 10 mM pH 7.4 with 0.02% sodium azide to achieve 1 mL of the mixture final volume. Sample blanks consisted of the samples (different concentrations) with sufficient volume of PBS to achieve 1 mL of the mixture final volume (intrinsic fluorescence of the samples). Positive control was prepared by mixing 500 µL BSA, 25/50 µL 5/10mM MGO and 475/450 µL PBS, as previously explained. Aminoguanidine (AG) was used as reference (1, 4 and 8 mM final concentrations) mixed with BSA, MGO and PBS. All stock solutions were prepared in PBS 10 mM pH 7.4 with 0.02% sodium azide. Eppendorf tubes were incubated at 37 °C for 7 days. Fluorescence measurements were performed at 340 and 420 nm of excitation and emission wavelengths, respectively, and inhibition percentages were calculated by taking positive control as 100% of AGEs formation.

Pancreatic lipase inhibition capacity was determined as described in Fernández-Fernández et al. [14]. Measurements were determined after 30 min incubation at 25 °C by fluorescence measurements at 360 nm and 460 nm of excitation and emission wavelengths, respectively.

2.3.6. TC7-Cell Culture and Sample Deposits

TC7-cells were routinely grown according to Benzaria et al. [21,22] with minor changes in 75 cm^2 sterile cell culture flasks in phenol red-DMEM culture medium. TC7-cells (passages 41-49) were seeded in sterile 12-well plates (3.5 cm^2/well; Nunc, VWR, Fontenay-sous-Bois, France) at a density of 2.5×10^5 cells/well (1 mL of cell suspension/well) then cultivated at 37 °C in controlled atmosphere (8% CO_2, 92% air, 100% relative humidity, RH) (Thermo Scientific 8000 incubator, Thermo Electron, St-Herblain, France) for 19–20 days to reach cell-confluence and obtain differentiated cells, the culture medium (phenol red-DMEM supplemented with 20% *v/v* heat-inactivated FBS, 1% *v/v* penicillin-streptomycin and 1% *v/v* MEM non-essential amino acids), being changed every 2 days. Cell confluence was assessed by transepithelial electrical resistance (TEER; Millicell®-ERS volt-ohm meter, Millipore, St-Quentin-en-Yvelines, France) measurements before deposing extract samples onto the cells. For TEER measurements, cells were grown in sterile Transwell plates with ThinCert inserts (3 µm pore size; 1.13 cm^2/well; Greiner Bio-one, VWR International, Fontenay-sous-Bois, France) at a density of 2.5×10^5 cells/well, obtaining TEER values of 750–800 Ω cm^{-2}. Cell confluency of the cell cultures was also checked by inverse phase microscope examination.

After washing using Phenol red-free DMEM, differentiated TC7-cells were incubated for 3 h or 22 h at 37 °C in controlled atmosphere (8% CO_2, 92% air, 100% HR) with 500 µL of extract mixture or control sample. Exposure times (3 h or 22 h) were chosen on the basis of previous experiments [22], and taking into account the open time necessary to prepare cell series. All cell seeding and sample deposit experiments were carried in sterile conditions under a laminar flow cabinet (PSM MSC Advantage, ThermoFisher Scientific, St-Herblain, France), using 0.2 µm filtrated media, sterile solutions and sterile plastic material (pipets, tips, flasks, plates, microplates, Eppendorf® and Falcon® tubes).

Ethanolic and aqueous extracts were assayed on TC7-cells for a range of lyophilized extract concentrations in the cell deposit medium. For this, a 56.6 mg of lyophilized ethanolic extract (Me) was solubilized into 1 mL of 80% Ethanol (Ethanol/Water 80:20, v/v), and 28.3 mg of lyophilized aqueous extract (Mac) into 1 mL of sterile distilled water, due to its solubility. Various concentrations of lyophilized extracts were then prepared ranging from 0.88 to 56.6 mg/mL in 80% Ethanol for Me, and ranging from 3.54 to 28.3 mg/mL in sterile distilled water for Mac. In the case of HP extracts, samples were solubilized in the range 0.88–14.2 mg/mL in 80% Ethanol for Me HP, and in the range 3.54–28.3 mg/mL in sterile distilled water for Mac HP. Mixtures of 100 µL of the latter extract solutions and 1.9 mL of Phenol red-free DMEM were then prepared for cell deposit in 12-well plates. A 500 µL of mixture per well was deposed onto TC7-cells in triplicate. Control samples were also prepared using Phenol red-free DMEM alone, or 1.9 mL Phenol red-free DMEM plus 100 µL of one of the following solutions: distilled water, 80% Ethanol, 0.1% Triton X100, or purified quercetin at 0.156–0.625 mg/mL in 80% Ethanol.

2.3.7. Determination of In Vitro TC7-Cell Membrane Integrity and Cell Metabolic Activity

After 3 h or 22 h of exposure time with Me or Mac extracts, or control samples, the apical TC7-culture supernatants were collected on ice to determine the lactate dehydrogenase (LDH) activity, i.e., LDH-release from cytosols in cellular apical media. TC7-cells were recovered for the MTT colorimetric determination (i.e., evaluation of cellular metabolic activity or cell viability), both as described by Benzaria et al. [22] with minor modifications. LDH-leakage outside TC7-cells was determined to evaluate cellular membrane damage after exposure to extracts, as an indicative of further cell death. Apical TC7 media were collected then four-fold diluted (1/4) in Phenol red-free DMEM. A 25 µL of the latter solutions were added to 96-well plates (8 replicates for each apical diluted medium). Then, 250 µL of pH 9.3 NAD reagent (1.65 mM NAD, 165 mM KCl, 54 mM L-lactic acid, 108 mM Tris, final) was added per well. LDH induced the lactate oxidation into pyruvate with the simultaneous reduction of NAD to NADH. NADH absorbance was therefore measured in plate wells at 340 nm and 37 °C over 10 min (Multiskan Spectrum microplate reader, Thermo Electron, Vintaa, Finland). LDH activity was expressed as the difference in absorbance values taken at 0 and 10 min. Results were the means of eight absorbance determinations for each apical cell medium. A positive control was included in the series, corresponding to high LDH release by exposure of TC7-cells to Triton® X-100 in Phenol red-free DMEM (0.005% final, v/v).

TC7-cells in plate wells were then recovered for MTT-assay to evaluate cell viability after 3 h or 22 h of exposure time to Me or Mac extracts at various concentrations, or to control samples. After washing with Phenol red-free DMEM, cells were incubated for 3 h with 500 µL/well of MTT reagent (0.15 mg/mL MTT in FBS-free Phenol red-free DMEM) at 37 °C. MTT is reduced into Formazan® by a succinate dehydrogenase in living cells. After removing MTT solution, Formazan® was recovered by cell-lysing for 30 min at 37 °C using 1000 µL DMSO per well. Amounts of 100 µL were then transferred into 96-well plates to measure Formazan® absorbance at 570 nm (Multiskan Spectrum microplate reader) (8 replicates for each apical cell lysate). The cell ability to reduce MTT provides an indication of mitochondrial integrity, and therefore of cell metabolic activity or cell viability. Results were expressed as the means of 8 absorbance determinations for each apical cell lysate sample.

For each series ("3 h or 22 h of exposure time"), data were pooled from 4 independent cell culture experiments involving different TC7-cell passages. For Me and Mac extracts, 3 to 4 independent cell culture experiments were carried out on different days, and 1 to 3 for Me HP and Mac HP, each with currently 2-3 apical cell media analyzed per studied extract concentration.

2.3.8. Marcela Bioactive Compounds' Uptake

Cellular uptake of marcela compounds was determined after TC7-cell exposure to Me or Mac extracts for 3 h or 22 h, in triplicate, as described in Section 2.3.6. Me and Me HP samples were deposed at 0.088–0.354 mg/mL extract onto TC7-cells; Mac and Mac HP samples, at 0.71–1.42 mg/mL extract. After cell incubation, the apical culture media were taken off and the plate-wells washed with DPBS+. TC7-cells were scratched with 500 μL of cold acidified methanol (methanol with 0.1% *v/v* acetic acid) and transferred into Eppendorf tubes for centrifugation at 10,000 rpm and 4 °C for 5 min. Methanolic supernatants of centrifugation were kept in brown vials at 4 °C until further analysis by RP-UHPLC as described in Section 2.3.3. Prior to sample injection (20 μL) in RP-UHPLC, supernatants were dried at 40 °C in a conventional stove and re-suspended in 250 μL of methanol.

2.3.9. Statistical Analysis

All experiments were performed in triplicate and cell studies were performed at least in three different passages. The statistical analysis was established by analysis of variance (ANOVA). Results were expressed as means $\pm$ standard deviation (SD) ($n = 3$). Tukey test was applied to determine significant differences between values ($p < 0.05$) using Infostat v. 2015 program. Different letters state significant differences when $p < 0.05$.

In the case of cell studies, the statistical analysis was carried out using Sigma Plot vs. 11.0 program: the pooled data were analyzed by one-way analysis of variance (ANOVA) with all pairwise multiple comparison procedure (Holm–Sidak test) and an overall significance level of 0.05.

3. Results and Discussion

3.1. Chemical Composition

Aqueous and ethanolic extracts resulted in a yield of 6.9 and 3.1% *w/w*, respectively. The results of the proximate analysis showed in marcela powder (raw material) high fiber and carbohydrates contents. Marcela powder contained per 100 g: 4.77 $\pm$ 0.02 g proteins, 4.52 $\pm$ 0.18 g lipids, 21.01 $\pm$ 0.79 g carbohydrates (without fiber), 57.22 $\pm$ 0.73 g fiber, 4.94 $\pm$ 0.05 g ashes, and 7.54 $\pm$ 0.13 g moisture.

As to RP-HPLC (Figure 1A,B), and RP-UHPLC results (Figure 1C,D), the extracts presented a typical chromatogram as previously reported by De Souza et al. [6] in *Achyrocline satureioides* preparations. According to De Souza et al. [6] the three prominent peaks correspond to quercetin, luteolin and 3-*O*-methylquercetin, in order of appearance in the chromatogram (Figure 1B). In the present study, retention times (RT) were lower because of using a higher flow (1 mL/min) than 0.6 mL/min. Furthermore, the last prominent peak at RT 19 min in Me chromatogram (Figure 1B) could correspond to achyrobichalcone according to Zorzi et al. [23]. Polyphenol profile of marcela extracts in the current study are also in agreement with those reported by Martínez-Busi et al. [24].

Chromatograms at 370 nm showed that Me extract presents mainly flavonoids such as 3-*O*-methylquercetin (30% in the extract as stated in relative area, RA), being in a close proportion to quercetin (22% RA) (Figure 1B). In contrast, Mac chromatogram at 290 nm presented mainly phenolic acids (over 65% RA, Figure 1C) compared to flavonoids in which quercetin (5% RA) is in lower proportion than 3-*O*-methylquercetin (7–8% RA) as shown by Mac chromatogram at 370 nm (Figure 1A). These results agree with those reported by Polydoro et al. [25] where the aqueous extract of *Achyrocline satureioides* presented the lowest contents of quercetin, luteolin and 3-*O*-methylquercetin. Moreover, they found higher concentrations of the latter compounds in the extract with higher proportion of ethanol (80%) with similar ratio of quercetin to 3-*O*-methylquercetin. In the present study, quercetin was quantified in Mac and Me extracts by RP-HPLC obtaining 1.98 $\pm$ 0.13 and 88.9 $\pm$ 6.36 mg of quercetin/g extract, respectively. Calculating from quercetin calibration curve, 3-*O*-methylquercetin was estimated to 3.3 $\pm$ 0.3 mg and 127.1 $\pm$ 9.1 mg/g extract for Mac and Me, respectively. The Me extract in the present study showed greater quercetin content than the marcela aqueous extracts prepared by maceration and ultrasound-assisted extraction reported by Guss et al. [26].

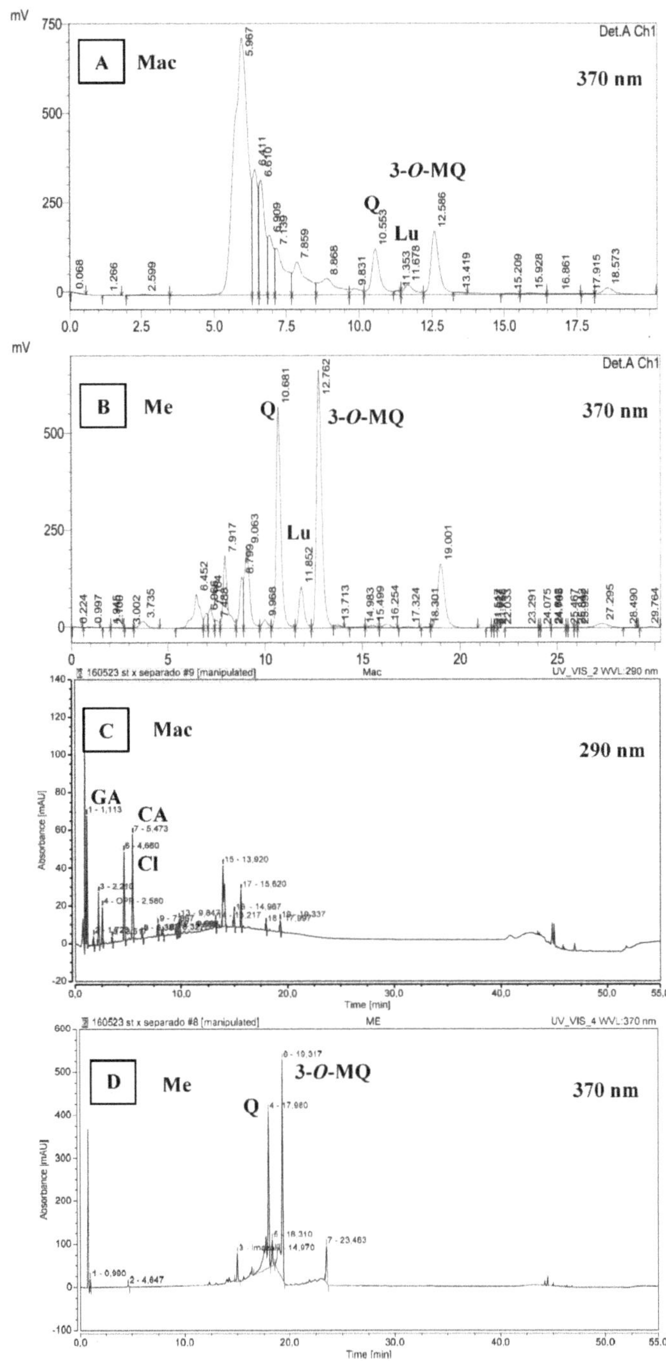

Figure 1. RP-HPLC chromatogram at 370 nm of samples aqueous extract (Mac 10 mg/mL) (**A**) and ethanolic extract (Me 1 mg/mL) (**B**). RP-UHPLC chromatogram at 290 nm of aqueous extract (Mac 2 mg/mL) (**C**) and at 370 nm of ethanolic extract (Me 1 mg/mL) (**D**). In order of appearance according to the retention times: GA, gallic acid; Cl, chlorogenic acid; CA, caffeic acid; Q, quercetin; Lu, luteolin; 3-O-MQ, 3-O-methylquercetin. [24]

In parallel, Me and Mac extracts were eluted by RP-UHPLC method (Figure 1C,D) in which, phenolic acids were identified and quantified. The composition in phenolic compounds is associated with the type of solvent used. Mac was characterized by the presence of gallic acid (RT 1.1 min), chlorogenic acid (RT 4.6 min) and caffeic acid (RT 5.4 min) with 11.9, 15.1 and 18.7% RA, respectively (Figure 1C) which corresponds to 3.76 ± 0.25 mg of gallic acid, 14.28 ± 0.01 mg of chlorogenic acid and 3.28 ± 0.62 mg of caffeic acid per g of Mac extract, in agreement with previous reports [27,28].

3.2. Antioxidant Capacity

As to total polyphenol content, aqueous extracts (Mac, Mac HP BL and Mac HP) presented lower content than ethanolic ones (Me, Me HP BL and Me HP) indicating ethanol favors polyphenols extraction (Table 1). For ABTS antioxidant capacity (Table 1), the tendency was different resulting Me as the best, followed by all the other extracts with no significant differences ($p > 0.05$). For ORAC-FL antioxidant capacity (Table 1), the highest antioxidant potential were Mac and Me, followed by ethanolic extracts Me HP BL and Me HP. When analyzing HP-assisted extracts, both aqueous and ethanolic HP extracts (Mac HP and Me HP) presented a lower polyphenol content by 3.2% and 7.6%, respectively (although non-significant, Table 1) when compared to their respective blanks (Mac HP BL and Me HP BL). Thus, high hydrostatic pressure did not significantly affect total polyphenol content nor antioxidant capacity when applied at the tested conditions to the crude extracts dispersed in phosphate buffer or 95% ethanol (Section 2.3.1).

Table 1. Results of total polyphenol content and antioxidant capacity by 2,20-azinobis-(3-ethylbenzothiazoline-6-sulfonic acid) diammonium salt (ABTS) and oxygen radical antioxidant capacity-fluorescein (ORAC-FL) methods of aqueous (Mac, Mac HP BL and Mac HP) and ethanolic (Me, Me HP BL and Me HP) extracts.

Samples	Total Polyphenol Content (mg GAE/g Extract)	ABTS (μmol TE/mg Extract)	ORAC-FL (μmol TE/mg Extract)
Mac	83.36 ± 6.69 [c]	2.71 ± 0.27 [a]	2.17 ± 0.10 [d]
Me	108.79 ± 16.47 [d]	4.19 ± 0.43 [b]	2.30 ± 0.07 [e]
Mac HP BL	44.07 ± 1.97 [a]	2.06 ± 0.17 [a]	0.34 ± 0.08 [a]
Mac HP	42.68 ± 3.55 [a]	2.30 ± 0.18 [a]	0.54 ± 0.04 [b]
Me HP BL	63.00 ± 3.12 [b]	1.84 ± 0.20 [a]	1.11 ± 0.08 [c]
Me HP	58.23 ± 4.51 [b]	1.88 ± 0.13 [a]	1.08 ± 0.07 [c]

Results are expressed as mean values $\pm$ SD ($n = 3$). ANOVA analysis was performed by column using Tukey test to state significant differences. Different letters indicate significant differences ($p < 0.05$) between values in the same column. Sample solutions were prepared in triplicate and assayed in triplicate. Marcela aqueous (Mac) and marcela ethanolic (Me) extracts. Marcela aqueous high pressure-assisted (Mac HP) and marcela ethanolic high pressure-assisted (Me HP) extracts. Blank of marcela aqueous HP (Mac HP BL) and marcela ethanolic HP (Me HP BL) extracts (Section 2.3.1).

Compared to other studies, these extracts presented similar total polyphenol content to the marcela extracts reported by Ferraro et al. [29] (23.0-112.6 mg GAE/g) with the highest polyphenol content, observing a correlation with antioxidant capacity determined by DPPH. In contrast, Guss et al. [26] reported greater polyphenol content and ABTS antioxidant capacity (i.e., lower IC_{50} value) of marcela maceration and ultrasound-assisted ethanolic extracts. In the current work, Me IC_{50} value was of 354 ± 25 μg/mL compared to 21.8 ± 0.8 and 21.3 ± 0.4 μg/mL for marcela maceration and ultrasound-assisted ethanolic extracts [26]. Marcela ethanolic extract (Me) presented higher total polyphenol content when compared to other medicinal herbs such as *Mentha x piperita* L., *Peumus boldus* Mol. and *Baccharis trimera* Iless. aqueous and ethanolic extracts [30], as well as antioxidant capacity. In contrast with Irazusta et al. [30] results, marcela aqueous extract showed lower antioxidant capacity than ethanolic extract. Antioxidant capacity of crude methanolic extracts of native Australian mint and common spearmint showed 398.5 ± 19.3 and 403.5 ± 14.8 μmol TE/g extract for ABTS, and 1727.2 ± 183.5 and 1551.1 ± 137.4 μmol TE/g extract for ORAC-FL, respectively [31], presenting lower antioxidant capacity than Mac and Me extracts in the present study (Table 1).

3.3. Antidiabetic Activities

α-amylase and α-glucosidase inhibition capacities were assessed (Figure 2A,B) as a strategy for post-prandial plasma glucose level regulation through delay/inhibition of complex carbohydrates hydrolysis during digestion, such as starch, causing lower glucose absorption [14]. For α-amylase inhibition (Figure 2A), acarbose and quercetin presented the highest inhibitions with IC_{50} values of 34.1 ± 0.8 and 2.4 ± 0.2 µg/mL, respectively. As to the extracts, aqueous extracts presented very low inhibition at the tested concentrations (up to 25 mg/mL, data not shown), in contrast with ethanolic extracts which showed IC_{50} values of 515 ± 44 (Me), 2900 ± 51 (Me HP BL) and 7974 ± 422 µg/mL (Me HP), demonstrating HP negatively affects α-amylase inhibition capacity. Moreover, quercetin seems to be one of the responsible for ethanolic extracts inhibition capacity, because of being one of the main compounds present in the latter extracts.

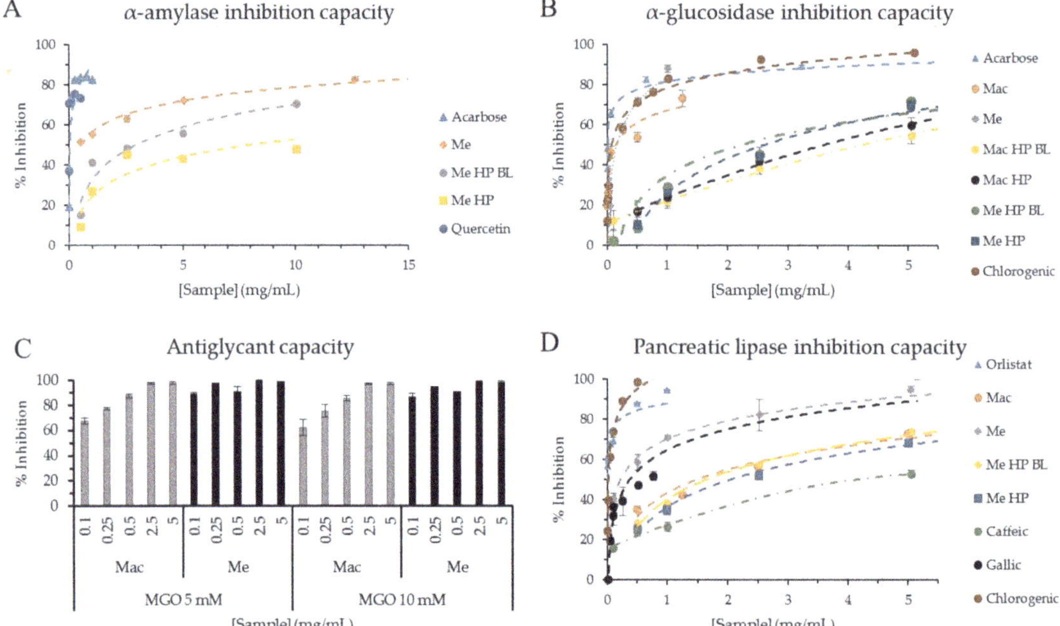

Figure 2. (**A,B**) Dose-response curves of α-amylase (**A**) and α-glucosidase (**B**) inhibition capacities expressed as % inhibition vs. sample concentration (mg/mL). (**C**) Inhibition capacity (%) of fluorescent AGEs formation with methylglyoxal (MGO) at 5 or 10 mM by different sample concentrations (mg/mL). (**D**) Dose-response curves of pancreatic lipase inhibition capacity expressed as % inhibition vs. sample concentration (mg/mL). Samples are: marcela aqueous (Mac) and ethanolic (Me) extracts, marcela aqueous HP (Mac HP) and ethanolic HP (Me HP) extracts. Blank of marcela aqueous HP (Mac HP BL) and ethanolic HP (Me HP BL) extracts. Acarbose was used as α-amylase and α-glucosidase inhibitory agent (**A,B**). Orlistat was used as lipase inhibitory agent (**D**). Quercetin, caffeic acid, gallic acid and chlorogenic acid were used as standards (**A,B,D**).

For α-glucosidase inhibition (Figure 2B), acarbose ($IC_{50} = 4.0 \pm 0.3$ µg/mL) and chlorogenic acid ($IC_{50} = 69.1 \pm 1.6$ µg/mL) presented the highest inhibition capacity (lowest IC_{50} value). Mac extract presented an IC_{50} value of 150.8 ± 54.0 µg/mL, and 157.6 ± 23.3 µg/mL was found for Me extract with no significant differences ($p > 0.05$). For Mac HP and its blank (Mac HP BL), IC_{50} values were 2973.1 ± 403.2 and 5392.0 ± 437.1 µg/mL respectively (significant difference for $p < 0.05$), stating bioactive compound extraction was favored by high hydrostatic pressure. In the case of Me HP and its blank (Me HP BL), IC_{50} values were 2587.3 ± 214.5 and 2211.1 ± 196.0 µg/mL, respectively, negatively affecting bioactivity by HP but with no significant differences ($p > 0.05$).

In accordance with the present work, quercetin has shown to possess more α-amylase inhibition capacity than acarbose [32]. Furthermore, other extracts from medicinal plants/herbs (Vietnamese and Amazonian plants, *Agrimonia asiatica*, species of *Myrcia* genus and *Euphorbia hirta* herbs) possessing phenolic acids (e.g., gallic acid) and flavonoids (quercetin and/or quercetin derivatives, and luteolin) like marcela extracts, have shown antidiabetic properties (α-amylase and α-glucosidase inhibition capacity) [33–36]. Particularly, *Euphorbia hirta* L. extract has shown to lower fast blood glucose level after 4 h and a significant reduction after 15 days treatment in streptozotocin-diabetic mice [36]. Guava (*Psidium guajava* L.) leaves possessing gallic, caffeic and chlorogenic acids, and quercetin, among others, have also shown antidiabetic properties [37]. The latter reports show the potential that marcela extracts could have as functional ingredients.

As another strategy for diabetes complications prevention/treatment, there is the inhibition of AGEs formation. Figure 2C showed maximum inhibition (close to 100%) of AGEs formation for Me extract at all the concentrations tested (0.1–5 mg/mL) in contrast with Mac extract that presented an increased inhibition trend with increasing concentration, although not significant ($p > 0.05$). This indicates that Me presents higher antiglycant capacity than Mac. Moreover, extracts did not present any differences when compared to methylglyoxal (MGO) at 5 and 10 mM. Inhibition capacity was not affected by MGO concentration at the tested concentrations (5 and 10 mM). Other medicinal herbs used as infusions, like marcela, have shown to inhibit AGEs formation such as *Mentha x piperita* L., *Peumus boldus Mol.* and *Baccharis trimera Iless.* [30] in a similar level as marcela extracts. Ethanolic extracts of ten common household condiments/herbs (*Allium sativum, Zingiber officinale, Thymus vulgaris, Petroselinum crispum, Murraya koenigii Spreng, Mentha piperita* L., *Curcuma longa* L., *Allium cepa* L., *Allium fistulosum* and *Coriandrum sativum* L.) showed correlation between total polyphenol content, antioxidant capacity and anti-glycant capacity [38], showing the same tendency when comparing marcela aqueous and ethanolic conventional extracts. Ethanolic extract (Me) showed higher total polyphenol content, antioxidant and anti-glycant capacity than aqueous extract (Mac).

3.4. Antiobesity Activity

Lipase inhibition capacity was determined (Figure 2D) as a strategy for post-prandial fat absorption control during digestion by delay/inhibition of triglycerides hydrolysis into free fatty acids, leading to lower fat absorption [14]. Mac extract presented an IC_{50} value of 1.471 ± 0.103 mg/mL and ethanolic extract (Me) 0.219 ± 0.028 mg/mL, the latter having no significant difference ($p > 0.05$) with Orlistat IC_{50} value (1.9 ± 0.2 µg/mL). Mac HP extract presented very little inhibition at the tested concentrations (0.1–10 mg/mL) and lower for its blank without pressure (Mac HP BL) (data not shown). Me HP extract presented an IC_{50} value of 2.025 ± 0.053 mg/mL and its blank without HP (Me HP BL) of 1.634 ± 0.038 mg/mL, the latter having no significant differences with Mac ($p > 0.05$). Considering all extracts, ethanolic extracts presented the best inhibition capacity, although HP negatively affected the inhibition capacity when compared to the blank (increased IC_{50} value of Me HP compared to Me HP BL) with significant differences ($p < 0.05$). Aqueous extracts seems to be more bioactive with applied high temperature (boiling extraction, Section 2.3.1) and for ethanolic extracts, it seems as if time was a key factor for bioactive compounds solvent extraction.

In parallel, polyphenol standards were tested finding IC_{50} values of 4.566 ± 0.231, 0.332 ± 0.032 and 0.012 ± 0.001 mg/mL for caffeic, gallic and chlorogenic acids, respectively, stating gallic and chlorogenic acids as the main responsible for Mac antiobesity activity. Chlorogenic acid presented no significant differences with Orlistat ($p > 0.05$), followed by gallic acid, with no significant differences with Me ($p > 0.05$), and by quercetin with an IC_{50} value of 1.105 ± 0.065 mg/mL (data not shown). In accordance with the present work, quercetin (25 µg/mL) has already been reported for inhibiting porcine pancreatic lipase by a 27.4% [39]. Caffeic acid presented the lowest inhibition capacity of the samples tested in the current work ($p < 0.05$).

3.5. Cell Studies

Results of cell membrane integrity through LDH-assay after 3 h of exposure time to Me extract (Figure 3A) showed a significant increase in LDH activity up to a maximum being of Triton level (positive control for cell membrane disruption), for 0.35 and 0.71 mg/mL of final extract in cell deposit media, which corresponds to 0.104 and 0.208 μM/mL quercetin in Me, respectively. From 0.71 to 2.83 mg/mL of final Me extract in cell deposit media, LDH activity decreased down to DMEM value (negative control) which could be explained by a solubility loss of Me constituents (initially soluble in 80% Ethanol) at the highest concentrations in the cell deposit medium (i.e., in DMEM) during incubation. Indeed, it was checked by absorbance measurement at 370 nm of Me deposit mixtures, as carried out before and after centrifugation (1200 rpm for 4 min, 30 °C), a decrease in absorbance by 13.6%, 21.4% and 39.4%, for the 0.71, 1.42 and 2.83 mg/mL extract concentrations, respectively, due to some precipitate formation. Such precipitate could correspond to the most hydrophobic compounds and/or marcela fibers contained in Me extract (initially soluble in 80% ethanol, but no more in DMEM, i.e., an aqueous dispersion of amino-acids, vitamins, salts and glucose). Such precipitate on cell monolayers could limit the access of harmful compounds to the cell membrane, and therefore membrane damage. We have checked that the deposit of Me extract at the highest concentrations after centrifugation to exclude insoluble compounds displayed similar LDH activity values (data not shown) than that obtained without centrifugation (Figure 3). Such precipitate was not observed for Mac extract at the studied concentrations.

For Mac extract, values of LDH activity after 3 h incubation at the tested concentrations (0.177–1.42 mg/mL in cell deposit media) were maintained below 25% of Triton® value indicating no noticeable cell-membrane damage compared to control samples (DMEM ± water or ethanol) (Figure 3A).

For Me HP sample, LDH activity increased with the extract concentration as for the non-HP processed sample but significantly less steeply, presenting a maximum at 0.71 mg/mL, close to Triton level, as for the non-HP processed Me. In contrast, Mac HP sample maintained cell membrane integrity at all tested concentrations (0.177–1.42 mg/mL in cell deposit media) such as quercetin solutions (0.026–103 μM/mL), and close to DMEM level (negative control). A lower extraction of polyphenols during HP aqueous extraction might be the reason for the maintenance of cell membrane integrity, supported by total polyphenol content and bioactivity results shown above.

Generally speaking, LDH activity was higher after 22 h than 3 h of exposure time to all extract samples, and especially in the case of Me and Me HP samples (Figure 3A,B). Me induced a marked increase in LDH activity being of Triton level from lower concentrations (0.088–0.35 mg/mL in the cell deposit medium) than after 3 h incubation, followed by a significant decrease in LDH activity at ≥1.42 mg/mL extract. As previously explained for 3 h incubation, a decrease in Me constituent solubility at the highest studied concentrations (1.42–2.83 mg/mL) in cell deposit media probably explained the observed decrease in LDH activity. Such decrease in LDH activity was associated to a positive level of cell metabolic activity as evaluated by MTT-assay (Figure 3C,D).

For Mac extract, LDH activity after 22 h incubation was ≥ to that observed after 3 h, remaining ≤31% of Triton level and showing no noticeable or little cell membrane damage. For Me HP, the cell membrane integrity loss was significantly higher than after 3 h incubation and close to that observed for the non-HP processed sample. Mac HP and quercetin standard showed low LDH activity, ≤29% of Triton level indicating no or little cell membrane damage.

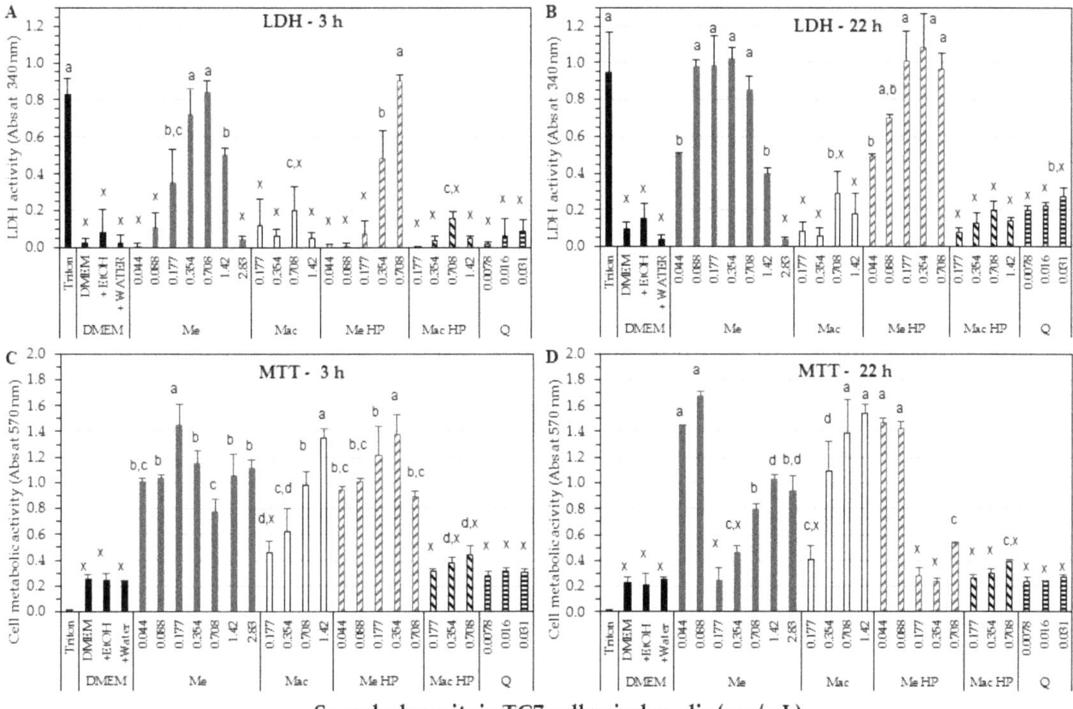

Sample deposits in TC7-cell apical media (mg/mL)

Figure 3. TC7-cell membrane integrity determined by LDH activity after 3 h (**A**) or 22 h (**B**) incubation, and TC7-cell viability assessed by MTT-assay after 3 h (**C**) or 22 h (**D**) incubation. Incubation of TC7-cells in the presence of marcela ethanolic extract (Me) or marcela aqueous extract (Mac) with or without HP-processing (HP), or purified quercetin (Q). Concentrations are expressed in mg/mL of marcela extracts or purified quercetin in the apical cell medium. Dulbecco's modified Eagle medium (DMEM) (±water or ethanol) was used as negative control and Triton as positive control. Bars and error bars represent the mean values and standard deviation, respectively. For each figure, the different letters on bars state significant differences for $p < 0.05$.

As for cell metabolic activity, MTT-assay after 3 h of exposure time (Figure 3C) showed significant increases in cell viability for Me and Mac samples comparing with control samples (DMEM ± water or ethanol), indicating some benefit effect of both extracts on TC7-cells. For Me extract, a high metabolic activity was maintained with increasing concentration with a maximum at 0.177 mg/mL extract in the cell deposit medium, corresponding to 0.052 μM/mL quercetin. For Mac extract, cell viability progressively increased with the extract concentration in cell deposit media reaching a maximal value for a higher concentration (1.42 mg/mL) compared to Me (0.177 mg/mL), suggesting different metabolic mechanisms for both extracts due to their composition. In contrast, purified quercetin deposed at 0.026 to 0.103 μM/mL (i.e., 0.0078 to 0.031 mg/mL) in apical cell media did not induced some significant improvement in cell viability compared to control samples (DMEM ± water or ethanol), and remained well below Me extract deposits containing similar quercetin amounts (0.088–0.354 mg/mL Me extract with 0.026–0.104 μM/mL quercetin). Me and Me HP extracts displayed similar MTT-profiles as a function of extract concentration, indicating no particular benefit or detrimental effect from HP-process. In the opposite, Mac HP presented the same tendency as Mac but with lower values remaining at the quercetin or DMEM level, possibly due to lower total polyphenol content (Table 1).

Higher increases in cell metabolic activity was observed (Figure 3D) after 22 h than 3 h of exposure to Me extract at its lowest studied concentrations (0.044–0.088 mg/mL in cell deposit media), but not at higher concentrations (namely 0.177–0.71 mg/mL) probably due to an excessive cell membrane damage as assessed by cellular LDH-leakage (Figure 3B). Figure 3D shows similar viability profiles for Me and Me HP samples after 22 h than 3 h of exposure, suggesting no detrimental effect from HP-process on ethanolic extraction. Higher increase in cell metabolic activity was observed after 22 h than 3 h of incubation with Mac extract in the range 0.354–1.42 mg/mL in apical cell media, due to the longer exposure time without noticeable cell membrane damage. In the case of Mac HP, the longer incubation time (22 h) did not improve the observed cell metabolic activity comparing with DMEM controls, possibly due to low polyphenolic content shown in Table 1.

Taking into account the whole results, the ethanolic extract displayed higher effects on TC7-cells than the aqueous extract at similar tested concentrations, with a dose-time dependence coupling mechanisms both inside the cells after constituent uptake (i.e., cell viability) and at the cell membrane surface (i.e., membrane integrity). It would be interesting to identify both kinds of active constituents.

Me and Me HP appears beneficial to TC7-cell metabolic activity at the lowest tested concentrations (0.044–0.177 mg/mL extract) and exposure time (3 h). However, higher concentrations (0.35–0.71 mg/mL extract) and exposure time (22 h) induced dramatic LDH-leakage. Indeed, a loss of cell-membrane integrity leads to further cell dysfunction then death. In contrast, Mac extract induced increased cell metabolic activity with increasing extract concentrations and exposure time, without noticeable loss of membrane integrity. However, the HP process led to a significant loss of its beneficial bioactive properties.

The fact that Me induced high levels of membrane damage and, simultaneously a high metabolic activity could result from a lag time between both mechanisms: membrane damage and decrease in mitochondrial activity; mitochondrial activity was still operating while the membrane started to be significantly damaged.

Purified quercetin deposed on TC7-cells at 0.026 to 0.103 µM/mL did not significantly increase cell metabolic activity as evaluated by MTT-assay, which does not highlight some prominent role of aglycone quercetin.

Previous reports of Polydoro et al. [25] showed ethanolic extracts (40 and 80% of ethanol) cytotoxicity assessed on Sertoli cells from Wistar rats at a concentration of 0.125 mg/mL with less than 80% of cell viability. Quercetin showed cytotoxicity [25] at high concentration (0.25 mg/mL or 0.827 µM/mL) which is higher than in the current study (0.026–0.103 µM/mL). Hence, the cytotoxicity observed by cell membrane disruption (LDH-assay) induced by ethanolic extracts in the present study might be displayed by other bioactive compounds than aglycone quercetin.

The RP-UHPLC analysis of TC7-cell methanolic extracts (Section 2.3.8) was carried out to detect the possible compound uptake by TC7-cells through their exposure to Me or Mac extracts compared to control DMEM. What was shown by RP-UHPLC chromatograms at 290 nm and/or 370 nm (Figure 4A–J) is outlined below.

A marked peak clearly appeared at 5.3 min on chromatograms at 290 nm, for most cell-extract samples after 3 h or 22 h incubation, included DMEM control samples. The latter peak detected at 290 nm but not at 370 nm, and that absorb in the UV 200–295 nm range with a maximum at 285 nm (Figure 4E), could correspond to aromatic aminoacids/peptides and protein material present in the living cells. A set of 5 to 6 'intermediate peaks' that could be also interpreted as cellular metabolites were detected in the 11.6–17.4 min range at 290 nm but not at 370 nm (Figure 4E,F,I,J) for some cell-extract samples included the DMEM control sample after 22 h. Consequently, looking at the chromatograms at 370 nm that exclude the latter peaks highlights the possible presence of flavonoids that absorb at 370 nm.

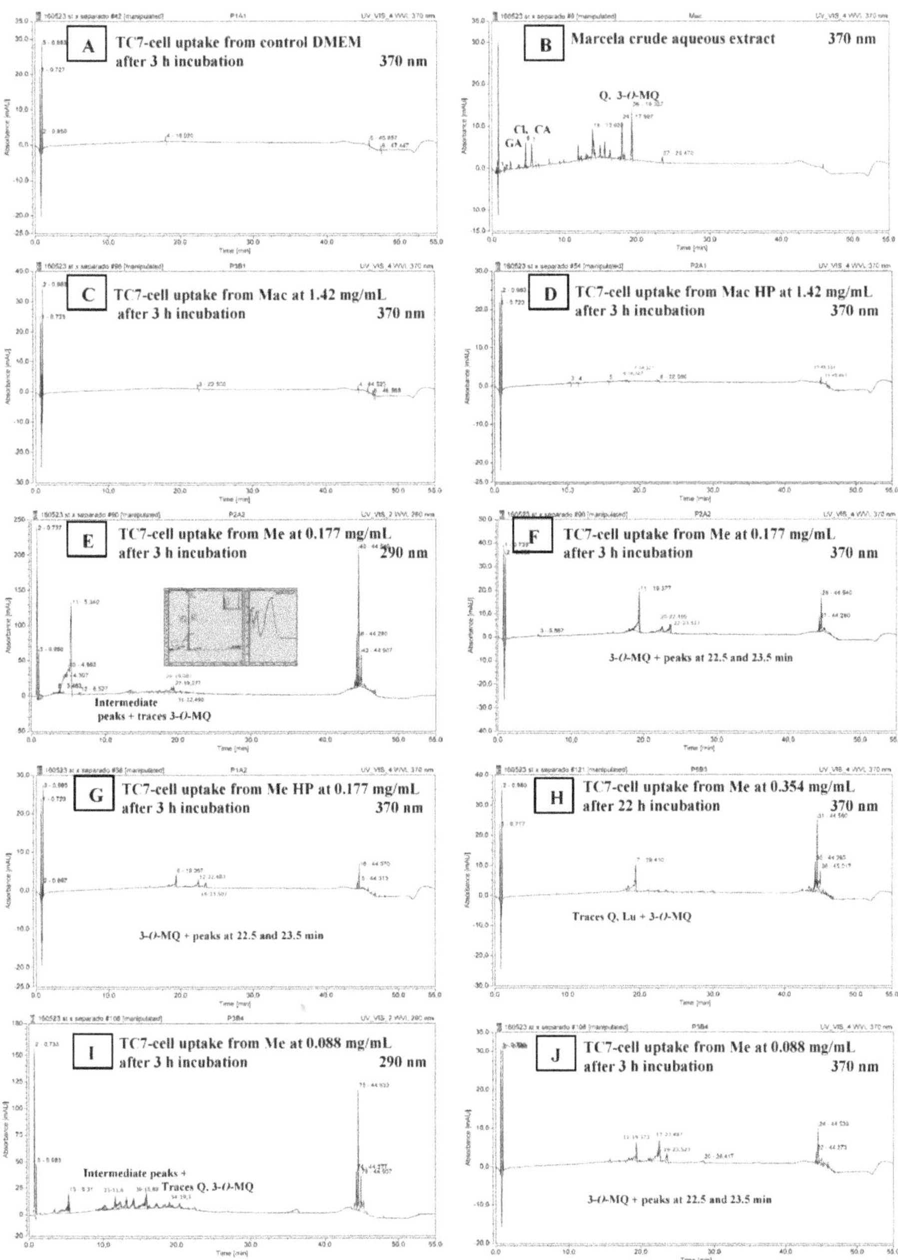

Figure 4. TC7-Cell uptake of marcela aqueous (Mac) and ethanolic (Me) extracts' compounds after 3 or 22 h of incubation. (**A**) TC7-cell methanolic extract (Section 2.3.8) for control DMEM. (**B**) Crude marcela aqueous extract (2 mg/mL) given for comparison. (**C–J**) TC7-cell methanolic extracts (Section 2.3.8) for Mac 1.42 mg/mL (**C**), Mac HP 1.42 mg/mL (**D**), Me 0.177 mg/mL and UV 200-295 nm spectrum for the 5.3 min peak allegedly amino-acids/protein material (**E**), Me 0.177 mg/mL (**F**), Me HP 0.177 mg/mL (**G**), Me 0.354 mg/mL (**H**), Me 0.088 mg/mL (**I,J**). Mac and Me concentrations in TC7-Cell apical media, incubation times, and elution wavelength are indicated. Figures can be amplified on the screen.

Chromatograms of Mac TC7-cell-methanolic extracts at 370 nm (Figure 4C,D) did not show the typical peaks corresponding to standards of gallic, chlorogenic and caffeic acids (visible at 290 nm, Figure 1C and 370 nm Figure 4B) which presented retention times of 1.1 min, 4.6 min, 5.4 min, respectively, nor the set of specific peaks visible in the crude aqueous extract in the 12-16.4 min RT range (Figure 4B).

In the opposite, the 3 peaks of Quercetin (18.0 min), Luteolin (18.3 min) and 3-*O*-methylquercetin (19.4 min) detected in Me crude extract at 290 nm (not shown) and more strongly at 370 nm (Figure 1D), were revealed in Me and Me HP cell-methanolic extracts analyzed at 370 nm comparing with the flat DMEM chromatogram (Figure 4A): indeed, traces for quercetin and luteolin at 18.1–18.4 min, plus a clear emerging peak at 19.4 min attributed to 3-*O*-methylquercetin were visible on Figure 4F–H,J. The lower ratio of quercetin to 3-*O*-methylquercetin found in Me TC7-cell-methanolic extracts comparing with the crude Me (Figure 1D) suggested that quercetin was poorly absorbed into TC7-cells, or further degraded/excreted in cell supernatants. Such a result was in accordance with the absence of visible peak on chromatograms at 370 nm, in the case of cell exposure to purified aglycone quercetin for 3 h or 22 h (not shown). It has been demonstrated that methylated flavonoids are better absorbed into Caco-2 cells and present a higher resistance to microsomal oxidation than their corresponding non-methylated aglycone forms [40,41] which could simply explain the present results. As indicative of the compound uptake by TC7-cells shown in Figure 4, and as estimated on the basis of chromatogram peak areas, the retained quercetin represented $0.44 \pm 0.08\%$, $0.24 \pm 0.01\%$ and $0.16 \pm 0.03\%$ of the quercetin present in Me deposits in the case of Figure 4F,H,J, respectively. Similarly estimated, the uptake of 3-*O*-methylquecetin represented $6.86 \pm 0.72\%$, $1.91 \pm 0.15\%$ and $4.91 \pm 0.43\%$ of the 3-*O*-methylquecetin contained in Me deposits in the case of Figure 4F,H,J, respectively. Small peaks at 22.5 and 23.5 min were also noticed after 3 h incubation (Figure 4F,G,J) suggesting the presence of newly formed derivatives at higher RT values (Figure 1B,D). More experiments are needed to achieve a quantitative evaluation of cellular uptake and thorough identification of the retained molecules.

By comparison, Mac and Mac HP samples, although deposed onto TC7-cells at higher extract concentrations (0.71–1.42 mg/mL) than Me and Me HP samples (0.088–0.354 mg/mL), revealed no peak or non-quantifiable traces on chromatograms at 370 nm in the RT range characteristic of flavonols (Figure 4C,D). These findings were in accordance with UV-Vis spectra of marcela crude extracts and TC7-cell methanolic extracts (Figure 5A,B). Indeed, while a main band (band 1) characteristic of flavonols [6,24,42] was observed at 350–365 nm in both crude Me (Figure 5A) and Me TC7-cell-methanolic extracts (Figure 5B), such a band was not found in crude Mac and Mac TC7-cell-methanolic extracts. This agrees with the fact that Mac contains low amounts of quercetin (2 mg/g aqueous extract) compared to Me (89 mg/g ethanolic extract), as well as much lower amounts of 3-*O*-methylquercetin (Figure 1). Mac TC7-cell-methanolic extracts displayed high UV absorption at 220 nm plus a broad peak with a maximum at 260–263 nm (Figure 5B). Such UV-bands also present for control DMEM could correspond to cellular material (hydrophobic amino acids/nucleic acids) solubilized by methanol during the cell-extraction step. The higher UV-light absorption in the 220–280 nm range observed for Mac and Me TC7-cell-methanolic extracts compared to control DMEM (Figure 5B) may be an indicative of enhanced TC7-cell metabolic activity induced by marcela extracts as demonstrated by MTT-assay (Figure 3C).

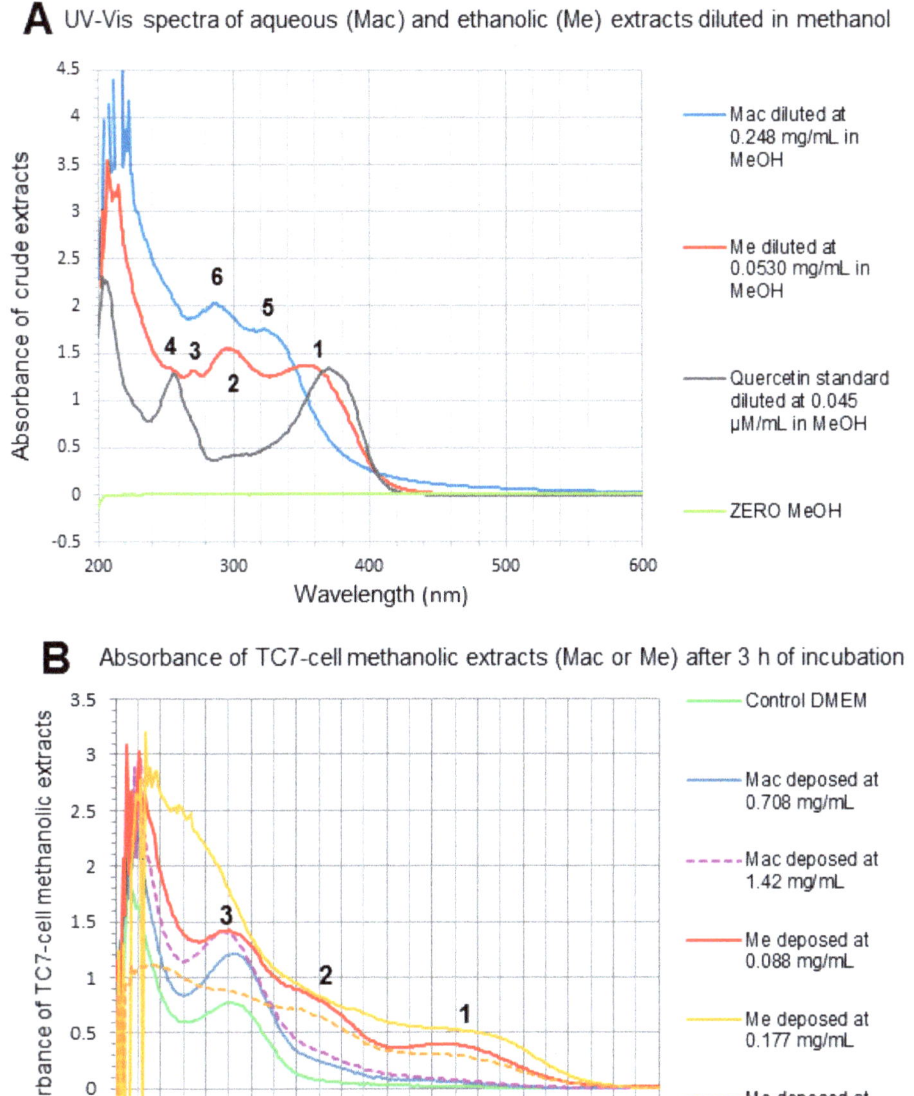

Figure 5. (**A**) UV-Vis spectra of crude aqueous extract (Mac), crude ethanolic extract (Me) and purified aglycone quercetin from Sigma, diluted in methanol at the indicated concentrations for absorbance measurement. Absorption maxima characteristic of: (1) B-ring absorption (band I) of flavonols (glycosylated Q, 3-*O*-MQ) and flavones (Lu); (2) hydroxycinnamic acid shoulder, flavanones, all phenolic compounds; (3) shoulder for most flavonols and flavones; (4) A-ring absorption (Band II) of flavonols, flavones; (5) hydroxycinnamic acids; (6) hydroxybenzoic acids and flavanols. (**B**) UV-Vis spectra of TC7-cell methanolic extracts after 3 h incubation with control DMEM, Mac or Me mixtures deposed at the indicated concentrations in TC7-cell apical media.

It is known that quercetin has highly variable and poor bioavailability, still quercetin aglycone intestinal absorption in Caco-2 cells occurs by passive diffusion and organic anion transporting polypeptide. In contrast, glycosylated form of quercetin are deglycosylated at the small intestine prior to absorption followed by quercetin metabolization through Phase II conjugation at the small intestine involving methylation, glucuronidation, and sulfation. Moreover, quercetin glucoside has been found to possess greater bioavailability due to the presence of the glucoside moiety when compared to quercetin aglycone [43].

Small intestine cell permeability for quercetin and luteolin has already been reported [40], making marcela extracts, mainly ethanolic extracts, rich in potentially bioavailable compounds. In addition, methylated flavones that show improved transport through biological membranes and increased metabolic stability compared to unmethylated flavones could present a greater oral bioavailability [40,41].

Still, bioaccessibility studies should be assessed in order to determine the stability/bioactivity of marcela bioactive compounds after digestion conditions and whether quercetin is still present in the bioaccessible fraction to be absorbed in the small intestine. Should this be the case, it would be reasonable to encapsulate the bioactive compounds into delivery systems such as liposomes or food emulsions to protect them from the extreme conditions of the gastro-intestinal tract and assess both delivery efficiency and cytotoxicity.

The incorporation of herbal extracts into traditional foods such as yogurts, cookies and meat sausages has been previously studied [44–47]; evidence suggests that food matrix and processing conditions must also be taken into account as factors that may influence bioaccessibility of the polyphenol compounds [48].

4. Conclusions

Achyrocline satureioides aqueous and ethanolic extracts presented different polyphenolic composition being characterized by phenolic acids and flavonoids, respectively. The extracts presented high polyphenol content and great antioxidant capacity determined by ABTS and ORAC-FL when compared to other medicinal plants, as well as antidiabetic (α-amylase, α-glucosidase and AGEs formation inhibition capacity) and antiobesity (pancreatic lipase inhibition capacity) activities. High hydrostatic pressure applied in the experimented conditions of pressure, pressurization duration and temperature did not prove to enhance marcela bioactive compounds extraction. Moreover, high hydrostatic pressure resulted in negative effects on some marcela bioactive properties. TC7-cell studies showed different tendencies for aqueous and ethanolic extracts as determined by LDH and MTT-assays, finding no cytotoxicity for Mac extracts at the tested concentrations (0.177–1.42 mg/mL of extract in apical cell media) compared to conventional ethanolic extracts that presented increased cell membrane disruption with increasing extract concentration. However, the lowest tested Me concentrations (0.044–0.177 mg/mL of extract in apical cell media) allowed high TC7-cell metabolic activity with limited cellular membrane damage. Cellular uptake studies revealed the presence of mainly 3-*O*-methylquercetin in Me and Me HP TC7-cell-methanolic extracts analyzed by RP-UHPLC at 370 nm, demonstrating the uptake of marcela bioactive flavonoids (mainly flavonols) into intestinal cell monolayers, in the particular case of ethanolic extracts. This suggests that marcela extracts present great potential as functional food ingredients for the prevention and/or treatment of chronic diseases.

Author Contributions: Conceptualization, A.M.F.-F. and E.D.; methodology, A.M.F.-F., E.D., A.M.-F. and T.L.-P.; formal analysis, A.M.F.-F. (chemical and biological assays), F.L. (biological assays) and I.M. (RP-UHPLC analysis); investigation, A.M.F.-F. and E.D.; data curation, A.M.F.-F. and E.D.; writing—original draft preparation, A.M.F.-F.; writing—review and editing, A.M.F.-F., E.D., H.H., P.L., T.L.-P. and A.M.-F.; supervision, E.D. and A.M.-F.; project administration, E.D., T.L.-P. and A.M.-F.; funding acquisition, E.D., T.L.-P. and A.M.-F. All authors have read and agreed to the published version of the manuscript.

Funding: This research was funded by the National Agency for Research and Innovation (Grant POS_NAC_2013_1_11655, ANII), Programa de Desarrollo de las Ciencias Básicas (PEDECIBA-

UDELAR), Comisión Sectorial de Investigación Científica (Project 2023—CSIC-i+d 2018- UDELAR), Campus France (Grant 185URYB150012), and by ECOS-Sud Committee funded project U08B01 (Procedimientos Innovadores y valorización de compuestos bioactivos destinados a la industria alimentaria, con particular atención en la industria láctea).

Conflicts of Interest: The authors declare no conflict of interest.

References

1. WHO. Noncommunicable Diseases. Available online: https://www.who.int/news-room/fact-sheets/detail/noncommunicable-diseases (accessed on 25 November 2019).
2. Arredondo, M.F.; Blasina, F.; Echeverry, C.; Morquio, A.; Ferreira, M.; Abin-Carriquiry, J.A.; Lafon, L.; Dajas, F. Cytoprotection by Achyrocline satureioides (Lam) D.C. and some of its main flavonoids against oxidative stress. *J. Ethnopharmacol.* **2004**, *91*, 13–20. [CrossRef]
3. Rivera, F.; Gervaz, E.; Sere, C.; Dajas, F. Toxicological studies of the aqueous extract from Achyrocline satureioides (Lam.) DC (Marcela). *J. Ethnopharmacol.* **2004**, *95*, 359–362. [CrossRef] [PubMed]
4. Retta, D.; Dellacassa, E.; Villamil, J.; Suárez, S.A.; Bandoni, A.L. Marcela, a promising medicinal and aromatic plant from Latin America: A review. *Ind. Crops Prod.* **2012**, *38*, 27–38. [CrossRef]
5. De Souza, P.O.; Bianchi, S.E.; Figueiró, F.; Heimfarth, L.; Moresco, K.S.; Gonçalves, R.M.; Hoppe, J.B.; Klein, C.P.; Salbego, G.; Pens Gelain, D.; et al. Anticancer activity of flavonoids isolated from Achyrocline satureioides in gliomas cell lines. *Toxicol. Vitr.* **2018**, *51*, 23–33. [CrossRef] [PubMed]
6. De Souza, K.C.B.; Schapoval, E.E.S.; Bassani, V.L. LC determination of flavonoids: Separation of quercetin, luteolin and 3-O-methylquercetin in Achyrocline satureioides preparations. *J. Pharm. Biomed. Anal.* **2002**, *28*, 771–777. [CrossRef]
7. Galanakis, C.M. Emerging technologies for the production of nutraceuticals from agricultural by-products: A viewpoint of opportunities and challenges. *Food Bioprod. Process.* **2013**, *91*, 575–579. [CrossRef]
8. Ignat, I.; Volf, I.; Popa, V.I. A critical review of methods for characterisation of polyphenolis compounds in fruits and vegetables. *Food Chem.* **2011**, *126*, 1821–1835. [CrossRef]
9. Deliza, R.; Rosenthal, A.; Abadio, F.B.D.; Silva, C.H.O.; Castillo, C. Application of high pressure technology in the fruit juice processing: Benefits perceived by consumers. *J. Food Eng.* **2005**, *67*, 241–246. [CrossRef]
10. Barba, F.J.; Zhu, Z.; Koubaa, M.; de Souza Sant'Ana, A.; Orlien, V. Green alternative methods for the extraction of antioxidant bioactive compounds from winery wastes and by-products: A review. *Trends Food Sci. Technol.* **2016**, *49*, 96–109. [CrossRef]
11. AOAC. *Official Methods of Analysis*, 16th ed.; Association of Official Analytical Chemists: Washington, DC, USA, 1999.
12. Reza, H.M.; Gias, Z.T.; Islam, P.; Sabnam, S.; Jain, P.; Hossain, M.H.; Alam, M.A. HPLC-DAD system-based phenolic content analysis and in vitro antioxidant activities of rice bran obtained from aush dhan (Oryza sativa) of Bangladesh. *J. Food Biochem.* **2015**, *39*, 462–470. [CrossRef]
13. Slinkard, K.; Singleton, V.L. Total Phenol Analysis: Automation and Comparison with Manual Methods. *Am. J. Enol. Vitic.* **1977**, *28*, 49–55.
14. Fernández-Fernández, A.M.; Iriondo-DeHond, A.; Dellacassa, E.; Medrano-Fernandez, A.; del Castillo, M.D. Assessment of antioxidant, antidiabetic, antiobesity, and anti-inflammatory properties of a Tannat winemaking by-product. *Eur. Food Res. Technol.* **2019**, *245*, 1539–1551. [CrossRef]
15. Re, R.; Pellegrini, N.; Proteggente, A.; Pannala, A.; Yang, M.; Rice-Evans, C. Antioxidant activity applying an improved ABTS radical cation decolorization assay. *Free Radic. Biol. Med.* **1999**, *26*, 1231–1237. [CrossRef]
16. Ou, B.; Hampsch-Woodill, M.; Prior, R.L. Development and Validation of an Improved Oxygen Radical Absorbance Capacity Assay Using Fluorescein as the Fluorescent Probe. *J. Agric. Food Chem.* **2001**, *49*, 4619–4626. [CrossRef] [PubMed]
17. Dávalos, A.; Bartolomé, B.; Gómez-Cordovés, C. Antioxidant properties of commercial grape juices and vinegars. *Food Chem.* **2005**, *93*, 325–330. [CrossRef]
18. Li, K.; Yao, F.; Du, J.; Deng, X.; Li, C. Persimmon Tannin Decreased the Glycemic Response through Decreasing the Digestibility of Starch and Inhibiting α-Amylase, α-Glucosidase, and Intestinal Glucose Uptake. *J. Agric. Food Chem.* **2018**, *66*, 1629–1637. [CrossRef]
19. Fernández-Fernández, A.M.; Iriondo-DeHond, A.; Nardin, T.; Larcher, R.; Dellacassa, E.; Medrano-Fernandez, A.; del Castillo, M.D. In Vitro Bioaccessibility of Extractable Compounds from Tannat Grape Skin Possessing Health Promoting Properties with Potential to Reduce the Risk of Diabetes. *Foods* **2020**, *9*, 1575. [CrossRef]
20. Starowicz, M.; Zieliński, H. Inhibition of Advanced Glycation End-Product Formation by High Antioxidant-Leveled Spices Commonly Used in European Cuisine. *Antioxidants* **2019**, *8*, 100. [CrossRef] [PubMed]
21. Benzaria, A.; Maresca, M.; Taieb, N.; Dumay, E. Interaction of curcumin with phosphocasein micelles processed or not by dynamic high-pressure. *Food Chem.* **2013**, *138*, 2327–2337. [CrossRef]
22. Benzaria, A.; Gràcia-Julià, A.; Picart-Palmade, L.; Hue, P.; Chevalier-Lucia, D.; Marti-Mestres, G.; Hodor, N.; Dumay, E. UHPH-processed O/W submicron emulsions stabilised with a lipid-based surfactant: Physicochemical characteristics and behaviour on in vitro TC7-cell monolayers and ex vivo pig's ear skin. *Colloids Surf. B Biointerfaces* **2014**, *116*, 237–246. [CrossRef] [PubMed]
23. Zorzi, G.K.; Caregnato, F.; Cláudio, J.; Moreira, F.; Ferreira Teixeira, H.; Luis, E.; Carvalho, S. Antioxidant Effect of Nanoemulsions Containing Extract of Achyrocline satureioides (Lam) D.C.—Asteraceae. *AAPS PharmSciTech* **2015**, *17*, 844–850. [CrossRef]

24. Martínez-Busi, M.; Arredondo, F.; González, D.; Echeverry, C.; Vega-Teijido, M.A.; Carvalho, D.; Rodríguez-Haralambides, A.; Rivera, F.; Dajas, F.; Abin-Carriquiry, J.A. Purification, structural elucidation, antioxidant capacity and neuroprotective potential of the main polyphenolic compounds contained in Achyrocline satureioides (Lam) D.C. (Compositae). *Bioorg. Med. Chem.* **2019**, *27*, 2579–2591. [CrossRef]

25. Polydoro, M.; de Souza, K.C.; Andrades, M.; Da Silva, E.; Bonatto, F.; Heydrich, J.; Dal-Pizzol, F.; Schapoval, E.E.; Bassani, V.; Moreira, J.C. Antioxidant, a pro-oxidant and cytotoxic effects of Achyrocline satureioides extracts. *Life Sci.* **2004**, *74*, 2815–2826. [CrossRef]

26. Guss, K.L.; Pavanni, S.; Prati, B.; Dazzi, L.; De Oliveira, J.P.; Nogueira, B.V.; Pereira, T.M.C.; Fronza, M.; Endringer, D.C.; Scherer, R. Ultrasound-assisted extraction of Achyrocline satureioides prevents contrast-induced nephropathy in mice. *Ultrason. Sonochem.* **2017**, *37*, 368–374. [CrossRef] [PubMed]

27. Ferraro, G.E.; Norbedo, C.; Coussio, J.D. Polyphenols from Achyrocline satureioides. *Phytochemistry* **1981**, *20*, 2053–2054. [CrossRef]

28. Grassi-Zampieron, R.; França, L.V.; Carollo, C.A.; do Carmo Vieira, M.; Oliveros-Bastidas, A.; de Siqueira, J.M. Comparative profiles of Achyrocline alata (Kunth) DC. and A. satureioides (Lam.) DC., Asteraceae, applying HPLC-DAD-MS. *Braz. J. Pharmacogn.* **2010**, *20*, 575–579. [CrossRef]

29. Ferraro, G.; Anesini, C.; Ouviña, A.; Retta, D.; Filip, R.; Gattuso, M.; Gattuso, S.; Hnatyszyn, O.; Bandoni, A. Total Phenolic Content and Antioxidant Activity of Extracts of Achyrocline satureoides Flowers from Different Zones in Argentina. *Lat. Am. J. Pharm.* **2008**, *27*, 626–628.

30. Irazusta, A.; Caccavello, R.; Panizzolo, L.; Gugliucci, A.; Medrano, A. The potential use of Mentha x piperita L., Peumus boldus Mol. and Baccharis trimera Iless. extracts as functional food ingredients. *Int. J. Food Nutr. Res.* **2018**, *2*, 14.

31. Tang, K.S.C.; Konczak, I.; Zhao, J. Identification and quantification of phenolics in Australian native mint (Mentha australis R. Br.). *Food Chem.* **2016**, *192*, 698–705. [CrossRef] [PubMed]

32. Nyambe-Silavwe, H.; Villa-Rodriguez, J.A.; Ifie, I.; Holmes, M.; Aydin, E.; Jensen, J.M.; Williamson, G. Inhibition of human α-amylase by dietary polyphenols. *J. Funct. Foods* **2015**, *19*, 723–732. [CrossRef]

33. Trinh, B.T.D.; Staerk, D.; Jäger, A.K. Screening for potential α-glucosidase and α-amylase inhibitory constituents from selected Vietnamese plants used to treat type 2 diabetes. *J. Ethnopharmacol.* **2016**, *186*, 189–195. [CrossRef]

34. Figueiredo-González, M.; Grosso, C.; Valentão, P.; Andrade, P.B. α-Glucosidase and α-amylase inhibitors from Myrcia spp.: A stronger alternative to acarbose? *J. Pharm. Biomed. Anal.* **2016**, *118*, 322–327. [CrossRef]

35. Kashchenko, N.I.; Olennikov, D.N. Phenolome of asian agrimony tea (agrimonia asiatica juz., rosaceae): LC-MS profile, α-glucosidase inhibitory potential and stability. *Foods* **2020**, *9*, 1348. [CrossRef]

36. Tran, N.; Tran, M.; Truong, H.; Le, L. Spray-Drying Microencapsulation of High Concentration of Bioactive Compounds Fragments from Euphorbia hirta L. Extract and Their Effect on Diabetes Mellitus. *Foods* **2020**, *9*, 881. [CrossRef] [PubMed]

37. Kumar, M.; Tomar, M.; Amarowicz, R.; Saurabh, V.; Nair, M.S.; Maheshwari, C.; Sasi, M.; Prajapati, U.; Hasan, M.; Singh, S.; et al. Guava (Psidium guajava L.) Leaves: Nutritional Composition, Phytochemical Profile, and Health-Promoting Bioactivities. *Foods* **2021**, *10*, 752. [CrossRef]

38. Ramkissoon, J.; Mahomoodally, M.; Ahmed, N.; Subratty, A. Antioxidant and anti-glycation activities correlates with phenolic composition of tropical medicinal herbs. *Asian Pac. J. Trop. Med.* **2013**, *6*, 561–569. [CrossRef]

39. Zheng, C.D.; Duan, Y.Q.; Gao, J.M.; Ruan, Z.G. Screening for Anti-lipase Properties of 37 Traditional Chinese Medicinal Herbs. *J. Chin. Med. Assoc.* **2010**, *73*, 319–324. [CrossRef]

40. Tian, X.-J.; Yang, X.-W.; Yang, X.; Wang, K. Studies of intestinal permeability of 36 flavonoids using Caco-2 cell monolayer model. *Int. J. Pharm.* **2009**, *367*, 58–64. [CrossRef]

41. Walle, T. Methylation of dietary flavones increases their metabolic stability and chemopreventive effects. *Int. J. Mol. Sci.* **2009**, *10*, 5002–5019. [CrossRef]

42. Santos, J.; Oliveira, M.B.P.P.; Ibáñez, E.; Herrero, M. Phenolic profile evolution of different ready-to-eat baby-leaf vegetables during storage. *J. Chromatogr. A* **2014**, *1327*, 118–131. [CrossRef] [PubMed]

43. Guo, Y.; Bruno, R.S. Endogenous and exogenous mediators of quercetin bioavailability. *J. Nutr. Biochem.* **2015**, *26*, 201–210. [CrossRef] [PubMed]

44. Najgebauer-Lejko, D.; Sady, M.; Grega, T.; Walczycka, M. The impact of tea supplementation on microflora, pH and antioxidant capacity of yoghurt. *Int. Dairy J.* **2011**, *21*, 568–574. [CrossRef]

45. Dabija, A.; Codină, G.G.; Ropciuc, S.; Gâtlan, A.M.; Rusu, L. Assessment of the antioxidant activity and quality attributes of yogurt enhanced with wild herbs extracts. *J. Food Qual.* **2018**, *2018*. [CrossRef]

46. Alirezalu, K.; Hesari, J.; Yaghoubi, M.; Khaneghah, A.M.; Alirezalu, A.; Pateiro, M.; Lorenzo, J.M. Combined effects of ε-polylysine and ε-polylysine nanoparticles with plant extracts on the shelf life and quality characteristics of nitrite-free frankfurter-type sausages. *Meat Sci.* **2021**, *172*, 3–5. [CrossRef] [PubMed]

47. Namal Senanayake, S.P.J. Green tea extract: Chemistry, antioxidant properties and food applications—A review. *J. Funct. Foods* **2013**, *5*, 1529–1541. [CrossRef]

48. Quirós-Sauceda, A.E.; Palafox-Carlos, H.; Sáyago-Ayerdi, S.G.; Ayala-Zavala, J.F.; Bello-Perez, L.A.; Álvarez-Parrilla, E.; De La Rosa, L.A.; González-Córdova, A.F.; González-Aguilar, G.A. Dietary fiber and phenolic compounds as functional ingredients: Interaction and possible effect after ingestion. *Food Funct.* **2014**, *5*, 1063–1072. [CrossRef] [PubMed]

Article

In Vitro Study of Two Edible Polygonoideae Plants: Phenolic Profile, Cytotoxicity, and Modulation of Keap1-Nrf2 Gene Expression

Marina Jovanović [1,*] , Dina Tenji [2] , Biljana Nikolić [3] , Tatjana Srdić-Rajić [4] , Emilija Svirčev [2] and Dragana Mitić-Ćulafić [3]

[1] Institute of General and Physical Chemistry, Studentski trg 12-14/V, 11000 Belgrade, Serbia
[2] Faculty of Sciences, University of Novi Sad, Trg Dositeja Obradovića 3, 21000 Novi Sad, Serbia; dinatenji@gmail.com (D.T.); emilija.svircev@dh.uns.ac.rs (E.S.)
[3] Faculty of Biology, University of Belgrade, Studentski trg 16, 11000 Belgrade, Serbia; biljanan@bio.bg.ac.rs (B.N.); mdragana@bio.bg.ac.rs (D.M.-Ć.)
[4] Institute of Oncology and Radiology of Serbia, Pasterova 14, 11000 Belgrade, Serbia; tsrdic@gmail.com
* Correspondence: marina.rajic.jovanovic@gmail.com; Tel.: +381-63-74-43-004

Citation: Jovanović, M.; Tenji, D.; Nikolić, B.; Srdić-Rajić, T.; Svirčev, E.; Mitić-Ćulafić, D. In Vitro Study of Two Edible Polygonoideae Plants: Phenolic Profile, Cytotoxicity, and Modulation of Keap1-Nrf2 Gene Expression. *Foods* **2021**, *10*, 811. https://doi.org/10.3390/foods10040811

Academic Editors: Francisca Rodrigues and Verica Dragović-Uzelac

Received: 26 February 2021
Accepted: 7 April 2021
Published: 9 April 2021

Publisher's Note: MDPI stays neutral with regard to jurisdictional claims in published maps and institutional affiliations.

Abstract: *Polygonum aviculare* and *Persicaria amphibia* (subfam. Polygonoideae) are used in traditional cuisines and folk medicine in various cultures. Previous studies indicated that phytochemicals obtained from Polygonoideae plants could sensitize chemoresistant cancer cells and enhance the efficacy of some cytostatics. Here, the cytotoxic properties of chemically characterized ethanol extracts obtained from *P. aviculare* and *P. amphibia*, individually and in combination with doxorubicin (D), were determined against hepatocarcinoma HepG2 cells. Phenolic composition, cell viability, cell cycle, apoptosis, and the expression of Keap1 and Nrf2 were examined by following methods: LC-MS/MS, LC-DAD-MS, MTT, flow cytometry, and qRT-PCR. Extracts were rich in dietary polyphenolics. Synergistic cytotoxicity was detected for extracts combined with D. The observed synergisms are linked to the interference with apoptosis, cell cycle, and expression of Keap1-Nrf2 genes involved in cytoprotection. The combined approach of extracts and D could emerge as a potential pathway of chemotherapy improvement.

Keywords: edible plants; Polygonoideae; phenolic profile; doxorubicin; apoptosis; cell cycle; Keap1-Nrf2 expression

1. Introduction

Widespread throughout Europe, Asia and the Americas, wild plants *Polygonum aviculare* and *Persicaria amphibia* (syn. *Polygonum amphibium*), subfamily Polygonoideae, are used in traditional cuisines and folk medicine in various cultures [1,2]. *P. aviculare*, known as the common knotweed, is edible and used as a Korean salad plant, an Australian honey plant, and a traditional Vietnam culinary herb [3–5]. In the USA, *P. amphibia*, popularly known as water smart weed, has been utilized in soft drink preparation [2]. Described as healing weeds, these plants are widely used as a home remedy to treat ailments such as stomach pains and diarrhea [1,2,6,7]. Concerning Serbia, *P. aviculare* is mainly used as an appetite stimulant [8]. Importantly, in the folk medicine of China and Austria *P. aviculare* and *P. amphibia* are employed to treat some types of cancer [9,10]. Chemical properties of these plants have been thoroughly investigated in recent decades and acquired data showed that their extracts are rich in flavonoids, sesquiterpenoids, and tannins, thus, justifying their ethnopharmacological use and contributing to their recognition in contemporary pharmacology [1,2,11,12]. Although a limited number of pharmacological studies regarding these herbs are available, some of them indicate that these plants and their active compounds could be used for the treatment of various diseases in clinical medicine, including diabetes and some types of cancer [13–15]. Furthermore, epidemiological evidence has demonstrated that a diet rich in natural bioactive compounds could decrease the risk of cancer development and could be used in chemoprevention. The discovery of plant-derived drugs has emerged as a potential pathway in the search for chemotherapeutics owing to the accepted assumption that plant medicaments are safer than their synthetic counterparts. In addition, toxic and other unfavorable effects of synthetic anticancer drugs have been widely noted [2,16,17]. Chemotherapy treatment with anthracycline drugs, such as doxorubicin (D), results in high hepatotoxicity. Apart from the numerous side effects, the medical application of D is also limited due to the frequent development of resistance in tumor cells [18,19]. Considering that D could rely on an increase in the free radical production to exhibit its effect, the reduction of antioxidant defense could initially make the cancerous cells susceptible to chemotherapeutics [20,21]. Importantly, numerous cancerous cells possess increased endogenous antioxidant defense due to the constitutive overexpression of the nuclear factor erythroid 2-related factor 2 (Nrf2) related to the disruption of Kelch-like ECH-associated protein 1 (Keap1) [19,22]. Keap1 acts as a negative regulator of Nrf2, and hence, it may act as a tumor suppressor in cancer cells. Nrf2 is the redox-sensitive transcription activator that regulates the expression of a large number of cytoprotective enzymes [23]. Thereby, Nrf2 has been proposed as a novel therapeutic target to overcome chemoresistance in various types of cancer, including hepatocellular carcinoma (HCC) [22]. Moreover, it has been observed that some phytochemicals have the potential to sensitize chemoresistant HCC through the suppression of Nrf2 [22].

Therefore, in this work, the phenolic profiles and cytotoxic properties of ethanol extracts of aerial parts of *P. aviculare* (POA) and *P. amphibia* (PEA) were explored, as well as their potential to modulate the response of human hepatocellular carcinoma cells (HepG2) to D, the most widely used cytostatic in HCC treatment [16]. The cytotoxic properties of extracts, alone and combined with D, were estimated by MTT assay and flow cytometric analysis. The potential of co-treatments of extracts and D to influence Nrf2 and Keap1 expression was assessed by qRT-PCR. Thus, this study, in a comprehensive manner, investigated cytotoxic properties of POA and PEA. Using both plant extracts and cytostatic D, it aimed to present the benefits of the combined approach in order to make an initial step in chemotherapy improvement.

2. Materials and Methods

2.1. Materials

Reference standards of the secondary metabolites used in LC-MS/MS analysis were obtained from Sigma–Aldrich Chem (Steinheim, Germany): 4-hydroxy-benzoic acid, 2,5-dihydroxybenzoic acid, vanillic acid, gallic acid, cinnamic acid, caffeic acid, trans-ferulic acid, 3,4-dimethoxycinnamic acid, D-(−)-quinic acid, umbelliferon, matairesinol, secoisolariciresinol, chlorogenic acid, predominantly trans, quercetin dihydrat, (+)-catechin hydrate, baicalein, genistein, daidzein, baicalin, syringic acid, p-coumaric acid (predominantly trans isomer), 2-hydroxycinnamic acid (predominantly trans), sinapic acid (predominantly trans isomer), scopoletin, (−)-epicatechin, quercetin-3-*O*-beta-D-glucoside, quercitrin-hydrate, (−)-epigallocatechin gallate; Roth/Carl Roth GmbH/Rotichrom®: protocatechuic acid, esculetin, apigenin, apigenin-7-*O*-glucoside, hyperoside, chrysoeriol, amentoflavone trihydrate, apiin; Chromadex (Santa Ana, CA, USA): kaempherol, kaempferol 3-*O*-glucoside, naringenin, isorhamnetin; Extrasynthese Genay Cedex France: luteolin, luteolin-7-*O*-glucoside, and from Fluka Chemie GmbH (Buchs, Switzerland): myricetin, vitexin, rutin, trihydrate. HPLC gradient-grade methanol was purchased from J. T. Baker (Deventer, The Netherlands) and p.a. formic acid from Merck (Darmstadt, Germany). Folin and Ciocalteu's Phenol Reagent (FC) was provided by Sigma–Aldrich, while sodium carbonate and aluminium(III) chloride were purchased from Centrohem (Stara Pazova, Serbia) and Kemika (Zagreb, Croatia), respectively. William's medium, fetal bovine serum (FBS), penicillin-streptomycin mixture, phosphate-buffered saline (PBS), trypsin from porcine pancreas, dimethyl sulfoxide (DMSO), protease inhibitor cocktails, Triton® X-100, and

3-(4,5-dimethylthiazol-2-yl)-2,5-diphenyltetrazolium bromide (MTT) were purchased from Sigma-Aldrich (Steinheim, Germany). Reagents for apoptosis and cell cycle assay were obtained from Invitrogen Life Technologies^{TM} (FITC-AnexinV, Binding puffers 2x, Rnase A Pure Link^{TM}, Waltham, MA, USA) and 7- amino actinomycin D was provided from Pharmingen^{TM} (Franklin Lakes, NJ, USA). Trizol reagent, Power SYBR green PCR master mix and specific primers for qRT-PCR were obtained from Invitrogen Life Technologies^{TM} (Carlsbad, CA, USA). Doxorubicin (D, Cas. No. 25316-40-9) was provided by Actavis, S.C. Sindan-Pharma S.R.L. (Bucureşti, Romania). All the other chemicals and reagents were purchased from local companies and were of a molecular biology grade.

2.2. Plant Material, Extracts Preparation, and Chemical Analysis

Aerial parts of *P. aviculare* and *P. amphibia* were collected at Vlasina Lake (N42°42′40.09″ E22°20′32.942″) in Serbia. Plant materials were identified and the voucher specimens were deposited at the Herbarium of Department of Biology and Ecology, Faculty of Natural Sciences, University of Novi Sad, Serbia (BUNS Herbarium; voucher numbers for *P. aviculare* and *P. amphibia* are 2-1669 and 2-1691, respectively).

Extracts were prepared by the maceration of air dried and powdered aboveground plant material (10 g) with 80% ethanol (100 mL) for 72 h under constant stirring at room temperature. Extracts were removed from plant material by filtration and after vacuum drying, yield of dry raw extracts were: 1.31 g/13.1% (*P. aviculare*) and 1.32 g/13.2% (*P. amphibia*). Raw extracts were suspended in water and purified by liquid–liquid extraction with petroleum ether, to remove chlorophyll and other ballasts. Defatting of the extracts with petroleum ether can lead to losses of the compounds of interest [24,25], so the petroleum ether layer was washed with methanol and methanol fraction pooled with water layer. After purification, herb extracts yield decreased for 1.1% (1.20 g; *P. aviculare*) and for 0.8% (1.23 g; *P. amphibia*). Vacuum dried (<45 °C) purified extracts were dissolved in DMSO to a final concentration of 100 mg/mL.

The phytochemical profiles of *P. aviculare* and *P. amphibia* were evaluated by measuring the total phenolic content (by means of Folin–Ciocalteu reagent under alkaline conditions) and total flavonoids content (based on aluminum–flavonoids complex formation). Detailed procedure of these two spectrophotometric methods was previously published by Beara et al. [26]. Quantitative LC-MS/MS analysis of selected 45 secondary metabolites was carried out according to the previously reported method [27]. Standard mixture (containing 45 phenolics) was double diluted with mobile phase solvents: A (0.05% aqueous formic acid): B (methanol), in 1:1 ratio, to obtain fifteen working standards (from 25,000 ng/mL to 1.53 ng/mL). Extracts were diluted also with solvents A:B (1:1) to a final concentration of 2 mg/mL. Samples and standards were analyzed using Agilent Technologies 1200 Series high-performance liquid chromatograph coupled with Agilent Technologies 6410A Triple Quad tandem mass spectrometer with electrospray ion source (ESI), and controlled by Agilent Technologies MassHunter Workstation software—Data Acquisition (ver. B.03.01). The detailed procedure and method validation were published previously [27] (Supplementary material).

Qualitative LC-DAD-MS analysis of extracts was performed on Agilent Technologies 1200 Series HPLC with DAD, coupled with Agilent Technologies 6410A Triple Quad tandem mass spectrometer with electrospray ion source, and controlled by Agilent Technologies MassHunter Workstation software—Data Acquisition (ver. B.03.01). Working solution of standard mixture-45 (1.56 µg/mL) and 5 µL of extracts (20 mg/mL diluted with A:B (1:1)) were injected into the system, with Zorbax Eclipse XDB-C18 (50 mm, 4.6 mm, 1.8 µm) rapid resolution column held at 50 °C. Mobile phase A (0.05% aqueous formic acid) and B (methanol) was delivered at flow rate of 0.8 mL/min in a gradient mode (0 min 20% B, 6.67 min 60% B, 8.33 min 100% B, 12.5 min 100% B, re-equilibration time 4 min). Eluted components were firstly recorded on diode array detector (DAD), full spectra in 190–700 nm range, chromatograms were acquired at 254 nm, 340 nm and 430 nm; and secondly on triple quadrupole mass spectrometer, using MS2Scan run mode (both, positive and negative ionization, m/z range of 120–1000 and fragmentor voltage of 80 V). ESI ion source parameters were as follows: nebulization gas (N_2) pressure 40 psi, drying gas (N_2) flow 9 L/min and temperature 350 °C, capillary voltage 4 kV.

2.3. Human Cell Line

The human cell line used in this study was hepatocellular carcinoma HepG2 (ATCC HB-8065, Manassas, VA, USA). HepG2 cells were grown in William's medium, with 15% fetal bovine serum, 1% penicillin/streptomycin, and 2 mM of L-glutamine. The cell line was maintained in an incubator at 37 °C with 5.0% CO_2 in a humidified atmosphere. The cells were sub-cultured at 90% confluence, twice a week, using 0.1% trypsin. Cell viability was determined by the trypan blue dye exclusion method. Cells in the logarithmic growth phase were used in all experiments.

2.4. Cytotoxicity and Drug Synergism Analysis

The cytotoxic effects of plant extracts and D, both as single compounds and in a mixture, were assessed by MTT assay, as described by Jovanović et al. [28]. HepG2 cells were seeded into 96-well plates at a density 2×10^4 cells/well and incubated overnight with 5% CO_2 at 37 °C. Further on, the cells were exposed to a series of two-fold dilutions of extracts and D in the ranges 4000–125 μg/mL and 22.8–0.712 μg/mL, respectively. To prepare mixtures of extracts and D, the highest concentrations of each substance were combined and subsequently diluted two-fold. This process was repeated until reaching 125 μg/mL and 0.712 μg/mL of extracts and D, respectively. After the incubation for 24 h, the medium with test substances was replaced with MTT (final concentration 0.5 mg/mL) and incubated for additional 3 h. At the end of incubation with MTT, the medium was removed, and the formazan crystals were dissolved in DMSO. The optical density was measured at 570 nm, using a micro-plate reading spectrophotometer (Multiskan FC, Thermo Scientific, Shanghai, China). Three independent experiments were conducted.

To evaluate the nature of interaction between extracts and D, combination index (CI) analysis was used, providing quantitative definition for the additive effect (CI = 1), synergism (CI < 1), and antagonism (CI > 1) in drug combinations [29]. The CI was calculated for IC_{25} and IC_{50} values of the mixtures, using the formula: $CI = C_A/IC_A + C_B/IC_B$, where C_A is the concentration of the first test substance in the binary mixture; IC_A is the concentration of the first test substance alone; C_B is the concentration of the second test substance in the binary mixture; and IC_B is the concentration of the second test substance alone.

2.5. Flow Cytometry Analysis of Apoptosis and Cell Cycle Phase Distribution

Apoptotic cell death and analysis of the cell cycle phase distribution were analyzed using a fluorescence-activated cell sorting flow cytometer (FACS) (Calibur Becton Dickinson, Heidelberg, Germany) and Cell Quest computer software, according to manufacturer's protocol. HepG2 (1×10^6 cells/well) was cultured with plant extracts with and without D. Concentrations of tested substances were selected in accordance with the results of the MTT assay. IC_{50} values of extracts and D, individually and combined, were tested. Apoptotic or necrotic cell death was assessed after the 24 h treatment. As described by Srdic-Rajic et al. [30], cells were harvested, washed with PBS, and stained with Annexin V FITC and 7- amino actinomycin D (7-AAD). In brief, Annexin V FITC binds to the exposed phosphatidylserine of the early apoptotic cells, whereas 7-AAD labels the late apoptotic/necrotic cells, containing damaged membrane. The numbers of viable (annexin V FITC $^-$ 7AAD$^-$), early apoptotic (annexin V FITC$^+$ 7AAD$^-$), and late apoptotic/necrotic (annexin V FITC$^+$ 7AAD$^+$) cells were determined.

The quantitative analysis of the proportion of cells in different cell cycle phases was performed after the treatment and incubation for 24 h. Cells were harvested and fixed with ice-cold 70% ethanol at −20 °C for 30 min. Subsequently, cells were resuspended in PBS containing propidium iodide and RNase A and incubated for 30 min at room temperature. The distribution of the cells was measured by FACS analysis, as previously described.

2.6. Real-Time Quantitative PCR (qRT-PCR) Analysis

In order to detect the expression pattern of Keap1 and Nrf2 genes in HepG2 cells, qRT-PCR analysis was conducted as described in Kaisarevic et al. [31], with minor modifications. For the experiment, the cells were seeded into 12-well plate (10^6 cells/well) and, after 24 h, exposed in duplicates to the selected concentrations of combined extracts and D. The selected concentrations for this assay were the ones that induced 25% inhibitions of cell survival (IC_{25}), considering that the test procedure requires high cell viability. After the 24 h treatment, the medium was removed, the cells were washed by PBS, and total RNA was extracted using trizol reagent according to supplier's instructions. The quality and quantity of RNA was determined spectrophotometrically by BioSpecnano (Schimadzu Corporation, Kyoto, Japan). Reverse transcription of each total RNA sample (2 µg) to cDNA was conducted using High-Capacity cDNA Reverse Transcription Kit with RNase inhibitor (Applied Biosystems). The reverse transcription reaction was conducted in the Veriti Thermal Cycler (Applied Biosystems), under the following incubation conditions: 10 min at 25 °C, 120 min at 37 °C, and 5 min at 85 °C. The expression level of Keap1 and Nrf2 were quantified by qPCR, which was conducted on Mastercycler® ep realplex (Eppendorf, Germany). Each PCR system contained cDNA (15 ng) and 500 nM of specific primers for the target mRNA, and the reaction was catalyzed by Power SYBR Green PCR Master Mix, according to the manufacturer's instruction. Cycling conditions were as follows: 50 °C for 2 min, 95 °C for 10 min, 40 cycles of 95 °C for 15 s, and 60 °C for 1 min. Keap1 and Nrf2 expression was detected with the amplification by 40 cycles. The following primers were used: 5′-GACAGCCTCTGACAACACAAC-3′ (forward for Keap1), 5′-GAAATCAAAGAACCTGTGGC-3′ (reverse for Keap1); 5′-CCTCAACTATAGCGATGCTGAATCT-3′ (forward for Nrf2), 5′-AGGAGTTGGGCATGAG TGAGTAG-3′ (reverse for Nrf2); 5′-AGAGCTACGAGCTGCCTGAC-3′ (forward for β-actin), 5′-AGCACTGTGTTGGCGTACAG-3′ (reverse for β-actin). Data were analyzed by GraphPad Prism software with β-actin as a reference gene, and its expression was not altered by any of the treatments. The relative expression levels of each target were calculated based on the cycle threshold (Ct) method, as described by Voelker et al. [32].

2.7. Statistical Analysis

The values obtained from the following tests, MTT assay, apoptosis, cell cycle and qRT-PCR, were analyzed by analysis of variance (One-way ANOVA, Dunnett's multiple comparisons test) using GraphPad Prism 6.0 (GraphPad Software Inc. San Diego, CA, USA). The level of statistical significance was defined as $p \leq 0.05$. To describe the type of pharmacokinetic interactions between extracts and D, the combination index (CI) was calculated, and data from MTT assay were employed. The values of CI being lower, equal, or higher than 1 (CI < 1, CI = 1, CI > 1) indicated the synergistic, additive, and antagonistic effect, respectively.

3. Results

3.1. Identification of Compounds in the Extracts

The results of spectrophotometric measurement of total phenolics and flavonoids content of the extracts were expressed as equivalents of gallic acid per g of dry extract (eq GA/g DE) and equivalents of quercetin per g of dry extract (eq Querc/g DE), respectively. They were determined to be 282.8 ± 73 mg eq GA/g DE and 306.9 ± 43 mg eq GA/1g DE, and 28.9 ± 0.5 mg eq Querc/1g DE and 38.5 ± 2.0 mg eq Querc/1g DE, for POA and PEA, respectively. The comparison of data concerning phenolics content of POA and PEA is presented in Table 1. The results of the LC-MS/MS analysis (Figure 1) showed that both extracts are rich in phenolic acids and flavonoids. POA is rich in quinic acid (8.72 mg/g DE), kaempherol-3-*O*-glucoside (1.33 mg/g DE), quercetin-3-*O*-glucoside (1.38 mg/g DE), and quercetin-3-*O*-galactoside (3.02 mg/g DE). PEA is characterized by a high content of aglycone, such as quercetin (5.50 mg/g DE) and a high content of quercetin derivatives: quercetin-3-*O*-galactoside (11.90 mg/g DE), quercetin-3-*O*-L-rhamnoside (9.79 mg/g DE), and quercetin-3-*O*-glucoside (1.49 mg/g DE). PEA is also rich in free gallic acid (3.49 mg/g DE) and epigallocatechin gallate (1.28 mg/g DE).

Table 1. Concentrations of phenolics found in *Polygonum aviculare* (POA) and *Persicaria amphibia* (PEA) ethanol extracts (expressed as µg of phenolics per gram of dry extract).

Class of Secondary Metabolites	Compound	No [a]	Rt [b] [min]	LoQ [c] [µg/g de]	Content [µg/g dw] [d] POA	PEA
Cyclohexanecarboxylic acids	Quinic acid	1	0.52	5.0	$(8.7 \pm 0.9)\,10^3$	$(8.8 \pm 0.9)\,10^2$
Hydroxybenzoic acids	Gallic acid	2	0.58	10	$(9.5 \pm 0.8)\,10^2$	$(3.5 \pm 0.3)\,10^3$
	Protocatechuic acid	3	0.79	2.0	$(2.3 \pm 0.2)\,10^2$	$(1.9 \pm 0.1)\,10^1$
	2,5-dihydroxybenzoic acid	5	1.03	3.5	$(3.0 \pm 0.2)\,10^1$	<LoQ
	p-Hydroxybenzoic acid	8	1.08	4.0	$(3.1 \pm 0.2)\,10^1$	$(4.8 \pm 0.3)\,10^1$
	Vanillic acid	12	1.24	50	$(2.6 \pm 0.8)\,10^1$	$(0.3 \pm 0.1)\,10^2$
	Syringic acid	13	1.31	20	$(1.5 \pm 0.3)\,10^2$	$(1.1 \pm 0.2)\,10^2$
Phenylpropanoids	Cinnamic acid	36	3.91	40	<LoQ	<LoQ
Hydroxycinnamic acids	Caffeic acid	11	1.18	3.0	$(2.2 \pm 0.2)\,10^1$	$(6.2 \pm 0.4)\,10^1$
	p-Coumaric acid	14	1.69	2.0	$(5.1 \pm 0.4)\,10^1$	$(3.7 \pm 0.3)\,10^1$
	Ferulic acid	17	1.90	5.0	$(2.8 \pm 0.3)\,10^1$	$(5.0 \pm 0.5)\,10^1$
	Sinapic acid	18	1.92	20	$(1.6 \pm 0.2)\,10^1$	$(3.9 \pm 0.4)\,10^1$
	o-coumaric acid	24	2.62	3.0	1.3 ± 0.1	<LoQ
	3,4-dimethoxycinnamic acid	31	2.99	25	<LoQ	<LoQ
Chlorogenic acids	5-*O*-caffeoylquinic acid	6	0.80	3.5	$(6.9 \pm 0.3)\,10^2$	11.0 ± 0.5
Flavan-3-ols	Catechin	4	0.74	25	$(7.0 \pm 0.7)\,10^2$	$(6.5 \pm 0.6)\,10^2$
	Epicatechin	10	0.95	30	$(8.8 \pm 0.9)\,10^1$	$(2.3 \pm 0.2)\,10^2$
Flavan-3-ol-derivatives	Epigallocatechin gallate	7	0.81	50	$(1.5 \pm 0.1)\,10^2$	$(1.3 \pm 0.1)\,10^3$
Coumarins	Esculetin	9	1.13	3.0	5.1 ± 0.3	9.1 ± 0.5
	Umbelliferone	15	1.73	5.0	<LoQ	<LoQ
	Scopoletin	16	1.77	3.5	1.0 ± 0.1	11.2 ± 0.9
Flavone glycosides	Luteolin-7-*O*-glucoside	20	2.13	2.5	$(11.0 \pm 0.3)\,10^{-1}$	3.1 ± 0.1
	Vitexin	19	1.90	2.0	3.2 ± 0.2	<LoQ
	Apiin	25	2.60	1.5	<LoQ	<LoQ
	Apigenin-7-*O*-glucoside	26	2.81	3.0	<LoQ	$(6.0 \pm 0.3)\,10^{-1}$
	Baicalin	32	3.4	10	<LoQ	<LoQ
Flavones	Luteolin	38	4.03	2.0	6.4 ± 0.3	12.6 ± 0.6
	Apigenin	39	4.71	5.0	$(8.5 \pm 0.6)\,10^1$	$(4.0 \pm 0.3)\,10^1$
	Baicalein	41	5.15	15	<LoQ	<LoQ

Table 1. *Cont.*

Class of Secondary Metabolites	Compound	No [a]	Rt [b] [min]	LoQ [c] [µg/g de]	Content [µg/g dw] [d]	
					POA	PEA
	Chrysoeriol	43	4.82	2.0	$(5.0 \pm 0.1) \, 10^{-1}$	$(8.0 \pm 0.4) \, 10^{-1}$
Biflavonoid	Amentoflavone	45	5.78	2.5	<LoQ	<LoQ
Flavonol-glycosides	Quercetin-3-*O*-galactoside	21	2.16	3.0	$(3.0 \pm 0.2) \, 10^3$	$(11.9 \pm 0.7) \, 10^3$
	Quercetin-3-*O*-rutinoside	22	2.33	1.5	$(26.2 \pm 0.8) \, 10^1$	$(20.5 \pm 0.6) \, 10^1$
	Quercetin-3-*O*-glucoside	23	2.25	2.0	$(13.8 \pm 0.4) \, 10^2$	$(14.9 \pm 0.4) \, 10^2$
	Quercetin-3-*O*-L-rhamnoside	28	2.75	1.5	$(1.6 \pm 0.1) \, 10^2$	$(9.8 \pm 0.6) \, 10^3$
	Kaempferol-3-*O*-glucoside	30	2.8	2.0	$(13.3 \pm 0.5) \, 10^2$	$(2.8 \pm 0.1) \, 10^1$
Flavonols	Myricetin	27	2.67	50	$(11.1 \pm 0.8) \, 10^1$	$(8.6 \pm 0.6) \, 10^2$
	Quercetin	35	3.74	50	$(0.4 \pm 0.1) \, 10^3$	$(5.5 \pm 1.6) \, 10^3$
	Kaempferol	40	4.55	3.0	$(12.0 \pm 0.8) \, 10^1$	$(13.4 \pm 0.9) \, 10^1$
	Isorhamnetin	42	4.79	10	5.1 ± 0.3	8.6 ± 0.5
Flavanones	Naringenin	37	3.87	3.5	6.5 ± 0.4	$(1.6 \pm 0.1) \, 10^1$
Isoflavones	Daidzein	33	3.43	5.0	<LoQ	<LoQ
	Genistein	44	4.12	3.0	<LoQ	<LoQ
Lignans	Secoisolariciresinol	29	2.90	25	<LoQ	<LoQ
	Matairesinol	34	3.66	50	<LoQ	<LoQ
TOTAL					18,782 (1.88%)	37,113 (3.71%)

[a] Numbers are used as labels on given chromatograms bellow. [b] From the method validation published in Orčić et al. [27]. [c] Calculated from the instrument quantification limit (Orčić et al. [27]) and sample dilution. [d] Results are given as the concentration (µg/g of dry extract) ± relative standard deviation of repeatability (as determined by method validation [27]). LoQ–limit of quantitation; the standard curves were provided in Supporting materials (Figure S1).

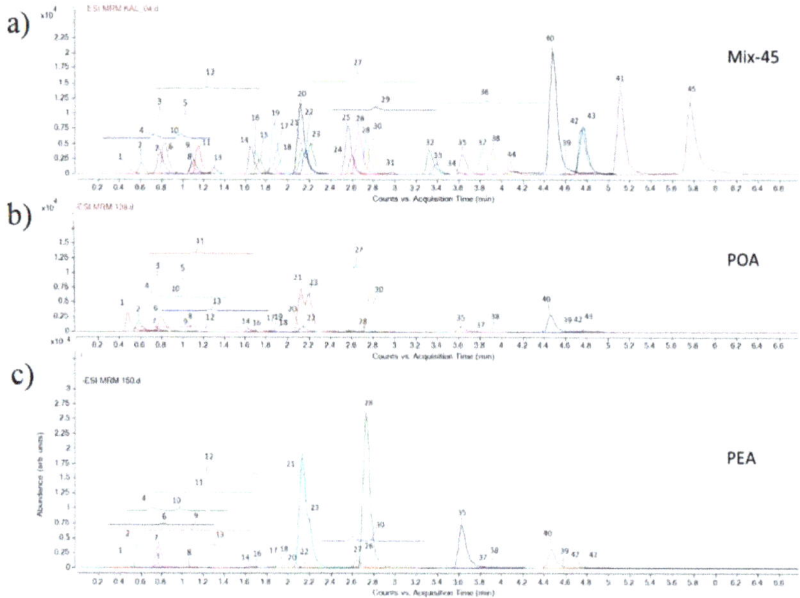

Figure 1. MRM chromatograms of standard compounds (3.125 µg/mL each standard), (**a**); of *P. aviculare* herb ethanol extract, POA, (**b**); and of *P. amphibia* herb ethanol extract, PEA, (**c**); 1: Quinic acid, 2: Gallic acid, 3: Protocatechuic acid, 4: Catechin, 5: 2,5-dihydroxybenzoic acid, 6: 5-*O*-caffeoylquinic acid, 7: Epigallocatechin gallate, 8: *p*-Hydroxybenzoic acid, 9: Esculetin, 10: Epicatechin, 11: Caffeic acid, 12: Vanillic acid, 13: Syringic acid, 14: *p*-Coumaric acid, 15: Umbelliferon, 16: Scopoletin, 17: Ferulic acid, 18: Sinapic acid, 19: Vitexin, 20: Luteolin-7-*O*-glucoside, 21: Quercetin-3-*O*-galactoside, 22: Rutin, 23: Quercetin-3-*O*-glucoside, 24: *o*-Coumaric acid, 25: Apiin, 26: Apigenin-7-*O*-glucoside, 27: Myricetin, 28: Quercetin-3-*O*-L-rhamnoside, 29: Secoisolariciresinol, 30: Kaempferol-3-*O*-glucoside, 31: 3,4-dimethoxycinnamic acid, 32: Baicalin, 33: Daidzein, 34: Matairesinol, 35: Quercetin, 36: Cinnamic acid, 37: Naringenin, 38: Luteolin, 39: Apigenin, 40: Kaempferol, 41: Baicalein, 42: Isorhamnetin, 43: Chrysoeriol, 44: Genistein, 45: Amentoflavone.

3.2. Cytotoxicity and Drug Synergism Analysis of Herbal Extracts and Doxorubicin

The evaluation of the cytotoxic effect of herbal extracts, alone and combined with D, was conducted on HepG2 cells. The IC_{25} and IC_{50} values of extracts and D, determined from the dose–response curves (Figure 2A–C), are presented in Table 2. Applied individually, PEA was more effective against HepG2 cells than POA. Applied in a mixture, in lower tested concentrations, POAD induced higher sensitivity of HepG2 cells than PEAD. To quantify the mode of interaction between tested substances, the combination index (CI) was calculated for IC_{25} and IC_{50} concentrations. Remarkable synergism for both mixtures, POAD (CI =0.62 and 0.13) and PEAD (CI = 0.89 and 0.39), was detected. Thus, the concentration required to inhibit cell viability for 25% and 50% for both agents in the mixtures has been remarkably reduced.

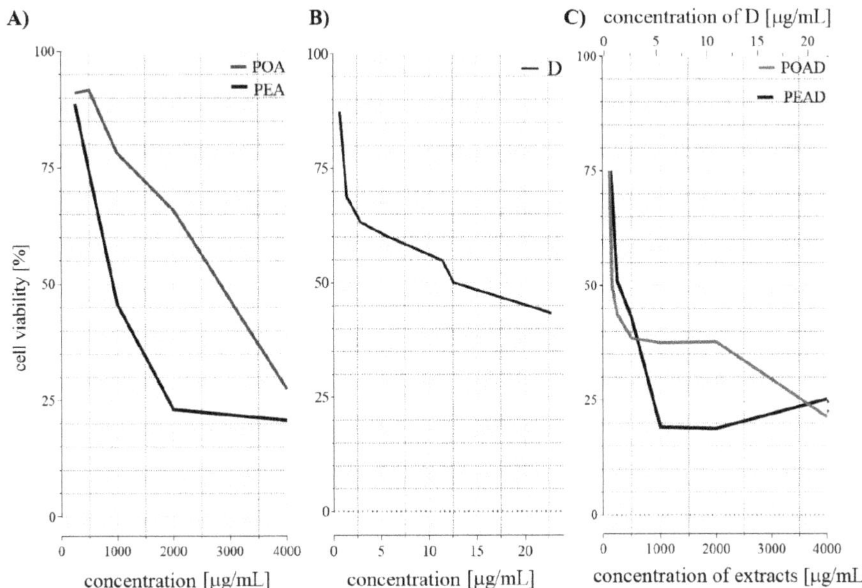

Figure 2. Inhibition rates of HepG2 cells treated with individual extracts (**A**); doxorubicin (**B**); and their combination (**C**) after 24 h.

Table 2. The cytotoxicity of herbal extracts and doxorubicin (D), either alone or in a two-drug combination on HepG2 cells.

	Individual Treatments			
	IC$_{25}$ *		IC$_{50}$ *	
D	1.3		12.56	
POA	1250		2800	
PEA	500		910	
	IC$_{25}$ * values of the co-treatments			
	POAD		PEAD	
POA	D		PEA	D
120	0.68		140	0.79
	IC$_{50}$ * values of the co-treatments			
	POAD		**PEAD**	
POA	D		PEA	D
160	**0.91**		**250**	**1.43**
	CI			
	POAD		PEAD	
IC$_{25}$	0.62		0.89	
IC$_{50}$	0.13		0.39	

* The concentrations are expressed in µg/mL. *Polygonum aviculare* ethanol extract (POA); *Persicaria amphibia* ethanol extract (PEA); Doxorubicin (D); Co-treatment of POA and D (POAD); Co-treatment of PEA and D (PEAD); Combination index (CI). The concentrations in bold, individually and combined, were used in flow cytometry analysis.

3.3. Effect of Herbal Extracts and Doxorubicin on Apoptosis and Cell Cycle

To determine whether the cytotoxicity of individual agents and their combinations is related to apoptosis and mediated by cell cycle arrest, the flow cytometry was applied. A significant increase in early apoptosis was determined after treatment with POA (21%). (Figure 3A,B). In addition, both concentrations of D, used for the preparation of combinations with POA and PEA, also increased early apoptosis of cancer cells (36% and 48.65%). An increase in both early and late apoptosis was observed when the PEA (29% and 38%, respectively) and POAD (26% and 17% respectively) were applied. However, in the case of PEAD co-treatment, only an increase in late apoptosis was detected (46.51%).

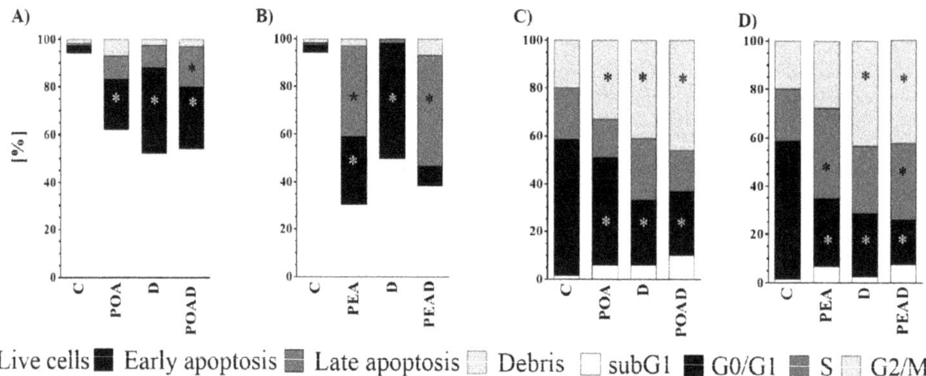

Figure 3. Analysis of apoptosis (**A,B**) and cell cycle arrest (**C,D**) in HepG2 cells after treatment with extracts, doxorubicin (**D**), and their combinations. An increase in the number of early and late apoptotic cells is always at the expense of living cells; the results are expressed as percentages compared to the untreated control * $p \leq 0.05$.

The analysis of the cell cycle phase distribution of HepG2 treated cells showed a significant cell cycle arrest in G2/M phase when individual treatment, POA (33%), and both concentration of D (41% and 43.45%), as well as co-treatments POAD (46%) and PEAD (42.69%) were applied (Figure 3C,D). Additionally, a significant increase in HepG2 cells in S phase was observed after individual treatment with PEA (37.4%) and co-treatment with PEAD (31.84%).

3.4. Effect of Herbal Extracts and Doxorubicin on Keap1 and Nrf2 Genes Expression

In malignant cells, alterations of the expression of Keap1 and Nrf2 genes are not rare. Here, the expression of Nrf2 and Keap1 was examined in HepG2 cells. Both co-treatments significantly increased Keap1 and simultaneously decreased Nrf2 gene expression (Figure 4).

4. Discussion

Polygonaceae species are rich sources of valuable secondary metabolites, mainly flavonoids. Data concerning chemical composition of *P. aviculare* is abundant, while *P. amphibia* has been less studied. Several studies gave important contributions in elucidating chemical composition of *P. aviculare* extracts [1,33–42], but Granica et al. [43–45], and Cai et al. [46] stands out. Granica et al. [43,44] focused on developing new standardization HPLC methods for *P. aviculare*. Taking into account the results of Granica [43], we have focused on a targeted search for given flavonol (myricetin (M), quercetin (Q), kaempferol (K), isorhamnetin (IR), kaempferide (KD)) glucuronides (U) and their acetylated derivatives (acU), as some of these tend to be the major compounds occurring in *P. aviculare* (Q-3-*O*-glucuronide or kaempherol-3-*O*-glucuronide). For HPLC separation conditions used in our work there is a pattern of elucidation order, as follows: MU, QU, MacU, KU, IRU, QacU1, QacU2, KacU1, IRacU1, KacU2, IRacU2, KDU, KDacU (Figures 5 and 6) This way of analysis revealed significant differences between *P. aviculare* and *P. amphibia* plant extracts.

Although both plant extracts contain myricetin-glucuronide and quercetin-glucuronide in significant amounts (not quantified), in PEA extracts the following compounds were not found: MacU, IRU, QacU1, QacU2, KacU1, IRacU1, KacU2, IRacU2, KDU, KDacU. Considering the scarce data on chemical composition of *P. amphibia* [34,35,42,47] this work, with quantitative and tentative qualitative HPLC analysis, creates a notable contribution.

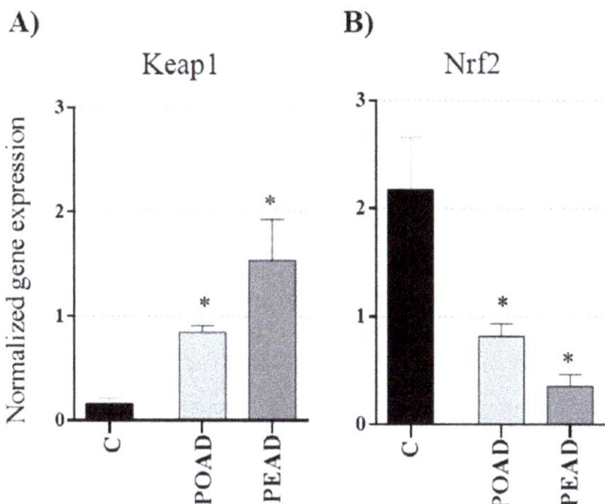

Figure 4. The effect of extracts and doxorubicin (D) combined (IC_{25}) on the expression of (**A**) Keap1 and (**B**) Nrf2 genes in HepG2 cells evaluated by the qRT-PCR, * $p \leq 0.05$.

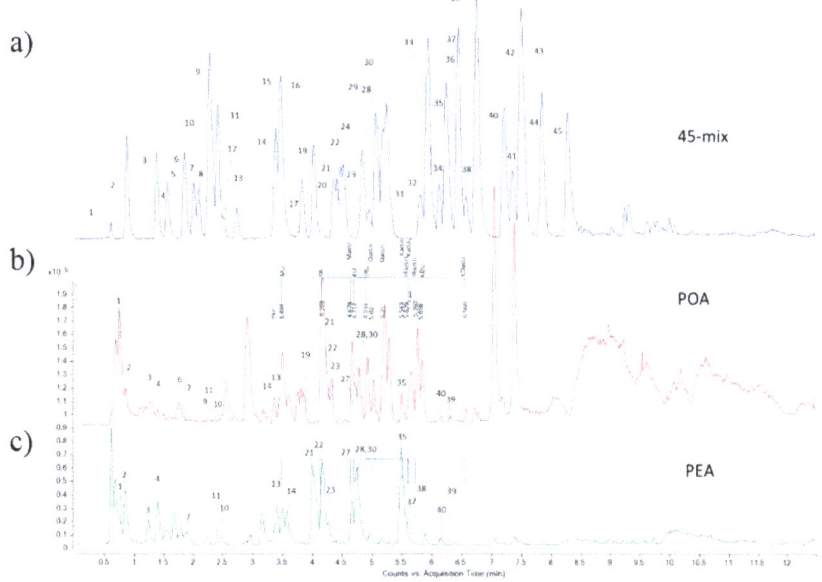

Figure 5. ESI BPC chromatograms, negative ionization mode (MS2Scan) of standard mix-45 (1.56 µg/mL, each compound), (**a**); of *Poligonum aviculare* ethanol herb extract, POA, (**b**); and *Persicaria amphibia* ethanol herb extract, PEA, (**c**); with labeled phenolics that were confirmed by quantitative LC-MS/MS analysis, and tentatively determined -glucuronides (U), and acetylglucuronides (acU) derivatives of Myricetin (M), Quercetin (Q), Kaempferol (K), Isorhamenetin (IR) and Kaempferide (KD), e.g., MU-myricetin-glucuronide, MacU-myricetinacetylglucuronide (Tables S1 and S2).

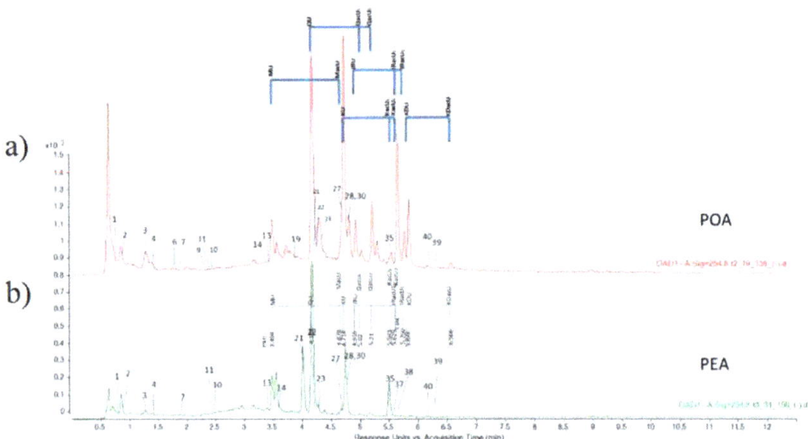

Figure 6. HPLC-DAD chromatograms (255 nm) of *Poligonum aviculare* ethanol herb extract, POA, (**a**); and *Persicaria amphibia* ethanol herb extract, PEA, (**b**); with labeled phenolics that were confirmed by quantitative LC-MS-MS analysis. Tentatively determined -glucuronides (U), and acetylglucuronides (acU) derivatives of Myricetin (M), Quercetin (Q), Kaempferol (K), Isorhamenetin (IR) and Kaempferide (KD) are also labeled, e.g., MU-myricetin-glucuronide, MacU-myricetinacetylglucuronide (Figure S2).

On the other hand, to overcome the problem of overall toxicity and resistance of cancer cells to chemotherapeutics, a combined approach that employs both commercial cytostatic and herbal extracts was subjected to the analysis. The benefit of this approach involves the induction of more diverse mechanisms of action that hinder the development of resistance and allow for the reduction of cytostatic doses. In this study, we examined in vitro cytotoxic properties of POA and PEA, alone and combined with commercial cytostatic D. Our investigation provided corroborative evidence that POA and PEA extract could potentiate D cytotoxicity in hepatocarcinoma (HepG2) cells. This result is in accordance with our previous findings [28], which demonstrated a synergistic interaction between *Polygonum maritimum* extract and D in HepG2 cells. Likewise, Ghazali et al. [48] demonstrated that herb extracts obtained from *Polygonum minus* have an antiproliferative effect on HepG2 cells. It has also been reported that the *Polygonum cuspidatum* extract has an antiproliferative effect on hepatocarcinoma cells Bel-7402 and Hepa 1–6 [49], whereas extracts obtained from *Polygonum glabrum* and *Polygonum orientale* exhibited the protective activity on normal hepatocytes in vivo [50,51].

It is well known that inhibition of cancer cell proliferation may be a result of a pro-apoptotic effect and a cell cycle disruption. Consequently, the effect of extracts and their combinations with D on apoptosis and cell cycle arrest were monitored. Herein, it was confirmed that apoptosis induction plays an important role in D-induced cell death of HepG2 cells. Also, all treatments, but particularly with POA, applied alone and in a combination, were capable of inducing early or late apoptosis. Similar to this finding, Habibi et al. [7] have reported that methanol extract of *P. aviculare* induced apoptosis in breast cancer MCF-7 cells. As for the impact of *Polygonum* spp. extracts on the molecular mechanism of apoptosis, up-regulation of the apoptotic gene *p53* and down-regulation of the anti-apoptotic *Bcl-2* gene were demonstrated [7]. The pro-apoptotic effect controlled by p53 is accompanied by cell cycle arrest in G2/M phase [52]. Further on, the arrest of the cell cycle in G2/M phase is a well-established feature of D [53]. Importantly, the G2/M checkpoint serves to prevent damaged cells from entering mitosis and proliferate. Not only D, but also the extracts, particularly POA alone and combined with D, increased the number of HepG2 cells in the G2/M phase. Likewise, recent studies have shown that various *Polygonum* spp. extracts and their active compounds induced the pro-apoptotic effect and arrested HepG2 cells in G2/M phase [54,55]. Moreover, this study demonstrated that PEA alone and combined with D induced cell cycle arrest in the S phase. The obtained results are in line with Ghazali et al. [48], implying that *Polygonum minus* extracts induced S phase cell cycle arrest in HepG2 cells. Therefore, the observed synergism between D and tested extracts in cancer cells might be attributed to the interference with pathways involved in the regulation of apoptosis and cell cycles.

Data reported in the literature indicates that pro-apoptotic effects and cell cycle arrest induced by plant extracts could be attributed to their chemical composition. Searching for possible active compounds among the main constituents of the tested extracts pointed to free gallic acid, as well as quercetin and its derivatives, since they have been well documented to possess cytotoxicity linked to pro-apoptotic effects and the ability to induce cell cycle arrest [56–58]. Thus, quercetin caused the cell cycle arrest in G2/M phase, which was followed by a decrease in cell numbers in the G0/G1 phase [58]. Furthermore, gallic acid induced cell cycle arrest in malignant cells, contributing to inhibition of cancerous cell proliferation [59]. Moreover, quercetin induced apoptosis in various cancer cells [60,61].

Beside the growth-inhibiting and apoptosis-inducing effects, phytochemicals are capable of modulating Nrf2 expression. Furthermore, due to the overexpression of Nrf2, malignant cells are frequently highly resistant to different chemotherapeutics [19,21]. Therefore, Nrf2 is an important pharmacological target of effective chemotherapy. This study demonstrated that both co-treatments decreased Nrf2 expression in HepG2 cells. As expected, this response was followed by increased Keap1 gene expression. Similar results were obtained for resveratrol, an active compound of *Polygonum cuspidatum*, which was shown to modulate Nrf2 expression in a concentration- and time-dependent way [62]. Comprehensively observed, Nrf2 has a functional link with numerous genes reported to play specific roles in the development of drug resistance. For instance, Nrf2 influences the regulation of phase II-detoxifying enzymes, antioxidant defense enzymes, and multidrug resistance-associated proteins 1-6 (MRP 1-6) [23,63]. Altogether, regulation of Nrf2 is responsible, at least partially, for chemotherapy resistance, indicating the importance of determining Nrf2 inhibitors, such as POAD and PEAD.

In conclusion, since synergistic cytotoxicity in hepatocarcinoma cells was observed, the combined approach that employs *Polygonum aviculare* and *Persicaria amphibia* ethanol extracts and cytostatic D, could serve as a good starting point in the search for hepatocarcinoma chemotherapy improvement.

Supplementary Materials: The following are available online at https://www.mdpi.com/article/10.3390/foods10040811/s1, Figure S1: Calibration curves of standards (34 out of 45) which presence was detected in analysed extracts, Figure S2: Extract Ion Chromatograms: -EIC: 493-,477-,461-,491-,535-,519-,475-,503-,533-,517-,317-,301-,285-,315-,299- indicating the significant differences in flavonol-glucronides composition in *P. aviculare* and *P. amphibia* species, Table S1: Identification of the main compounds from *Polygonum aviculare* ethanol extracts (POA) by HPLC-DAD-MS, Table S2 Identification of the main compounds from *Persicaria amphibia* ethanol extracts (PEA) by HPLC-DAD-MS.

Author Contributions: Conceptualization, D.M.-Ć., B.N., and M.J.; methodology, T.S.-R., E.S., D.T., and M.J.; validation, B.N.; formal analysis, T.S.-R., E.S., D.T., and M.J.; investigation, M.J., D.M.-Ć., and B.N.; resources, D.M.-Ć.; data curation, E.S., T.S.-R., and D.T.; writing—original draft preparation, M.J.; writing—review and editing, B.N., D.M.-Ć., and E.S.; visualization, M.J.; supervision, D.M.-Ć., B.N., and T.S.-R.; project administration, D.M.-Ć.; funding acquisition, D.M.-Ć., and M.J.; All authors have read and agreed to the published version of the manuscript.

Funding: This work was supported by Ministry of Education, Science, and Technological Development of Republic of Serbia (451-03-68/2020-14/200178 and 200051).

Data Availability Statement: The data presented in this study are available on request from the corresponding author.

Acknowledgments: The authors are grateful to Bojana Žegura, National Institute of Biology, Slovenia, for providing HepG2 cells for this work and to Goran Anačkov, University of Novi Sad—Faculty of Sciences, Serbia, for identifying plant material.

Conflicts of Interest: The authors declare no conflict of interest.

References

1. Nugroho, A.; Kim, E.J.; Choi, J.S.; Park, H.J. Simultaneous quantification and peroxynitrite-scavenging activities of flavonoids in *Polygonum aviculare* L. herb. *J. Pharm. Biomed. Anal.* **2014**, *89*, 93–98. [CrossRef] [PubMed]
2. Özbay, H.; Alim, A. Antimicrobial activity of some water plants from the northeastern Anatolian region of Turkey. *Molecules* **2009**, *14*, 321–328. [CrossRef]
3. Chon, S.U.; Heo, B.G.; Park, Y.S.; Cho, J.Y.; Gorinstein, S. Characteristics of the leaf parts of some traditional Korean salad plants used for food. *J. Sci. Food Agric.* **2008**, *88*, 1963–1968. [CrossRef]
4. Costea, M.; Tardif, F.J. The biology of Canadian weeds. 131. *Polygonum aviculare* L. *Can. J. Plant Sci.* **2005**, *85*, 481–506. [CrossRef]
5. Thu, N.N.; Sakurai, C.; Uto, H.; Van Chuyen, N.; Do, T.K.; Yamamoto, S.; Ohmori, R.; Kondo, K. The polyphenol content and antioxidant activities of the main edible vegetables in northern Vietnam. *J. Nutr. Sci. Vitaminol.* **2004**, *50*, 203–210. [CrossRef]
6. Shikov, A.N.; Pozharitskaya, O.N.; Makarov, V.G.; Wagner, H.; Verpoorte, R.; Heinrich, M. Medicinal plants of the Russian Pharmacopoeia; their history and applications. *J. Ethnopharmacol.* **2014**, *154*, 481–536. [CrossRef]
7. Habibi, R.M.; Mohammadi, R.A.; Delazar, A.; Halabian, R.; Soleimani, R.J.; Mehdipour, A.; Bagheri, M.; Jahanian-Najafabadi, A. Effects of Polygonum aviculare herbal extract on proliferation and apoptotic gene expression of MCF-7. *Daru J. Pharm. Sci.* **2011**, *19*, 326.
8. Zlatković, B.K.; Bogosavljević, S.S.; Radivojević, A.R.; Pavlović, M.A. Traditional use of the native medicinal plant resource of Mt. Rtanj (Eastern Serbia): Ethnobotanical evaluation and comparison. *J. Ethnopharmacol.* **2014**, *151*, 704–713. [CrossRef] [PubMed]
9. Bolotova, Y.V. Aquatic plants of the Far East of Russia: A review on their use in medicine, pharmacological activity. *BJMS* **2015**, *14*, 9–13. [CrossRef]
10. Ravipati, A.S.; Zhang, L.; Koyyalamudi, S.R.; Jeong, S.C.; Reddy, N.; Bartlett, J.; Smith, P.T.; Shanmugam, K.; Münch, G.; Wu, M.J.; et al. Antioxidant and anti-inflammatory activities of selected Chinese medicinal plants and their relation with antioxidant content. *BMC Complement. Altern. Med.* **2012**, *12*, 173. [CrossRef] [PubMed]
11. Hsu, C.Y. Antioxidant activity of extract from *Polygonum aviculare* L. *Biol. Res.* **2006**, *39*, 281–288. [CrossRef] [PubMed]
12. Prota, N.; Mumm, R.; Bouwmeester, H.J.; Jongsma, M.A. Comparison of the chemical composition of three species of smart-weed (genus Persicaria) with a focus on drimane sesquiterpenoids. *Phytochemistry* **2014**, *108*, 129–136. [CrossRef]
13. Orbán-Gyapai, O.; Lajter, I.; Hohmann, J.; Jakab, G.; Vasas, A. Xanthine oxidase inhibitory activity of extracts prepared from *Polygonaceae* species. *Phytother. Res.* **2015**, *29*, 459–465. [CrossRef] [PubMed]
14. Sung, Y.Y.; Yoon, T.; Yang, W.K.; Kim, S.J.; Kim, D.S.; Kim, H.K. The antiobesity effect of *Polygonum aviculare* L. ethanol extract in high-fat diet-induced obese mice. *Evid. Based Complement. Alternat. Med.* **2013**, *2013*, 626397. [CrossRef] [PubMed]
15. Dong, X.; Fu, J.; Yin, X.; Li, X.; Wang, B.; Cao, S.; Zhang, J.; Zhang, H.; Zhao, Y.; Ni, J. Pharmacological and other Bioactivities of the Genus Polygonum—A Review. *Trop. J. Pharm. Res.* **2014**, *13*, 1749–1759.
16. Zhu, A.X. Systemic therapy of advanced hepatocellular carcinoma: How hopeful should we be? *Oncologist* **2006**, *11*, 790–800. [CrossRef]
17. Tu, D.G.; Chyau, C.C.; Chen, S.Y.; Chu, H.L.; Wang, S.C.; Duh, P.D. Antiproliferative Effect and Mediation of Apoptosis in Human Hepatoma HepG2 Cells Induced by Djulis Husk and Its Bioactive Compounds. *Foods* **2020**, *9*, 1514. [CrossRef]
18. Eid, S.Y.; El-Readi, M.Z.; Wink, M. Synergism of three-drug combinations of sanguinarine and other plant secondary metab-olites with digitonin and doxorubicin in multi-drug resistant cancer cells. *Phytomedicine* **2012**, *19*, 1288–1297. [CrossRef]
19. Gao, A.M.; Ke, Z.P.; Shi, F.; Sun, G.C.; Chen, H. Chrysin enhances sensitivity of BEL-7402/ADM cells to doxorubicin by suppressing PI3K/Akt/Nrf2 and ERK/Nrf2 pathway. *Chem. Biol. Interact.* **2013**, *206*, 100–108. [CrossRef]
20. Shafa, M.H.; Jalal, R.; Kosari, N.; Rahmani, F. Efficacy of metformin in mediating cellular uptake and inducing apoptosis activity of doxorubicin. *Regul. Toxicol. Pharmacol.* **2018**, *99*, 200–212. [CrossRef]
21. Lee, Y.J.; Lee, D.M.; Lee, S.H. Nrf2 expression and apoptosis in quercetin-treated malignant mesothelioma cells. *Mol. Cells* **2015**, *38*, 416. [CrossRef]
22. Raghunath, A.; Sundarraj, K.; Arfuso, F.; Sethi, G.; Perumal, E. Dysregulation of Nrf2 in hepatocellular carcinoma: Role in cancer progression and chemoresistance. *Cancers* **2018**, *10*, 481. [CrossRef]
23. Dong, L.; Han, X.; Tao, X.; Xu, L.; Xu, Y.; Fang, L.; Yin, L.; Qi, Y.; Li, H.; Peng, J. Protection by the total flavonoids from *Rosa laevigata* Michx fruit against lipopolysaccharide-induced liver injury in mice via modulation of FXR signaling. *Foods* **2018**, *7*, 88. [CrossRef]
24. Šibul, F.S.; Orčić, D.Z.; Svirčev, E.; Mimica-Dukić, N.M. Optimization of extraction conditions for secondary biomolecules from various plant species. *Hem. Ind.* **2016**, *70*, 473–483. [CrossRef]
25. Zhang, C.; Liu, D.; Wu, L.; Zhang, J.; Li, X.; Wu, W. Chemical characterization and antioxidant properties of ethanolic extract and its fractions from sweet potato (*Ipomoea batatas* L.) leaves. *Foods* **2020**, *9*, 15. [CrossRef] [PubMed]
26. Beara, I.N.; Lesjak, M.M.; Jovin, E.Đ.; Balog, K.J.; Anackov, G.T.; Orcic, D.Z.; Mimica-Dukic, N.M. Plantain (*Plantago* L.) species as novel sources of flavonoid antioxidants. *J. Agric. Food Chem.* **2009**, *57*, 9268–9273. [CrossRef] [PubMed]

27. Orčić, D.; Francišković, M.; Bekvalac, K.; Svirčev, E.; Beara, I.; Lesjak, M.; Mimica-Dukić, N. Quantitative Determination of Plant Phenolics in Urtica dioica Extracts by High-Performance Liquid Chromatography Coupled with Tandem Mass Spec-trometric Detection. *Food Chem.* **2014**, *143*, 48–53. [CrossRef]

28. Jovanović, M.; Srdić-Rajić, T.; Svirčev, E.; Jasnić, N.; Nikolić, B.; Bojić, S.; Stević, T.; Knežević-Vukčević, J.; Mitić-Ćulafić, D. Evaluation of anticancer and antimicrobial activities of the Polygonum maritimum ethanol extract. *Arch. Biol. Sci.* **2018**, *70*, 665–673. [CrossRef]

29. Vasilijević, B.; Knežević-Vukčević, J.; Mitić-Ćulafić, D.; Orčić, D.; Francišković, M.; Srdic-Rajic, T.; Jovanović, M.; Nikolić, B. Chemical characterization, antioxidant, genotoxic and in vitro cytotoxic activity assessment of *Juniperus communis* var. sax-atilis. *Food Chem. Toxicol.* **2018**, *112*, 118–125. [CrossRef] [PubMed]

30. Srdic-Rajic, T.; Nikolic, K.; Cavic, M.; Djokic, I.; Gemovic, B.; Perovic, V.; Veljkovic, N. Rilmenidine suppresses proliferation and promotes apoptosis via the mitochondrial pathway in human leukemic K562 cells. *Eur. J. Pharm. Sci.* **2016**, *81*, 172–180. [CrossRef] [PubMed]

31. Kaisarevic, S.; Dakic, V.; Hrubik, J.; Glisic, B.; Lübcke-von Varel, U.; Pogrmic-Majkic, K.; Fa, S.; Teodorovic, I.; Brack, W.; Kovacevic, R. Differential expression of CYP1A1 and CYP1A2 genes in H4IIE rat hepatoma cells exposed to TCDD and PAHs. *Environ. Toxicol. Pharmacol.* **2015**, *39*, 358–368. [CrossRef]

32. Voelker, D.; Vess, C.; Tillmann, M.; Nagel, R.; Otto, G.W.; Geisler, R.; Schirmer, K.; Scholz, S. Differential gene expression as a toxicant-sensitive endpoint in zebrafish embryos and larvae. *Aquat. Toxicol.* **2007**, *81*, 355–364. [CrossRef]

33. Fu, Q.; Liu, S.; Wang, H.; Chen, S. Simultaneous determination of eight flavonoids in *Polygonum aviculare* L. by RP-HPLC-UV. *J. Chin. Pharm. Sci.* **2014**, *23*, 170–176. [CrossRef]

34. Smolarz, H.D. Flavonoid glycosides in nine *Polygonum* L. taxons. *Acta Soc. Bot. Pol.* **2002**, *71*, 29–33. [CrossRef]

35. Smolarz, H.D. Comparative study on the free flavonoid aglycones in herbs of different species of *Polygonum* L. *Acta Pol. Pharm.* **2002**, *59*, 145–148. [PubMed]

36. Cong, H.J.; Zhang, S.W.; Zhang, C.; Huang, Y.J.; Xuan, L.J. A novel dimeric procyanidin glucoside from *Polygonum aviculare*. *Chin. Chem. Lett.* **2012**, *23*, 820–822. [CrossRef]

37. Yunuskhodzhaeva, N.A.; Eshbakova, K.A.; Abdullabekova, V.N. Flavonoid composition of the herb *Polygonum aviculare*. *Chem. Nat. Compd.* **2010**, *46*, 803–804. [CrossRef]

38. Avula, B.; Joshi, V.C.; Wang, Y.H. Simultaneous Identification and Quantification of Anthraquinones, Polydatin, and Resveratrol in *Polygonum multiflorum*, Various Polygonum Species, and Dietary Supplements by Liquid Chromatography and Microscopic Study of Polygonum Species. *J. AOAC Int.* **2007**, *90*, 1532–1538. [CrossRef]

39. Kawasaki, M.; Kanomata, T.; Yoshitama, K. Flavonoids in the Leaves of Twenty-Eight Polygonaceous Plants. *Bot. Mag. Shokubutsu-gaku-zasshi* **1986**, *99*, 63–74. [CrossRef]

40. Al-Hazimi, H.M.A.; Haque, S.N. A new naphthoquinone from *Polygonum aviculare*. *Nat. Prod. Lett.* **2002**, *16*, 115–118. [CrossRef]

41. Kim, H.J.; Woo, E.R.; Park, H.A. Novel Lignan and Flavonoids from *Polygonum aviculare*. *J. Nat. Prod.* **1994**, *57*, 581–586. [CrossRef]

42. Nikolaeva, G.G.; Lavrent'Eva, M.V.; Nikolaeva, I.G. Phenolic compounds from several Polygonum species. *Chem. Nat. Compd.* **2009**, *45*, 735–736. [CrossRef]

43. Granica, S. Quantitative and qualitative investigations of pharmacopoeial plant material polygoni avicularis herba by UHPLC-CAD and UHPLC-ESI-MS methods. *Phytochem. Anal.* **2015**, *26*, 374–382. [CrossRef] [PubMed]

44. Granica, S.; Piwowarski, J.P.; Popławska, M.; Jakubowska, M.; Borzym, J.; Kiss, A.K. Novel insight into qualitative standardization of *Polygoni avicularis* herba (Ph. Eur.). *J. Pharm. Biomed. Anal.* **2013**, *72*, 216–222. [CrossRef] [PubMed]

45. Granica, S.; Czerwińska, M.E.; Zyżyńska-Granica, B.; Kiss, A.K. Antioxidant and anti-inflammatory flavonol glu-curonides from *Polygonum aviculare* L. *Fitoterapia* **2013**, *91*, 180–188. [CrossRef] [PubMed]

46. Cai, Y.; Wu, L.; Lin, X.; Hu, X.; Wang, L. Phenolic profiles and screening of potential α-glucosidase inhibitors from *Polygonum aviculare* L. leaves using ultra-filtration combined with HPLC-ESI-qTOF-MS/MS and molecular docking analysis. *Ind. Crop. Prod.* **2020**, *154*, 112673. [CrossRef]

47. Smolarz, H.D.; Budzianowski, J.; Bogucka-Kocka, A.; Kocki, J.; Mendyk, E. Flavonoid glucuronides with anti-leukaemic activity from *Polygonum amphibium* L. *Phytochem. Anal.* **2008**, *19*, 506–513. [CrossRef] [PubMed]

48. Ghazali, M.; Alfazari, M.; Al-Naqeb, G.; Krishnan Selvarajan, K.; Hazizul Hasan, M.; Adam, A. Apoptosis induction by Polygonum minus is related to antioxidant capacity, alterations in expression of apoptotic-related genes, and S-phase cell cycle arrest in HepG2 cell line. *BioMed. Res. Int.* **2014**, *2014*, 539607.

49. Hu, B.; An, H.M.; Shen, K.P.; Song, H.Y.; Deng, S. Polygonum cuspidatum extract induces anoikis in hepatocarcinoma cells associated with generation of reactive oxygen species and downregulation of focal adhesion kinase. *Evid.-Based Compl. Alt.* **2012**, *2012*, 607675. [CrossRef] [PubMed]

50. Raja, S.; Ramya, I.A. Comprehensive review on Polygonum glabrum. *IJOP* **2017**, *8*, 457–467.

51. Chiu, Y.J.; Chou, S.C.; Chiu, C.S.; Kao, C.P.; Wu, K.C.; Chen, C.J.; Tsai, J.C.; Peng, W.H. Hepatoprotective effect of the ethanol extract of *Polygonum orientale* on carbon tetrachloride-induced acute liver injury in mice. *J. Food Drug Anal.* **2017**, *26*, 369–379. [CrossRef] [PubMed]

52. Stark, G.R.; Taylor, W.R. Analyzing the G2/M checkpoint. In *Checkpoint Controls and Cancer*; Humana Press: Totowa, NJ, USA, 2004; pp. 51–82.

53. Lee, T.; Lau, T.; Ng, I. Doxorubicin-induced apoptosis and chemosensitivity in hepatoma cell lines. *Cancer Chemother. Pharmacol.* **2002**, *49*, 78–86. [CrossRef]

54. Shieh, D.E.; Chen, Y.Y.; Yen, M.H.; Chiang, L.C.; Lin, C.C. Emodin-induced apoptosis through p53-dependent pathway in human hepatoma cells. *Life Sci.* **2004**, *74*, 2279–2290. [CrossRef]

55. Lee, B.H.; Huang, Y.Y.; Wu, S.C. Hepatoprotective activity of fresh Polygonum multiflorum against HepG2 hepatocarcinoma cell proliferation. *J. Food Drug Anal.* **2011**, *19*, 26–32.

56. Ow, Y.Y.; Stupans, I. Gallic acid and gallic acid derivatives: Effects on drug metabolizing enzymes. *Curr. Drug Metab.* **2003**, *4*, 241–248. [CrossRef]

57. Badhani, B.; Sharma, N.; Kakkar, R. Gallic acid: A versatile antioxidant with promising therapeutic and industrial applications. *RSC Adv.* **2015**, *5*, 27540–27557. [CrossRef]

58. Kim, S.R.; Lee, E.Y.; Kim, D.J.; Kim, H.J.; Park, H.R. Quercetin inhibits cell survival and metastatic ability via the EMT-mediated pathway in oral squamous cell carcinoma. *Molecules* **2020**, *25*, 757. [CrossRef] [PubMed]

59. Pinmai, K.; Chunlaratthanabhorn, S.; Ngamkitidechakul, C.; Soonthornchareon, N.; Hahnvajanawong, C. Synergistic growth inhibitory effects of Phyllanthus emblica and Terminalia bellerica extracts with conventional cytotoxic agents: Doxorubicin and cisplatin against human hepatocellular carcinoma and lung cancer cells. *World J. Gastroenterol.* **2008**, *14*, 1491. [CrossRef] [PubMed]

60. Gopalakrishnan, A.; Kong, A.N. Anticarcinogenesis by dietary phytochemicals: Cytoprotection by Nrf2 in normal cells and cytotoxicity by modulation of transcription factors NF-κB and AP-1 in abnormal cancer cells. *Food Chem. Toxicol.* **2008**, *46*, 1257–1270. [CrossRef] [PubMed]

61. Ramos, S. Cancer chemoprevention and chemotherapy: Dietary polyphenols and signalling pathways. *Mol. Nutr. Food Res.* **2008**, *52*, 507–526. [CrossRef] [PubMed]

62. Hybertson, B.M.; Gao, B.; Bose, S.K.; McCord, J.M. Oxidative stress in health and disease: The therapeutic potential of Nrf2 activation. *Mol. Asp. Med.* **2011**, *32*, 234–246. [CrossRef] [PubMed]

63. Wang, X.J.; Sun, Z.; Villeneuve, N.F.; Zhang, S.; Zhao, F.; Li, Y.; Chen, W.; Yi, X.; Zheng, W.; Wondrak, G.T.; et al. Nrf2 enhances resistance of cancer cells to chemotherapeutic drugs, the dark side of Nrf2. *Carcinogenesis* **2008**, *29*, 1235–1243. [CrossRef] [PubMed]

Article

Quality and Antioxidant Properties of Cold-Pressed Oil from Blanched and Microwave-Pretreated Pomegranate Seed

Tafadzwa Kaseke [1,2], Umezuruike Linus Opara [2,*] and Olaniyi Amos Fawole [2,3,*]

[1] Department of Food Science, Faculty of AgriSciences, Stellenbosch University, Private Bag X1, Stellenbosch 7602, South Africa; tafakaseqe@gmail.com
[2] Faculty of AgriSciences, Africa Institute for Postharvest Technology, South African Research Chair in Postharvest Technology, Stellenbosch University, Private Bag X1, Stellenbosch 7602, South Africa
[3] Department of Botany and Plant Biotechnology, Faculty of Science, University of Johannesburg, P.O. Box 524, Johannesburg 2006, South Africa
* Correspondence: opara@sun.ac.za (U.L.O.); olaniyi@sun.ac.za (O.A.F.)

Abstract: The present research studied the influence of blanching and microwave pretreatment of seeds on the quality of pomegranate seed oil (PSO) extracted by cold pressing. Pomegranate seeds (cv. Acco) were independently blanched (95 ± 2 °C/3 min) and microwave heated (261 W/102 s) before cold pressing. The quality of the extracted oil was evaluated with respect to oxidation indices, refractive index, yellowness index, total carotenoids content, total phenolic content, flavor compounds, fatty acid composition, and 2.2-diphenyl-1-picryl hydrazyl (DPPH) and 2.2-azino-bis (3-ethylbenzothiazoline-6-sulfonic acid) (ABTS) radical scavenging capacity. Blanching and microwave pretreatments of seeds before pressing enhanced oil yield, total phenolic content, flavor compounds, and DPPH and ABTS radical scavenging capacity. Although the levels of oxidation indices, including the peroxide value, free fatty acids, acid value, ρ-anisidine value, and total oxidation value, also increased, and the oil quality conformed to the requirements of the Codex Alimentarius Commission (CODEX STAN 19-1981) standard for cold-pressed vegetable oils. On the other hand, blanching and microwave heating of seeds decreased the pomegranate seed oil's yellowness index, whilst the refractive index was not significantly ($p > 0.05$) affected. Even though both blanching and microwave pretreatment of seeds added value to the cold-pressed PSO, the oil extracted from blanched seeds exhibited lower oxidation indices. Regarding fatty acids, microwave pretreatment of seeds before cold pressing significantly increased palmitic acid, oleic acid, and linoleic acid, whilst it decreased the level of punicic acid. On the contrary, blanching of seeds did not significantly affect the fatty acid composition of PSO, indicating that the nutritional quality of the oil was not significantly affected. Therefore, blanching of seeds is an appropriate and valuable step that could be incorporated into the mechanical processing of PSO.

Keywords: pomegranate seed; oil; pretreatment; cold pressing; total phenolic content; antiradical activity

Citation: Kaseke, T.; Opara, U.L.; Fawole, O.A. Quality and Antioxidant Properties of Cold-Pressed Oil from Blanched and Microwave-Pretreated Pomegranate Seed. *Foods* **2021**, *10*, 712. https://doi.org/10.3390/foods10040712

Academic Editors: Francisca Rodrigues and Cristina Delerue-Matos

Received: 27 February 2021
Accepted: 18 March 2021
Published: 26 March 2021

Publisher's Note: MDPI stays neutral with regard to jurisdictional claims in published maps and institutional affiliations.

1. Introduction

Pomegranates (*Punica granatum* L.) are consumed as fresh fruits and processed into products such as jam, juice, jelly, wine, and dried snacks [1]. In addition to increased production volumes, the inconvenience associated with fresh pomegranate consumption due to the fruit complexity has promoted the fruit's processing into these ready to eat and convenient products [2]. Moreover, the consumption of the fruit is related to its medicinal properties. The fruit's pharmacological value can be traced back to ancient times when the fruit was used as a traditional medicine to treat different ailments [3].

The literature has shown that every part of the fruit contains compounds with health benefits. The juice and peels contain punicalagins, hydrolyzable tannins, anthocyanins, and ellagic acid [4]. Pomegranate seeds, one of the waste products from the processing of the fruit, serve as a rich source of oil (12–20%) high in tocopherols, polyphenols, sterols, and punicic acid [5]. It has been demonstrated that these bioactive phytochemicals are implicated in pomegranate seed oil's chemopreventive activities such as anti-mutagenicity, antihypertension, antioxidative potential, and reduction in liver injury [6]. In line with this, pomegranate seed can be considered for value-added products. Further, processing the seeds into specialty oil is a profitable alternative to managing the postharvest waste from pomegranate fruit processing. Pomegranate seed oil can be extracted from the seeds using various techniques such as cold pressing and solvent, supercritical carbon dioxide, and ultrasound-assisted aqueous enzymatic extraction [7,8]. Prior research has indicated that the extraction technique is a major determinant of seed oil quality [9].

Seed oil extraction by cold pressing is the most preferred by processors and consumers because of the low production costs and high concentration of bioactive compounds such as essential fatty acids, tocopherols, phenols, carotenoids, and phytosterols in the oil [10]. The retention of antioxidant compounds may provide cold-pressed oils with acceptable oxidative stability and better health properties [11]. Cold-pressed oils are obtained mechanically using either a hydraulic or screw press without the application of heat, solvents, or chemical treatments, which makes the process environmentally friendly and the extracted oil safer for human consumption [12]. Therefore, there is a growing demand for cold-pressed oil, such as cold-pressed pomegranate seed oil. The maximum temperature of cold-pressed oil should not exceed 50 °C [13,14]. The cold-pressed oil may be physically purified through filtration, sedimentation, or centrifugation processes, which do not degrade the oil quality [10].

Despite the many advantages of cold pressing, the low-cost and sustainable oil extraction technique suffers from low oil yield due to a significant amount of oil that remains trapped in the pressed meal, which has hindered its development and commercial viability [15]. Nonetheless, this may be improved by blanching or microwave heating the oil-bearing seeds before pressing. According to Kaseke et al. [9], blanching seeds improved the pomegranate seed oil yield and bioactive compounds recovery with ethanol. Moreover, seed blanching is a novel technique that presents a sustainable strategy capable of improving seed oil quality whilst significantly reducing the oil extraction time and energy consumption during cold pressing [16]. Blanching significantly changes the seed matrix's structural integrity by disintegrating the cell walls and membranes, which may enhance the extractability of the intracellular material by cold pressing [17]. Nevertheless, microwave pretreatment is the commonly used technique to improve oil yield and bioactive compounds recovery in cold-pressed oils [18–21], due to its uniform energy delivery, high thermal conductivity to the interior of the material, energy saving, and precise process control [22]. Although the influence of seed pretreatment on the oil recovery efficiency of mechanical pressing has been studied, comparative studies on seed pretreatment techniques' potential to improve the quality of cold-pressed oil are limited.

In this regard, the current study aimed to investigate the effect of blanching and microwave heating pomegranate seeds on the quality and functional properties of oil extracted by cold pressing.

2. Materials and Methods

2.1. Plant Material

'Acco' pomegranates were harvested in April during the 2019 season from Blydeverwacht Farm (33°48′0″S, 19°53′0″E) in Western Cape Province, South Africa, at the commercial maturity stage (total soluble solids: 14.02–16.61°Brix). The seeds were separated from the peels, membranes, and juice before they were thoroughly cleaned with tap water.

2.2. Sample Preparation and Pretreatments

2.2.1. Blanching

Freshly extracted and clean pomegranate seeds (PS) were blanched in a water bath (Scientific, South Africa) at 95 ± 2 °C for 3 min [9]. After blanching, samples were cooled promptly in an ice water bath, drained off, and then oven dried at 55 ± 2 °C to 10% (*w/w*) moisture content. The thermogravimetric technique was applied to measure the moisture content using a moisture analyzer set at 100 °C (KERN, DBS60-3, Balingen, Germany).

2.2.2. Microwave Pretreatment

Fifty grams of oven-dried PS were uniformly spread in a glass Petri dish (190 mm in diameter) inside a calibrated domestic microwave oven (Model: DMO 351, Defy Appliances, Cape Town, South Africa) with a nominal power of 900 W. The microwave oven was calibrated following the method described by Rekas et al. [23]. The samples were microwave heated at 2450 MHz and 261 W for 102 s [24]. The microwave-heated seeds were cooled at ambient temperature (25–27 °C) and thoroughly mixed to uniform samples. The moisture content of the seeds after microwave heating was adjusted to 10%.

2.3. Cold Pressing

PS (250 g) were pressed using a single-screw press (Farmet UNO, Ceska Skalice, Czech Republic) equipped with a 10 mm diameter die. The capacity of the expeller press is about 8–12 kg seed/h. The press head was heated to 60 ± 5 °C before oil pressing using a removable heating element, and the temperature of the outflowing oil was 50 ± 5 °C. Temperature was measured using a type-K thermocouple connected to a digital temperature sensor (KIMO Instruments, Wilmington, NC, USA). The pressed oil was centrifuged at 4000 rpm for 15 min (Centrifuge 5810R, Eppendorf, Horsholm, Germany) to remove solid particles. Pomegranate seed oil (PSO) extraction yield was defined as gram per hundred gram of pomegranate seed (g/100 g seed). The oil samples were packed in brown bottles and stored at −20 °C before further analyses to minimize oxidation.

2.4. Determination of Pomegranate Seed Oil Quality Indices

2.4.1. Yellowness and Refractive Index

The refractive index was evaluated at 25 °C using a calibrated Abbe 5 refractometer (Bellingham + Stanley, Kent, United Kingdom). The yellowness index (YI) was calculated from the PSO color properties, including lightness (L*) and yellowness (b*) values, which were measured using a calibrated Chroma meter CR-410 (Konica Minolta, INC, Tokyo, Japan).

$$YI = \frac{142.86b^*}{L^*} \qquad (1)$$

2.4.2. Oxidation Indices

Free fatty acids (FFA) and acid value (AV) were measured following the AOCS standard [25]. The modified ferrous oxidation-xylenol orange (FOX) method was used to determine peroxide value (PV) [26]. ρ-Anisidine value (AnV) was determined according to [25]. Total oxidation (TOTOX) value was calculated from the PV and AnV using Equation (2).

$$TOTOX = 2PV + AnV \qquad (2)$$

2.5. Determination of Bioactive Compounds and Antiradical Activity

2.5.1. Total Carotenoids Content and Total Phenolic Content

The method described by Ranjith et al. [27] was used to determine total carotenoids content (TCC). PSO (0.2 g) was dissolved in hexane (5 mL) and 0.5 mL sodium chloride (NaCl) (0.5%, *w/w*) and thoroughly vortexed before being centrifuged (Centrifuge 5810R, Eppendorf, Horsholm, Germany) at 4000 rpm for 5 min. The absorbance was measured at 460 nm using a UV spectrophotometer (Spectrum Instruments, United Scientific, Cape

Town, South Africa), and the results were expressed as mgβ-carotene/g of PSO. The Folin–Ciocalteu method was applied to evaluate the total phenolic content (TPC) [28]. Briefly, 200 μL of PSO methanol extracts, 250 μL of the Folin–Ciocalteau reagent, 750 μL of 2% (*w/v*) sodium carbonate, and 3 mL of distilled water were sequentially mixed, and the mixtures were vortexed and incubated in the dark for 40 min. The absorbances were measured at 760 nm using a UV spectrophotometer (Spectrum Instruments, United Scientific, Cape Town, South Africa), and the results were reported as milligram gallic acid equivalent per g PSO (mg GAE/g PSO).

2.5.2. Antiradical Activity

The PSO antiradical activity was evaluated using 2.2-azino-bis (3-ethylbenzothiazoline-6-sulfonic acid) (ABTS) and 2.2-diphenyl-1-picryl hydrazyl (DPPH) assays. Briefly, ABTS radical cation (ABTS$^+$) stock solution, prepared by mixing equal volumes of 2.2-azino-bis (3-ethylbenzothiazoline-6-sulfonic acid) (ABTS) solution (7.4 mM) and potassium persulfate solution (2.6 mM), was kept in the dark for 12–16 h. The absorbance was adjusted to 0.7 ± 0.02 at 750 nm after the incubation period, using 80% (*v/v*) methanol [29]. Three hundred microliters of methanol PSO extracts were mixed with 300 μL of the freshly prepared ABTS$^+$ solution and the samples were incubated for 10 min in the dark. The absorbances of the samples were measured at 750 nm using a microplate reader (Thermo Fisher Scientific, Shanghai, China). The results were reported as mmol Trolox/g of PSO.

The method described by Siano et al. [30] was used to determine the DPPH radical scavenging capacity of PSO. Aliquots of 200 μL of methanol PSO extracts were added to 2.5 mL of 0.04% (*w/v*) DPPH in 80% (*v/v*) methanol and vortexed before incubation in the dark for 60 min. The absorbance was measured using a UV spectrophotometer (Spectrum Instruments, United Scientific, Cape Town, South Africa) at 517 nm. Results were expressed as mmol Trolox/g of PSO.

2.6. Fatty Acid Composition

The gas chromatography-mass spectrometry (GC-MS) method was used to determine the PSO fatty acid composition [31]. PSO (0.1 g), 2.0 mL hexane, 50 μL heptadecanoic acid (1000 ppm, internal standard), and 1.0 mL of 20% (*v/v*) H_2SO_4 in methanol were sequentially mixed, vortexed, and incubated at 80 °C for 1 h in an oven. To the cooled samples, 3 mL of saturated NaCl was added, and the mixture was further vortexed before centrifugation (Centrifuge 5810R, Eppendorf, Horsholm, Germany) at 4000 rpm for 3 min. The supernatant (hexane extract) was analyzed using a gas chromatograph (6890N, Agilent Technologies, Palo Alto, CA, USA) coupled to a flame ionization detector (FID). The fatty acid methyl esters were separated on a polar RT-2560 (100 m, 0.25 mm ID, 0.20 μm film thickness) (Restek, Bellefonte, PA, USA) capillary column and helium (1 mL/min) was used as the carrier gas. The sample (1 μL) was injected in a 5:1 split ratio and at 240 °C. The oven temperature was programmed as 60 °C/min and increased to 120 °C at a rate of 8 °C/min, then to 245 °C at 1.5 °C/min, and finally to 250 °C at 20 °C/min for 2 min. Gas-chromatographic peaks of FAME were identified by comparison with a commercial mixture of standards, and the NIST library was used to identify the pomegranate seed oil fatty acids profiles. The relative content (%) of each fatty acid was calculated by dividing the peak area of each fatty acid by the total peak area of all the fatty acids identified.

2.7. Determination of Volatile Compounds

Volatile compounds were analyzed by HS-SPME-GC-MS [32]. One thousand microliters of oil samples was put in 20 mL SPME vials and 1 μL was injected into the SPME-GC-MS system. Separation was performed on a gas chromatograph (6890N, Agilent technologies network) coupled to an Agilent technologies inert XL EI/CI Mass Selective Detector (MSD) (5975B, Agilent Technologies Inc., Palo Alto, CA). The GC-MS system was coupled to a CTC Analytics PAL autosampler. Separation of the oil volatiles was performed on a ZBWaxPlus (30 m, 0.25 mm ID, 0.25 μm film thickness) capillary column. Helium was used as the carrier gas at a flow rate of 1 mL/min. The injector temperature was maintained at 250 °C. Injection was performed in splitless mode. The oven temperature was programmed as follows: 35 °C for 5 min, followed by a ramping rate of 5 °C/min until 50 °C and held for 3 min, ramped again at a rate of 5 °C/min until 120 °C and held for 3 min, and finally ramped up to 240 °C at a rate of 10 °C/min for 3 min. The MSD was operated in a full scan mode, and the source and quad temperatures were maintained at 230 °C and 150 °C, respectively. The transfer line temperature was maintained at 250 °C. The mass spectrometer was operated under electron impact (EI) mode at an ionization energy of 70 eV, scanning from 25 to 650 m/z. Compound identification was based on mass spectral data of samples with the standard NIST and Wiley Library and with the comparison of retention indices. The relative content (%) of each volatile compound was calculated by dividing the peak area of each component by the total peak area of all the compounds identified.

2.8. Statistical Analysis

The results of all the studied variables are presented as mean ± SD (standard deviation). One-way analysis of variance (ANOVA) was carried out using Statistica software (Statistical v13, TIBC, Palo Alto, CA 94304, USA) and the mean values were separated according to Duncan's multiple range test. Graphical presentations were made using Microsoft Excel (Version: 16.0.13029.20344, Microsoft Cooperation, Washington, DC, USA).

3. Results and Discussion

3.1. Oil Yield

Oil yield is an essential factor in maximizing the gross income for seed oil processers. In this regard, pretreatment of the oleaginous material is crucial in promoting oil release from the seed matrix. The results in Figure 1 show that blanching and microwave heating of seeds enhanced the PSO yield (55 and 91%, respectively), with blanched seeds exhibiting a significantly higher oil yield (6.12%) than microwave-heated seeds (4.97%). The initial yield of PSO from unpretreated seeds was 3.20%. The finding that blanching and microwave heating of seeds significantly improved the PSO yield could be attributed to altering the pomegranate seed cellular structures, which increased the permeability of the cell walls and mass transfer of lipids during pressing [9,19]. Prior research has also reported significant enhancement of cold press oil extraction efficiency by thermal seed pretreatment [18,33,34]. For instance, seed microwave pretreatment doubled the yield of cold-pressed black cumin seed oil [35]. In a recent study, Lee et al. [36] observed a 2.3- to 2.4-fold increase in cold-pressed perilla seed oil yield after steam pretreatment of the seeds. On the other hand, in the absence of seed pretreatment, which provides cell disintegration, the permeability of the pomegranate seed to oil could have been limited, hence the lower oil yield from unpretreated seeds (Figure 1) [37]. The oil yield in the current study is 1.1- to 2.2-fold lower than the one reported by Khoddami et al. [7], a fact that could be explained by differences in seed variety, moisture content, oil press equipment, fruit maturity, and geographical location, among other factors.

3.2. Yellowness and Refractive Index

The color and appearance of foods, including seed oil, constitute the first set of sensory attributes and therefore affect the consumer perception of quality. The color of food may be attributed to natural pigments or biochemical or chemical products developed during processing such as seed thermal pretreatment [38]. The effect of processing on food product color can be determined by measuring the YI. Table 1 depicts the changes in the PSO yellowness index due to seed blanching and microwave heating. Blanching and microwave heating of PS significantly decreased the oil YI by 7%. With respect to PSO from blanched seeds, the decrease in the YI could be ascribed to the reduction in total carotenoids content due to the conversion of trans-carotenoids, which are the usual configuration, to cis-isomers, hence decreasing the oil yellowness (Figure 2a) [39]. According to Kha et al. [40], the extensive conjugated and trans-configured double bond system in carotenoids absorbs light in the visible region and provides foods such as seed oil with color. The pomegranate seed oil YI ranged from 37.30 to 40.21 and was lower than the one observed by Khoddami et al. [7] from cold-pressed oil (81.15–91.55) of different pomegranate cultivars.

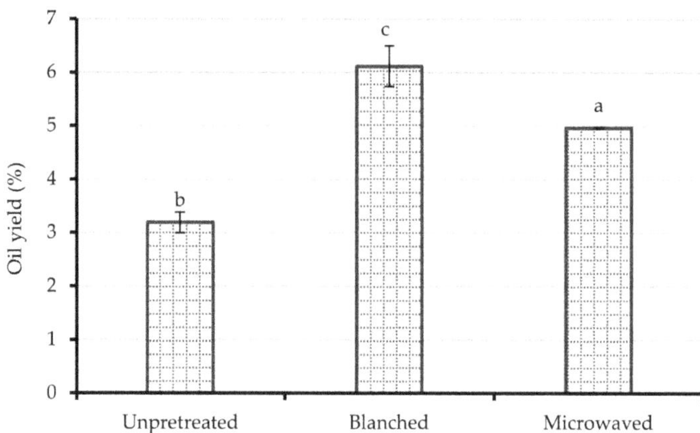

Figure 1. Oil yield of cold-pressed pomegranate seed oil from unpretreated, blanched (95 ± 2 °C/3 min), and microwave-heated (261 W for 102 s) seeds. Columns followed by different letters are significantly different ($p < 0.05$) according to Duncan's multiple range test. Vertical bars indicate the standard deviation of the mean.

Table 1. Physicochemical properties of cold-pressed pomegranate seed oil (PSO) from unpretreated, blanched (95 ± 2 °C/3 min), and microwave-heated (261 W for 102 s) seeds.

Treatment	RI	YI	FFA	AV	PV	AnV	TOTOX
Unpretreated	1.5197 ± 0.00 [a]	40.21 ± 0.03 [a]	0.60 ± 0.04 [b]	1.19 ± 0.09 [b]	0.73 ± 0.02 [a]	1.97 ± 0.15 [b]	3.42 ± 0.14 [a]
Blanched	1.5194 ± 0.00 [a]	37.30 ± 0.08 [b]	0.64 ± 0.10 [ab]	1.28 ± 0.19 [ab]	0.81 ± 0.01 [b]	2.71 ± 0.43 [ab]	4.33 ± 0.05 [c]
Microwaved	1.5195 ± 0.00 [a]	37.48 ± 0.09 [b]	0.92 ± 0.09 [a]	1.83 ± 0.18 [a]	0.86 ± 0.02 [b]	3.02 ± 0.08 [a]	4.74 ± 0.09 [b]

Means ± standard deviation of analysis ($n = 3$). Different superscript letters in the same column indicate significant difference ($p < 0.05$) according to Duncan's multiple range test. RI = refractive index (25 °C), YI = yellowness index, FFA = free fatty acid as punicic acid (%), AV = acid value (mg KOH/g PSO), PV = peroxide value (meqO$_2$/kg PSO, meqO$_2$/kg = milli-equivalents of active oxygen per kg), AnV = ρ-anisidine value, TOTOX = total oxidation value.

The refractive index is often applied to identify and characterize food materials, including seed oil. The relationships between RI and fatty acid chain length as well as the degree of unsaturation have been reported [41]. The refractive index of the oil did not significantly ($p > 0.05$) change after seed pretreatment, regardless of the significant change in fatty acid content after seed microwave pretreatment. In this sense, the interpretation of RI results in the present study should be made with caution. The RI narrowly ranged between 1.5194 and 1.5197, values typical of PSO and indicative of its high unsaturation (Table 1) [41]. These values agree with those reported by Costa et al. [42] (1.5091–1.5177) from cold-pressed PSO.

3.3. Peroxide Value, Free Fatty Acids, Acid Value, ρ-Anisidine, and Total Oxidation Value

The PV is used as the quantity measurement for peroxides, which are intermediate products of lipid oxidation. The PV test is a good way to determine the amount of primary oxidation products in freshly extracted seed oil. The PV of cold-pressed PSO from unpretreated seeds was relatively low (0.73 meqO$_2$/kg PSO). After seed blanching and microwave heating, the PV significantly ($p < 0.05$) increased by 11 and 18%, although no significant differences were observed between the PV of oils extracted from microwave-heated and blanched seeds (Table 1). Nevertheless, the PVs (0.73–0.86 meqO$_2$/kg PSO) from all oil samples were far below the level (15 meqO$_2$/kg oil) established by the World Health Organization (WHO) under the Codex Alimentarius Commission, indicating that the oils were of good quality and acceptable at the international market [43]. The lower values of peroxides in the oil samples may result from the lower extraction temperatures during oil pressing.

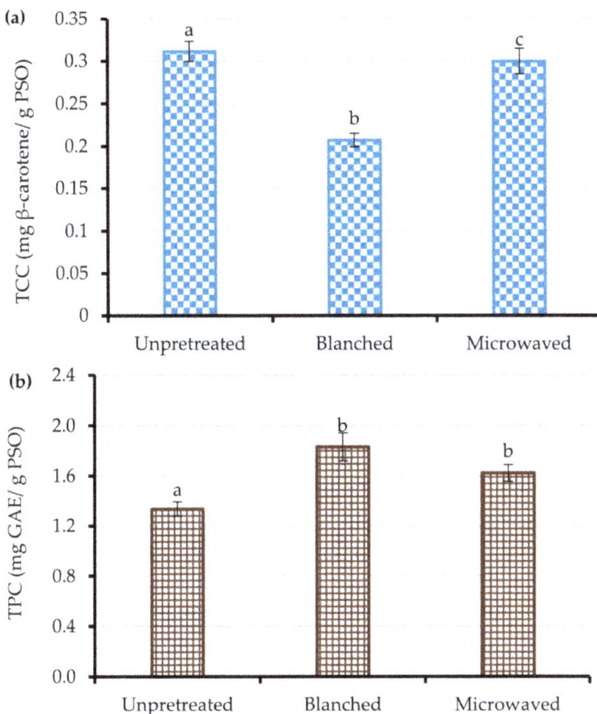

Figure 2. (**a**) Total carotenoids content (TCC) and (**b**) total phenolic content (TPC) of cold-pressed pomegranate seed oil (PSO) from unpretreated, blanched (95 ± 2 °C/3 min), and microwave-heated (261 W for 102 s) seeds. Columns followed by different letters are significantly different ($p < 0.05$) according to Duncan's multiple range test. Vertical bars indicate the standard deviation of the mean.

Free fatty acids and the acid value may be used to indicate lipase activity in the seed oil [7]. In this sense, higher FFA and AV in seed oil indicate a higher magnitude of hydrolytic deterioration and lower-quality oil product. As shown in Table 1, PSO from blanched and microwave-heated seeds had a relatively higher FFA and AV than PSO from unpretreated seeds. The increase in FFA and AV after seed blanching and microwave heating ranged between 7 and 54%. According to the quality requirements as recommended by the Codex Alimentarius Commission, cold-pressed oils should have a maximum of 4.0 mg KOH/g oil of AV [43]. Regardless of the increase after seed blanching and microwave heating, the AVs (1.19–1.83 mg KOH/g PSO) were within the standardized requirements (Table 1). The FFA (0.60–0.92%) in the present study were lower than those reported in previous studies. For instance, Khoddami et al. [7] reported FFA values of cold-pressed PSO ranging from 0.65 to 1.39%, which were 1.1- to 1.5-fold higher than our results.

The ρ-anisidine value measures the aldehyde and ketonic breakdown products of peroxides. These secondary products of oxidation are responsible for the development of rancidity in oils and fats. As shown in Table 1, the AnV of PSO from unpretreated, blanched, and microwave-heated seeds were 1.97, 2.71, and 3.02, respectively, which were 6 to 7 times lower than those reported by Costa et al. [42]. The result that microwave heating of seeds significantly increased the AV of PSO by 53% while blanching had an insignificant effect on AV indicates the difference in the pretreatment methods' mode of action. Despite the unavailability of an internationally recognized seed oil quality standard on AnV, there is a general agreement among researchers that for seed oil to be still acceptable, the AnV should be less than 10 [42,44].

The total oxidation value of PSO was determined using the PV and AnV values, representing the information for primary and secondary oxidation products. Therefore, the TOTOX value indicates both the oxidation history and further oxidation potential of the oils [45]. The changes in TOTOX values due to pomegranate seed pretreatment are shown in Table 1. The TOTOX value for PSO from unpretreated seeds was 3.42, which significantly increased to 4.33 and 4.74 after seed blanching and microwave heating, respectively. The results suggest that blanching and microwave heating of seeds could have promoted lipase enzyme activity and hydrolytic oxidation of the oil. The literature has reported increased activity of lipolytic enzymes on fat and oil in damaged cells [46].

3.4. Total Carotenoids Content, Total Phenolic Content, and Antiradical Activity

While the consumption of foods rich in carotenoids has been strongly linked to the reduction in incidences of diseases such as cancers, cardiovascular diseases, age-related macular degeneration, and cataracts, these thermolabile antioxidant compounds might be affected by processing [6]. According to Figure 2a, TCC significantly decreased (32%) after pomegranate seed blanching. Nevertheless, it was not significantly ($p > 0.05$) affected by seed microwave heating. The decrease in TCC after seed blanching could be explained by the breakdown of carotenoid molecules through isomerization and thermal oxidation [20]. These values were higher than TCC values reported in previous studies. For example, Costa et al. [42] reported TCC values ranging between 0.010 and 0.015 mg β-carotene/g PSO. Moreover, other studies failed to detect carotenoids in PSO [5]. The variation in the results could be due to dissimilarities in cultivars, fruit maturity, geographical location, and oil extraction process, among other factors [47]. It should also be highlighted that the absorbance in the spectrophotometric method might be increased by compounds other than carotenoids, which are active in the carotenoids' spectral range (400–500 nm) [48]. The TCC from other fruit seeds such as passion fruit and sour cherry ranged between 0.01 and 1.20 mg β-carotene/g oil [49,50]. The large disparity in the TCC of oil from different fruit seeds could reflect differences in the sensitivity of the methods of analysis, and it is, therefore, suggested that TCC calculated from the sum of individual carotenoids could be more reliable.

The total phenolic contents of PSO from unpretreated, blanched, and microwave-heated seeds are presented in Figure 2b. Blanching and microwave heating of pomegranate seeds significantly improved the TPC of cold-pressed oil by 21 and 37%, respectively. The findings suggest that blanching and microwaving of seeds facilitated the dissociation of glycosylated and esterified phenolic compounds, enhancing the amount of free phenolic compounds available for extraction [51]. The results coincide with Mazaheri et al. [20] and Lee et al. [36], who also reported improvement in TPC of cold-pressed black cumin and perilla seed oils after seed microwave and steam pretreatments, respectively. The levels of TPC from blanched and microwaved seeds did not significantly differ ($p > 0.05$). Given the potential bioactivity of phenolic compounds and the possible application of PSO as a functional food, enhancement of TPC after seed pretreatment was a desirable development. While the study of Zaouay et al. [52] reported TPC ranging from 0.03 to 0.07 mgGAE/g PSO, the TPC values in the current study varied from 1.33 to 1.83 mgGAE/g PSO (Figure 2b). Among other factors, the observed variation could be due to the selective nature of solvent extraction towards the phenolic compounds, hence the lower TPC values compared to the cold-pressed oil. On the contrary, Khoddami et al. [7] cold pressed oil from three different pomegranate cultivars and reported TPC values ranging from 8.52 to 10.44 mgGAE/g PSO that were 5.7 to 6.4 times higher than our results. These dissimilarities highlight the importance of preharvest and processing factors consideration in PSO processing and quality.

The antiradical radical activity of PSO was determined using the DPPH and ABTS assays. The DPPH radical scavenging activity of the cold-pressed PSO from unpretreated, blanched, and microwaved seeds is given in Figure 3a. While blanching seeds significantly improved the DPPH radical scavenging activity of the oil by 37%, microwave heating did not significantly ($p > 0.05$) change the DPPH radical scavenging activity of the cold-pressed PSO. Despite the insignificant effect of seed microwave pretreatment on the DPPH radical scavenging activity of the oil, previous studies on purslane and rape seed have shown increased DPPH radical scavenging activity in cold-pressed oil after seed microwave heating [53,54]. It is nevertheless noteworthy that seed physical and cellular structures that vary among different types of seeds and cultivars play a vital role in the efficiency of seed pretreatment, cold pressing, and recovery of the antioxidant compounds. Considering that antioxidant properties of oil have a major effect on its oxidative stability behavior, the PSO from blanched seeds might exhibit better stability and improved shelf life. The ABTS radical scavenging activity of the oil samples ranged between 10.95 and 11.55 mmol Trolox/g PSO, with oil from microwaved seeds exhibiting significantly higher ABTS scavenging activity than oil from blanched and unpretreated seeds (3 and 5%, respectively). The variation in the oil samples' (microwaved and blanched seeds) DPPH and ABTS radicals scavenging suggests that the antioxidant compounds react differently with the different radicals, due to factors such as synergism [55]. The high ABTS scavenging activity (10.95–11.55 mmol Trolox/g PSO) in the present study could be attributed to the high levels of phenols in the cold-pressed oils and their synergistic effect with other antioxidant compounds such as tocopherols, which are also abundantly found in PSO [55,56].

3.5. Fatty Acid Composition

Fatty acid composition is one of the most critical quality characteristics of oilseeds, considering that the suitability of the oil for food, nutraceutical, or pharmaceutical applications may be governed by the type of fatty acids. Table 2 shows the fatty acid composition of cold-pressed PSO from unpretreated, blanched, and microwaved seeds. Chromatograms of FAMES for the treatments are presented in Supplementary Figure S1. Ten different types of fatty acids were identified in PSO, with palmitic acid, oleic acid, linoleic acid, and punicic acid being the primary fatty acids and representing 7.73–9.22%, 9.53–10.48%, 15.93–17.11%, and 54.12–58.32% of the total composition, respectively. Other fatty acids identified but in minor quantities (0.06–4.32%) were arachidic acid, stearic acid, heneicosanoic acid, docosanoic acid, docosenoic acid, and linolenic acid. Generally, thermal pretreatment of oilseeds may alter the fatty acids composition due to the sensitivity of polyunsaturated fatty acids [33]. While microwave heating of seeds significantly decreased punicic acid by 7%, blanching did not significantly ($p > 0.05$) affect the fatty acid. In a similar study, Ozcan et al. [21] observed a 14 and 11% decrease in punicic acid after pomegranate seed roasting (150 °C for 10 min) and microwave heating (750 W for 7.5 min), respectively. Considering that punicic acid is implicated in most PSO biochemical properties, the decrease in punicic acid after microwave heating of the seeds in the current study was not desirable. Although punicic acid has been reported in other seeds such as bitter gourd [6], pomegranate seed remains the major source of this bioactive lipid. Compared to the literature, the levels of punicic acid (54.12–58.32%) in the current study are comparable to those reported by Costa et al. [42] (55.24–60.62%) from cold-pressed PSO. Nevertheless, some previous studies on cold-pressed PSO reported values that were higher (75.23–78.23%) than in the present study [7,57]. The dissimilarities in the punicic acid content could be ascribed to variation in processing techniques and pomegranate cultivars, among other factors. Linoleic acid and γ-linolenic acid, essential fatty acids, significantly increased by 1.1- and 3.4-fold after microwaving the pomegranate seed, whilst the blanching of the seeds did not significantly change the respective fatty acids. This indicates differences in microwaving and blanching modes of action and their impact on the fatty acids. Owing to the absence of appropriate enzymes, the human body cannot synthesize these essential fatty acids, and therefore their maximum extraction from oilseeds is essential [58]. Oleic acid, the major monosaturated fatty acid in PSO, insignificantly varied from 9.68% to 9.53% and 10.48% after blanching and microwave heating the seeds, respectively (Table 2). Although the concentration of palmitic acid and arachidic acid, the main saturated fatty acids, increased between 1.4 and 19% and 42 and 43%, respectively, after seed blanching and microwave heating, the levels of stearic acid, heneicosanoic acid, and docosanoic acid were not significantly changed by seed pretreatment. The insignificant effect of seed thermal pretreatment on some fatty acids has also been reported in prior research [19].

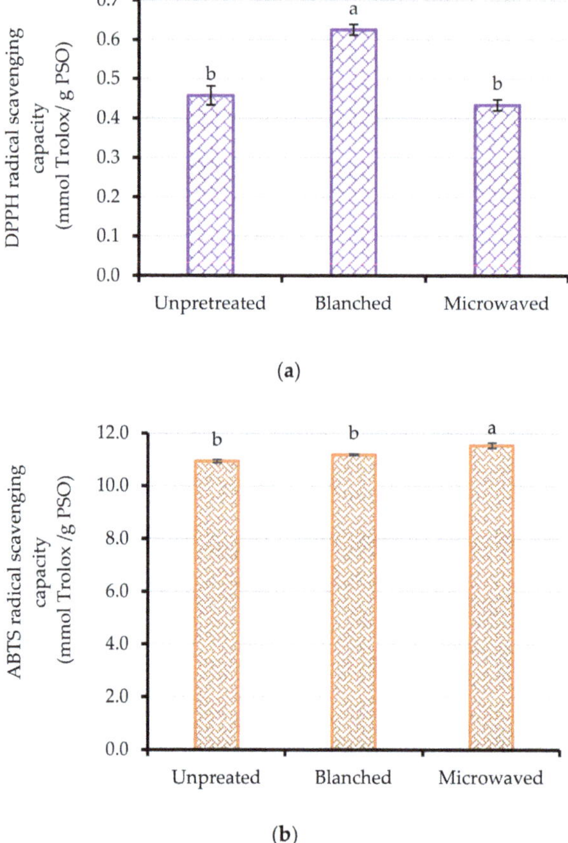

Figure 3. (**a**) DPPH and (**b**) ABTS radical scavenging capacity of cold-pressed pomegranate seed oil (PSO) from unpretreated, blanched (95 ± 2 °C/3 min), and microwave-heated (261 W for 102 s) seeds. Columns followed by different letters are significantly different ($p < 0.05$) according to Duncan's multiple range test. Vertical bars indicate the standard deviation of the mean.

Regarding total saturated fatty acids (SFA), blanching and microwave heating of seeds significantly increased the SFA by 9 and 18%, respectively. A 9% decrease in total monosaturated fatty acids (MUFA) was observed in PSO from blanched seeds, and this could be due to the significant decrease (7-fold) in docosenoic acid. No significant ($p > 0.05$) variation in MUFA of PSO pressed from microwaved seeds was observed. The total polyunsaturated fatty acids (PUFA) of oil from unpretreated seeds were 74.32% (Table 2). After seed microwave pretreatment, the level significantly decreased by 6%, whilst it insignificantly decreased in PSO from blanched seeds, indicating increased heat penetration and oxidation of polyunsaturated fatty acids during seed microwave heating (Table 1). The MUFA/PUFA index, which could be used as an indicator of the PSO stability to oxidation, among other factors [59], did not significantly vary after seed pretreatment. The finding implies that seed pretreatment did not affect the balance between the monosaturated and polyunsaturated fatty acids. However, the unsaturated fatty acids (UFA) to SFA index decreased after seed pretreatment, which could be explained by the significant increase in SFA after seed pretreatment.

Table 2. Fatty acid composition of cold-pressed pomegranate seed oil from unpretreated, blanched (95 $\pm$ 2 °C/3 min), and microwave-heated (261 W/102 s) seeds.

Treatment	Unpretreated	Blanched	Microwaved
SFA			
Palmitic (C16:0)	7.73 $\pm$ 0.30 [b]	7.84 $\pm$ 0.14 [b]	9.22 $\pm$ 0.22 [a]
Stearic (C18:0)	2.53 $\pm$ 0.04 [a]	2.53 $\pm$ 0.05 [a]	2.64 $\pm$ 0.05 [a]
Arachidic (20:0)	3.02 $\pm$ 0.16 [b]	4.32 $\pm$ 0.20 [a]	4.28 $\pm$ 0.18 [a]
Heneicosanoic acid (C21:0)	1.04 $\pm$ 0.21 [a]	1.14 $\pm$ 0.10 [a]	0.88 $\pm$ 0.07 [a]
Docosanoic acid (C22:0)	0.77 $\pm$ 0.11 [a]	0.61 $\pm$ 0.04 [a]	0.80 $\pm$ 0.02 [a]
$\sum$SFA	15.09 $\pm$ 0.50 [b]	16.44 $\pm$ 0.19 [a]	17.81 $\pm$ 0.30 [c]
MUFA			
Oleic (C18:1 *n*-9 *cis*)	9.68 $\pm$ 0.12 [b]	9.53 $\pm$ 0.24 [b]	10.48 $\pm$ 0.17 [a]
Docosenoic acid (C22:1)	0.92 $\pm$ 0.24 [a]	0.13 $\pm$ 0.01 [b]	0.25 $\pm$ 0.05 [b]
$\sum$MUFA	10.60 $\pm$ 0.15 [a]	9.65 $\pm$ 0.25 [b]	10.72 $\pm$ 0.22 [a]
PUFA			
Linoleic (C18:2 *n*-6 *cis*)	15.93 $\pm$ 0.49 [b]	16.00 $\pm$ 0.23 [ab]	17.11 $\pm$ 0.18 [a]
γ-Linolenic (C18:3 *n*-6)	0.07 $\pm$ 0.01 [b]	0.06 $\pm$ 0.00 [b]	0.24 $\pm$ 0.00 [a]
Punicic (*cis*-9 *trans*-11 *cis*-13 C18:3)	58.32 $\pm$ 0.87 [a]	57.85 $\pm$ 0.62 [a]	54.12 $\pm$ 0.36 [b]
$\sum$PUFA	74.32 $\pm$ 0.50 [a]	73.91 $\pm$ 0.41 [a]	71.47 $\pm$ 0.52 [b]
$\sum$MUFA/$\sum$PUFA index	0.14 $\pm$ 0.002 [ab]	0.13 $\pm$ 0.004 [b]	0.15 $\pm$ 0.004 [a]
$\sum$UFA/$\sum$SFA index	5.64 $\pm$ 0.22 [a]	5.09 $\pm$ 0.07 [b]	4.62 $\pm$ 0.09 [b]

Values (mean $\pm$ SD, n = 3) in the same row and followed by different superscript letters are significantly different ($p < 0.05$) according to Duncan's multiple range test, SFA = saturated fatty acids, MUFA = monounsaturated fatty acids, PUFA = polyunsaturated fatty acids, UFA = unsaturated fatty acids, $\sum$ = sum of SFA, MUFA, or PUFA.

3.6. Volatile Compounds

The results of volatile compounds of cold-pressed PSO from unpretreated, blanched, and microwave-heated seeds are presented in Table 3. A typical chromatogram of volatiles from the investigated pomegranate seed oil is presented in Supplementary Figure S2. Volatile compounds that can be perceived by humans have a greater influence on PSO flavor. These are the primary volatile flavor substances and constitute the characteristic flavor of PSO. The PSO samples showed varied volatile compounds belonging to the following chemical classes: alcohols, aldehydes, ketones, esters, carboxylic acids, and hydrocarbons. The groups of volatile compounds were comparable to the findings of Costa et al. [42] and Dun et al. [60] from cold-pressed pomegranate seed and peanut oils, respectively.

Esters, which are derived from the esterification of free fatty acids and alcohols, occur naturally in many fruits and enhance their flavors. Pentyl pentanoate, the only ester observed in the oil samples, was significantly higher in oil from blanched and microwaved seeds (10- and 1.5-fold, respectively) than in unpretreated seeds, suggesting that blanching and microwaving the seeds can enhance the oil flavor [61]. Ren et al. [62] also reported the enhancement of ester compounds in rapeseed oil after microwave pretreatment of the seeds. In addition, blanching and microwave heating of seeds may induce heterocyclic compounds through the Maillard reaction, which enhances the positive flavors. Furan and its derivatives belong to heterocyclic compounds and correlate with the flavor of foods [62].

In the present study, 2.5-dimethyltetrahydrofuran was significantly higher (69%) in PSO from microwaved seeds when compared to unpretreated seeds.

Table 3. Volatile compounds of cold-pressed pomegranate seed oil from unpretreated, blanched (95 ± 2 °C/3 min), and microwave-heated (261 W/102 s) seeds.

Treatment	Unpretreated	Blanched	Microwaved
Alcohols			
Cycloheptanol	0.77 ± 0.05 [b]	1.58 ± 0.08 [a]	1.32 ± 0.11 [a]
Ethanol	1.84 ± 0.30 [a]	1.74 ± 0.11 [a]	1.26 ± 0.08 [a]
Pentanol	10.82 ± 0.76 [b]	13.70 ± 0.84 [a]	10.92 ± 0.42 [b]
Hexanol	0.95 ± 0.10 [b]	1.83 ± 0.14 [a]	1.47 ± 0.17 [a]
Butanol	0.43 ± 0.01 [b]	0.59 ± 0.02 [a]	0.47 ± 0.00 [b]
Octanol	0.51 ± 0.02 [a]	0.52 ± 0.03 [a]	0.43 ± 0.05 [a]
Heptanol	0.80 ± 0.16 [a]	1.13 ± 0.08 [a]	ND
Aldehydes			
Hexanal	0.74 ± 0.01 [b]	1.30 ± 0.09 [a]	1.27 ± 0.08 [a]
3-Methylbutanal	1.36 ± 0.08 [b]	6.47 ± 0.96 [a]	6.71 ± 0.55 [a]
2-Heptenal	3.67 ± 0.65 [b]	5.71 ± 0.11 [a]	4.78 ± 0.35 [ab]
Nonanal	0.78 ± 0.00 [b]	1.04 ± 0.07 [a]	1.05 ± 0.01 [a]
2.4-*trans.trans*-Nonadienal	2.71 ± 0.34 [a]	2.53 ± 0.16 [a]	3.48 ± 0.50 [a]
2.4 Nonadienal	3.05 ± 0.34 [a]	2.72 ± 0.10 [a]	2.97 ± 0.01 [a]
Benzaldehyde	0.77 ± 0.08 [a]	0.64 ± 0.03 [a]	0.62 ± 0.04 [a]
Pentanal	10.99 ± 0.09 [a]	13.71 ± 0.83	11.87 ± 0.55 [a]
Trans-2-hexenal	ND	0.59 ± 0.01 [b]	0.50 ± 0.01 [a]
Benzene acetaldehyde	ND	1.97 ± 0.28 [b]	0.93 ± 0.17 [a]
Ketones			
2-Propanone	6.00 ± 0.40 [a]	7.01 ± 0.15 [a]	6.44 ± 0.29 [a]
5-Butyltetrahydro-2-furanone	0.22 ± 0.04	ND	ND
5-Butyl-5 H-furan-2-one	0.28 ± 0.03	ND	ND
Carboxylic acids			
Hexanoic acid	0.71 ± 0.03 [a]	0.69 ± 0.03 [a]	0.78 ± 0.04 [a]
Acetic acid	3.23 ± 0.54 [a]	3.46 ± 0.15 [a]	4.11 ± 0.08 [a]
Pentanoic acid	7.15 ± 1.19 [a]	ND	6.02 ± 0.13 [a]
Formic acid	ND	0.27 ± 0.01 [a]	0.29 ± 0.02 [a]
Butanoic acid	0.50 ± 0.02 [b]	0.62 ± 0.02 [a]	ND
Esters			
Pentyl pentanoate	0.69 ± 0.07 [b]	6.81 ± 0.24 [a]	1.05 ± 0.07 [b]
Furans			
2-Pentylfuran	ND	0.21 ± 0.00	ND
2-Butylfuran	ND	0.36± 0.04 [a]	0.37 ± 0.01 [a]
2.5-Dimethyltetrahydrofuran	0.22 ± 0.02 [a]	ND	0.35 ± 0.05 [b]
Hydrocarbons			
trans-alpha-Bergamotene	0.74 ± 0.09 [a]	0.32± 0.01 [b]	0.36 ± 0.03 [b]
2.3-Dimethyl-1-pentene	0.39 ± 0.06 [a]	0.51 ± 0.05 [a]	ND
Limonene	0.52 ± 0.09 [b]	0.13 ± 0.01 [a]	ND
Others			
Trichloromethane	2.43 ± 0.23 [a]	2.35 ±0.44 [a]	1.57 ± 0.09 [a]

Means ± standard deviation of analysis (n = 3). Different superscript letters in the same row indicate significant difference ($p < 0.05$) according to Duncan's multiple range test. ND = non-detected.

Moreover, other furans including 2-pentylfuran and 2-butylfuran were only detected in oil from blanched and microwaved seeds. This phenomenon indicates that the flavor of PSO may be improved by blanching and microwave pretreatment of seeds. Pentanol, the primary alcohol observed, was 25 to 27% higher in PSO from blanched seeds than microwave-heated and unpretreated seeds (Table 3). Likewise, butanol and cycloheptanol manifested higher in PSO extracts from blanched than microwaved and unpretreated seeds. Other alcohol compounds observed in lower concentrations such as ethanol and octanol were not significantly affected by seed blanching and microwave heating. Alcohols have also been reported in previous studies as important contributors to seed oil flavor [63].

Among the aldehydes, pentanal was the major compound observed in the cold-pressed PSO and was significantly higher in PSO extracts from blanched and microwave-heated seeds than unpretreated seeds. Pentanal is characterized by a nutty and fruity flavor and has been naturally found in other seed oils such as sesame, olive, and peanut [63]. Other compounds including hexanal, 3-methylbutanal, 2-heptenal, and nonanal were also significantly higher in oil extracts from blanched and microwaved pomegranate seeds (Table 2). Aldehydes in seed oil are primarily produced either through the lipoxygenase pathway during oilseed cell fragmentation or automatic oxidation of the oil during production [64]. Hexanal is a typical oxidation volatile and has been commonly used as a quality indicator for lipid oxidation in seed oils. It is characterized by green, oily, and fruity odors [60]. Its level has been positively correlated with rancid taste. As shown in Table 3, blanching and microwave pretreatment of seeds may promote the oxidative degradation of the oil. Our results, therefore, indicate higher oxidation liability of oil from pretreated pomegranate seeds compared with unpretreated seeds. The PV, AV, and AnV results found in this study support these findings (Table 1). While 2-propanone did not significantly differ in all oil samples, other ketones such as 5-butyltetrahydro-2-furanone and 5-butyl-5H-furan-2-one were only detected in oil from pretreated seeds. Saturated fatty acids including hexanoic acid, acetic acid, pentanoic acid, formic acid, and butanoic acid did not significantly ($p > 0.05$) vary among the PSO samples. These free fatty acids, which are linked to sour and pungent sensations synonymous with sensory defects, could have been produced from the oxidation of their respective aldehydes [65]. It can be stated that, although seed blanching and microwave heating may augment the positive flavor of cold-pressed PSO, they may also promote the development of undesirable flavors.

4. Conclusions

In the current study, the effect of blanching and microwave pretreatment of seeds on the quality of cold-pressed PSO was investigated. Blanching and microwave pretreatment of seeds prior to pressing improved oil yield, total phenolic content, flavor compounds, and DPPH and ABTS radical scavenging capacity. The findings are desirable to pomegranate seed oil processors and consumers along the value chain, given that cold pressing is also a greener and safer technology compared to the use of solvents such as hexane. The levels of oxidation indices including the peroxide value, free fatty acids, acid value, ρ-anisidine value, and total oxidation value also increased. Nevertheless, the oil quality conformed to the requirements of the Codex Alimentarius Commission standard (CODEX STAN 19-1981) on cold-pressed vegetable oils.

On the other hand, blanching and microwave heating of seeds decreased the pomegranate seed oil's yellowness index, whilst the refractive index was not significantly affected. Although both blanching and microwave pretreatment of seeds added value to the cold-pressed PSO, the oil extracted from blanched seeds exhibited lower oxidation indices. The finding affirms that the processing technique is one of the important seed oil quality determinants. Microwave pretreatment of seeds before cold pressing significantly increased palmitic acid, oleic acid, and linoleic acid, whilst it decreased the level of punicic acid, highlighting increased heat penetration and oxidation of the conjugated fatty acid. On the contrary, blanching of seeds did not significantly affect the fatty acid composition of PSO, an indication that the nutritional quality of the oil was not significantly affected. In conclusion, blanching of seeds is a practical step that could be incorporated into mechanical production of PSO.

Supplementary Materials: The following are available online at https://www.mdpi.com/article/10.3390/foods10040712/s1, Figure S1: Chromatograms of FAMES of oil from unpretreated, blanched and microwave heated pomegranate seed, Figure S2: Typical chromatograms of volatiles from the pomegranate seed oil.

Author Contributions: Conceptualization, T.K., O.A.F., and U.L.O.; formal analysis, T.K.; funding acquisition, O.A.F. and U.L.O.; investigation, T.K.; methodology, T.K.; supervision, O.A.F. and U.L.O.; validation, O.A.F. and U.L.O.; visualization, T.K.; writing—original draft, T.K.; writing—review and editing, T.K., O.A.F., and U.L.O. All authors have read and agreed to the published version of the manuscript.

Funding: This research was funded by the National Research Foundation of South Africa grant number 64813 and the APC was partly funded by Stellenbosch University.

Data Availability Statement: The data presented in this study are available on request from the corresponding author. Some of the data is contained within Supplementary Materials.

Acknowledgments: This work is based on the research supported wholly or in part by the National Research Foundation of South Africa (Grant Numbers: 64813). The opinions, findings, and conclusions or recommendations expressed are those of the author(s) alone, and the NRF accepts no liability whatsoever in this regard. The authors are grateful to the Cape Peninsula University of Technology, Agrifood Technology Station in South Africa for providing the oil press that was used to cold press the pomegranate seed oil.

Conflicts of Interest: The authors declare that they have no conflict of interest.

References

1. Rojo Gutiérrez, E.; Carrasco Molinar, O.; Tirado Gallegos, J.M.; Levario Gómez, A.; Chávez González, M.L.; Baeza Jiménez, R.; Buenrostro Figueroa, J.J. Evaluation of green extraction processes, lipid composition and antioxidant activity of pomegranate seed oil. *J. Food Meas. Charact.* **2021**, *101*, 1–10.
2. Fawole, O.A.; Opara, U.L. Developmental changes in maturity indices of pomegranate fruit: A descriptive review. *Sci. Hortic.* **2013**, *159*, 152–161. [CrossRef]
3. Moga, M.; Dimienescu, O.; Bălan, A.; Dima, L.; Toma, S.; Bîgiu, N.; Blidaru, A. Pharmacological and therapeutic properties of *Punica granatum* phytochemicals: Possible roles in breast cancer. *Molecules* **2021**, *26*, 1054. [CrossRef]
4. Chaves, F.M.; Pavan, I.C.B.; Da Silva, L.G.S.; De Freitas, L.B.; Rostagno, M.A.; Antunes, A.E.C.; Bezerra, R.M.N.; Simabuco, F.M. Pomegranate juice and peel extracts are able to inhibit proliferation, migration and colony formation of prostate cancer cell lines and modulate the Akt/mTOR/S6K signaling pathway. *Plant Foods Hum. Nutr.* **2020**, *75*, 54–62. [CrossRef] [PubMed]
5. Fernandes, L.; Pereira, J.A.; Lopez-Cortes, I.; Salazar, D.M.; Ramalhosa, E.; Casal, S. Fatty acid, vitamin E and sterols compo sition of seed oils from nine different pomegranate (*Punica granatum* L.) cultivars grown in Spain. *J. Food Compost. Anal.* **2015**, *39*, 13–22. [CrossRef]
6. Aruna, P.; Venkataramanamma, D.; Singh, A.K.; Singh, R. Health benefits of punicic acid: A review. *Compr. Rev. Food Sci. Food Saf.* **2015**, *15*, 16–27. [CrossRef]
7. Khoddami, A.; Bin, Y.; Man, C.; Roberts, T.H. Physicochemical properties, and fatty acid profile of seed oils from pomegranate (*Punica granatum* L.) extracted by cold pressing. *Eur. J. Lipid Sci. Tech.* **2014**, *116*, 553–562. [CrossRef]
8. Aruna, P.; Manohar, B.; Singh, R.P. Processing of pomegranate seed waste and mass transfer studies of extraction of pome-granate seed oil. *J. Food Process. Preserv.* **2018**, *42*, 1–11. [CrossRef]
9. Kaseke, T.; Opara, U.L.; Fawole, O.A. Effect of blanching pomegranate seed on physicochemical attributes, bioactive com-pounds, and antioxidant activity of extracted oil. *Molecules* **2020**, *25*, 2554. [CrossRef]
10. Chew, S.C. Cold-pressed rapeseed (*Brassicanapus*) oil: Chemistry and functionality. *Food Res. Int.* **2020**, *131*, 1–13. [CrossRef]
11. Parry, J.; Yu, L. Fatty acid content and antioxidant properties of cold-pressed black raspberry seed oil and meal. *J. Food Sci.* **2006**, *69*, FCT189–FCT193. [CrossRef]
12. Teh, S.-S.; Birch, J. Physicochemical and quality characteristics of cold-pressed hemp, flax, and canola seed oils. *J. Food Compos. Anal.* **2013**, *30*, 26–31. [CrossRef]
13. Östbring, K.; Malmqvist, E.; Nilsson, K.; Rosenlind, I.; Rayner, M. The effects of oil extraction methods on recovery yield and emulsifying properties of proteins from rapeseed meal and press cake. *Foods* **2020**, *9*, 19. [CrossRef] [PubMed]
14. Rabrenović, B.B.; Dimić, E.B.; Novaković, M.M.; Tešević, V.V.; Basić, Z.N. The most important bioactive components of cold pressed oil from different pumpkin (*Cucurbita pepo* L.) seeds. *LWT* **2014**, *55*, 521–527. [CrossRef]
15. Durdevic, S.; Milovanović, S.; Šavikin, K.; Ristić, M.; Menković, N.; Pljevljakušić, D.; Petrović, S.; Bogdanović, A.B. Improvement of supercritical CO_2 and n-hexane extraction of wild growing pomegranate seed oil by microwave pretreatment. *Ind. Crop. Prod.* **2017**, *104*, 21–27. [CrossRef]
16. Koubaa, M.; Mhemdi, H.; Barba, F.J.; Roohinejad, S.; Greiner, R.; Vorobiev, E. Oilseed treatment by ultrasounds and microwaves to improve oil yield and quality: An overview. *Food Res. Int.* **2016**, *85*, 59–66. [CrossRef]
17. Wang, H.-O.; Fu, Q.-Q.; Chen, S.-J.; Hu, Z.-C.; Xie, H.-X. Effect of hot-water blanching pretreatment on drying characteristics and product qualities for the novel integrated freeze-drying of apple slices. *J. Food Qual.* **2018**, *2018*, 1–12. [CrossRef]
18. Uquiche, E.; Jeréz, M.; Ortíz, J. Effect of pretreatment with microwaves on mechanical extraction yield and quality of vegetable oil from Chilean hazelnuts (*Gevuina avellana Mol*). *Innov. Food Sci. Emerg. Technol.* **2008**, *9*, 495–500. [CrossRef]

19. Wroniak, M.; Rekas, A.; Siger, A.; Janowicz, M. Microwave pretreatment effects on the changes in seeds microstructure, chemical composition, and oxidative stability of rapeseed oil. *LWT Food Sci. Technol* **2016**, *68*, 634–641. [CrossRef]
20. Mazaheri, Y.; Torbati, M.; Azadmard-Damirchi, S.; Savage, G.P. Effect of roasting and microwave pre-treatments of *Nigella sativa* L. seeds on lipase activity and the quality of the oil. *Food Chem.* **2019**, *274*, 480–486. [CrossRef]
21. Özcan, M.M.; Al-Juhaimi, F.Y.; Ahmed, I.A.M.; Osman, M.A.; Gassem, M.A. Effect of different microwave power setting on quality of chia seed oil obtained in a cold press. *Food Chem.* **2019**, *278*, 190–196. [CrossRef] [PubMed]
22. Soni, A.; Smith, J.; Thompson, A.; Brightwell, G. Microwave-induced thermal sterilization- A review on history, technical progress, advantages, and challenges as compared to the conventional methods. *Trends Food Sci. Technol.* **2020**, *97*, 433–442. [CrossRef]
23. Rekas, A.; Scibisz, I.; Siger, A.; Wroniak, M. The effect of microwave pretreatment of seeds on the stability and degradation kinetics of phenolics compounds in rapeseed oil during long time storage. *Food Chem.* **2017**, *222*, 43–54. [CrossRef]
24. Kaseke, T.; Opara, U.L.; Fawole, O.A. Effect of microwave pretreatment of seeds on the quality and antioxidant capacity of pomegranate seed oil. *Foods* **2020**, *9*, 1287. [CrossRef] [PubMed]
25. AOCS. *Official Methods and Recommended Practices of the American Oil Chemists' Society*; Firestone, D., Ed.; AOCS Press: Champaign, IL, USA, 2003.
26. Cruz, R.M.S.; Khmelinski, I.; Vieira, M.C. *Methods in Food Analysis*, 1st ed.; Taylor and Francis Group: Abingdon, UK, 2016; pp. 140–190.
27. Ranjith, A.; Kumar, K.S.; Venugopalan, V.V.; Arumughan, C.; Sawhney, R.C.; Singh, V. Fatty acids, tocols, and carotenoids in pulp oil of three sea buckthorn species (*Hippophae rhamnoides, H. salicifolia, and H. tibetana*) grown in the Indian Himalayas. *J. Am. Oil Chem. Soc.* **2006**, *83*, 359–364. [CrossRef]
28. Kale, S.; Matthau's, B.; Aljuhaimi, F.; Ahmed, I.S.M.; Özcan, M.M.; Ghafoor, K.; Babiker, E.E.; Osman, M.A.; Gassem, M.A.; Allah, H.A.S. A comparative study of the properties of 10 variety melon seeds and seed oils. *J. Food Process Preserv.* **2020**, *44*, 1–10. [CrossRef]
29. Uluata, S.; Özdemir, N. Antioxidant activities and oxidative stabilities of some unconventional oilseeds. *J. Am. Oil Chem. Soc.* **2011**, *89*, 551–559. [CrossRef]
30. Siano, F.; Straccia, M.C.; Paolucci, M.; Fasulo, G.; Boscaino, F.; Volpe, M.G. Physico-chemical properties and fatty acid composition of pomegranate, cherry, and pumpkin seed oils. *J. Sci. Food Agric.* **2015**, *96*, 1730–1735. [CrossRef]
31. Mphahlele, R.R.; Fawole, O.A.; Makunga, N.P.; Opara, U.L. Functional properties of pomegranate fruit parts: Influence of packaging systems and storage time. *J. Food Meas. Charact.* **2017**, *11*, 2233–2246. [CrossRef]
32. Wang, L.; Chen, Z.; Han, B.; Wu, W.; Zhao, Q.; Wei, C.; Liu, W. Comprehensive analysis of volatile compounds in cold-pressed safflower seed oil from Xinjiang, China. *Food Sci. Nutr.* **2019**, *8*, 903–914. [CrossRef] [PubMed]
33. Azadmard-Damirchi, S.; Habibi-Nodeh, F.; Hesari, J.; Nemati, J.; Achachlouei, B.F. Effect of pretreatment with microwaves on oxidative stability and nutraceuticals content of oil from rapeseed. *Food Chem.* **2010**, *121*, 1211–1215. [CrossRef]
34. Gaber, M.A.F.M.; Knoerzer, K.; Mansour, M.P.; Trujillo, F.J.; Juliano, P.; Shrestha, P. Improved canola oil expeller extraction using a pilot-scale continuous flow microwave system for pre-treatment of seeds and flaked seeds. *J. Food Eng.* **2020**, *284*, 110053. [CrossRef]
35. Bakhshabadi, H.; Mirzaei, H.; Ghodsvali, A.; Jafari, S.M.; Ziaiifar, A.M.; Farzaneh, V. The effect of microwave pretreatment on some physico-chemical properties and bioactivity of black cumin seeds' oil. *Ind. Crop. Prod.* **2017**, *97*, 1–9. [CrossRef]
36. Lee, K.-Y.; Rahman, M.S.; Kim, A.-N.; Jeong, E.-J.; Kim, B.-G.; Lee, M.-H.; Kim, H.-J.; Choi, S.-G. Effect of superheated steam treatment on yield, physicochemical properties, and volatile profiles of perilla seed oil. *LWT* **2021**, *135*, 110240. [CrossRef]
37. Fathi-Achachlouei, B.; Azadmard-Damirchi, S.; Zahedi, Y.; Shaddel, R. Microwave pretreatment as a promising strategy for increment of nutraceutical content and extraction yield of oil from milk thistle seed. *Ind. Crop. Prod.* **2019**, *128*, 527–533. [CrossRef]
38. Pathare, P.B.; Opara, U.L.; Al-Said, F.A.-J. Colour measurement and analysis in fresh and processed foods: A review. *Food Bioprocess. Technol.* **2013**, *6*, 36–60. [CrossRef]
39. Khoo, H.; Nagendra Prasad, K.; Kong, K.; Jiang, Y.; Ismail, A. Carotenoids, and their isomers: Colour pigments in fruits and vegetables. *Molecules* **2011**, *16*, 1710–1738. [CrossRef] [PubMed]
40. Kha, T.C.; Nguyen, M.H.; Roach, P.D.; Stathopoulous, C.E. Effect of galic aril microwave processing conditions on oil extraction efficiency and β-carotene and lycopene content. *J. Food Eng.* **2013**, *117*, 486–491. [CrossRef]
41. Davis, J.P.; Sweigart, D.S.; Price, K.M.; Dean, L.L.; Sanders, T.H. Refractive index and density measurements of peanut oil for determining oleic and linoleic acid contents. *J. Am. Oil Chem. Soc.* **2013**, *90*, 199–206. [CrossRef]
42. Costa, A.; Silva, L.; Torres, A. Chemical composition of commercial cold-pressed pomegranate (*Punica granatum*) seed oil from Turkey and Israel, and the use of bioactive compounds for samples' origin preliminary discrimination. *J. Food Compos. Anal.* **2019**, *75*, 8–16. [CrossRef]
43. Codex Alimentarius. Standard for Named Vegetable Oils-Codex Stan 210-1999 Standard for Named vegetable oils-Codex Stan 210-1999. Codex Alimentarius. 1999. Available online: http://www.fao.org/fao-whocodexalimentarius/codex-texts/list-standards (accessed on 7 September 2020).
44. Van der Merwe, G.H.; du Plessis, L.M.; Taylor, J.R.N. Changes in chemical quality indices during long-term storage of palm-olein oil under heated storage and transport-type conditions. *J. Sci. Food Agric.* **2003**, *84*, 52–58. [CrossRef]
45. Koohikamali, S.; Alam, M.S. Improvement in nutritional quality and thermal stability of palm olein blended with macadamia oil for deep-fat frying application. *J. Food Sci. Technol.* **2019**, *56*, 5063–5073. [CrossRef] [PubMed]

46. Moghimi, M.; Farzaneh, V.; Bakhshabadi, H. The effect of ultrasound pretreatment on some selected physicochemical properties of black cumin (*Nigella Sativa*). *Nutrire* **2018**, *43*, 18. [CrossRef]

47. Kaseke, T.; Opara, U.L.; Fawole, O.A. Fatty acid composition, bioactive phytochemicals, antioxidant properties and oxidative stability of edible fruit seed oil: Effect of preharvest and processing factors. *Heliyon* **2020**, *6*, 1–15. [CrossRef] [PubMed]

48. Corbu, A.R.; Rotaru, A.; Nour, V. Edible vegetable oils enriched with carotenoids extracted from by-products of sea buckthorn (*Hippophae rhamnoides* ssp. *sinensis*): The investigation of some characteristic properties, oxidative stability, and the effect on thermal behaviour. *J. Therm. Anal. Calorim.* **2020**, *142*, 735–747. [CrossRef]

49. De Santana, F.C.; Shinagawa, F.B.; Araujo, E.D.S.; Costa, A.M.; Mancini-Filho, J. Chemical composition and antioxidant capacity of Brazilian passiflora seed oils. *J. Food Sci.* **2015**, *80*, C2647–C2654. [CrossRef]

50. Górnaś, P.; Rudzińska, M.; Raczyk, M.; Mišina, I.; Soliven, A.; Segliņa, D. Composition of bioactive compounds in kernel oils recovered from sour cherry (*Prunus cerasus* L.) by-products: Impact of the cultivar on potential applications. *Ind. Crop. Prod.* **2016**, *82*, 44–50. [CrossRef]

51. Kumar, N.; Goel, N. Phenolic acids: Natural versatile molecules with promising therapeutic applications. *Biotechnol. Rep.* **2019**, *24*, e00370. [CrossRef]

52. Zaouay, F.; Brahem, M.; Boussaa, F.; Haddada, F.M.; Tounsi, M.S.; Mars, M. Effects of fruit cracking and maturity stage on quality attributes and fatty acid composition of pomegranate seed oils. *Int. J. Fruit Sci.* **2020**, *20*, S1959–S1968. [CrossRef]

53. Yang, M.; Huang, F.; Liu, C.; Zheng, C.; Zhou, Q.; Wang, H. Influence of microwave treatment of rapeseed on minor compo nents content and oxidative stability of oil. *Food Bioprocess. Technol.* **2013**, *6*, 3206–3216. [CrossRef]

54. Delfan-Hosseini, S.; Nayebzadeh, K.; Mirmoghtadaie, L.; Kavosi, M.; Hosseini, S.M. Effect of extraction process on composition, oxidative stability, and rheological properties of purslane seed oil. *Food Chem.* **2017**, *222*, 61–66. [CrossRef]

55. Choe, E.; Min, D.B. Mechanisms, and factors for edible oil oxidation. *Compr. Rev. Food Sci. Food Saf.* **2006**, *5*, 169–186. [CrossRef]

56. De Melo, I.L.P.; De Carvalho, E.B.T.; Silva, A.M.D.O.E.; Yoshime, L.T.; Sattler, J.A.G.; Pavan, R.T.; Mancini-Filho, J. Characterization of constituents, quality, and stability of pomegranate seed oil (*Punica granatum* L.). *Food Sci. Technol.* **2016**, *36*, 132–139. [CrossRef]

57. Silva, L.D.O.; Ranquine, L.G.; Monteiro, M.; Torres, A.G. Pomegranate (*Punica granatum* L.) seed oil enriched with conjugated linolenic acid (cLnA), phenolic compounds and tocopherols: Improved extraction of a specialty oil by supercritical CO_2. *J. Supercrit. Fluids* **2019**, *147*, 126–137. [CrossRef]

58. Orsavova, J.; Misurcova, L.; Ambrozova, J.V.; Vicha, R.; Mlcek, J. Fatty acids composition of vegetable oils and its contribution to dietary energy intake and dependence of cardiovascular mortality on dietary intake of fatty acids. *Int. J. Mol. Sci.* **2015**, *16*, 12871–12890. [CrossRef] [PubMed]

59. Prabakaran, M.; Lee, K.-J.; An, Y.; Kwon, C.; Kim, S.; Yang, Y.; Ahmad, A.; Kim, S.-H.; Chung, I.-M. Changes in soybean (*Glycine max* L.) flour fatty-acid content based on storage temperature and duration. *Molecules* **2018**, *23*, 2713. [CrossRef] [PubMed]

60. Dun, Q.; Yao, L.; Deng, Z.; Li, H.; Li, J.; Fan, Y.; Zhang, B. Effects of hot and cold-pressed processes on volatile compounds of peanut oil and corresponding analysis of characteristic flavour components. *LWT Food Sci. Technol.* **2019**, *112*, 1–9. [CrossRef]

61. Jiang, X.; Wu, S.; Zhou, Z.; Akoh, C.C. Physicochemical properties and volatile profiles of cold-pressed *Trichosanthes kirilowii maxim* seed oils. *Int. J. Food Prop.* **2016**, *19*, 1765–1775. [CrossRef]

62. Ren, X.; Wang, L.; Xu, B.; Wei, B.; Liu, Y.; Zhou, C.; Ma, H.; Wang, Z. Influence of microwave pretreatment on the flavour attributes and oxidative stability of cold-pressed rapeseed oil. *Dry. Technol.* **2019**, *37*, 397–408. [CrossRef]

63. Wei, C.Q.; Liu, W.Y.; Xi, W.P.; Cao, D.; Zhang, H.J.; Ding, M.; Chen, L.; Xu, Y.Y.; Huang, K.X. Comparison of volatile compounds of hot-pressed, cold-pressed, and solvent-extracted flaxseed oil analyzed by SPME-GC/MS combined with electronic nose: Major volatiles can be used as markers to distinguish differently processed oils. *Eur. J. Lipid Sci. Technol.* **2015**, *117*, 320–330. [CrossRef]

64. Zhou, Y.; Fan, W.; Chu, F.; Wang, C.; Pei, D. Identification of volatile oxidation compounds as potential markers of walnut oil quality. *J. Food Sci.* **2018**, *83*, 2745–2752. [CrossRef] [PubMed]

65. Xu, L.; Yu, X.; Li, M.; Chen, J.; Wang, X. Monitoring oxidative stability and changes in key volatile compounds in edible oils during ambient storage through HS-SPME/GC–MS. *Int. J. Food Prop.* **2017**, *20*, S2926–S2938. [CrossRef]

 *foods*

Article

Biological Effect of Different Spinach Extracts in Comparison with the Individual Components of the Phytocomplex

Laura Arru [1,2,*,†], Francesca Mussi [1,3,†], Luca Forti [1] and Annamaria Buschini [3,4,*]

1   Department of Life Sciences, University of Modena and Reggio Emilia, 41125 Modena, Italy; francesca.mussi@unipr.it (F.M.); luca.forti@unimore.it (L.F.)
2   International Center BIOGEST-SITEIA, University of Modena and Reggio Emilia, 42122 Reggio Emilia, Italy
3   Department of Chemistry, Life Sciences and Environmental Sustainability, University of Parma, 43124 Parma, Italy
4   International Center COMT, University of Parma, 43124 Parma, Italy
*   Correspondence: laura.arru@unimore.it (L.A.); annamaria.buschini@unipr.it (A.B.); Tel.: +39-0522-52-2016 (L.A.); +39-0521-90-5607 (A.B.)
†   The two authors contributed equally to the work.

Abstract: The Mediterranean-style diet is rich in fruit and vegetables and has a great impact on the prevention of major chronic diseases, such as cardiovascular diseases and cancer. In this work we investigated the ability of spinach extracts obtained by different extraction methods and of the single main components of the phytocomplex, alone or mixed, to modulate proliferation, antioxidant defense, and genotoxicity of HT29 human colorectal cells. Spinach extracts show dose-dependent activity, increasing the level of intracellular endogenous reactive oxygen species (ROS) when tested at higher doses. In the presence of oxidative stress, the activity is related to the oxidizing agent involved (H_2O_2 or menadione) and by the extraction method. The single components of the phytocomplex, alone or mixed, do not alter the intracellular endogenous level of ROS but again, in the presence of an oxidative insult, the modulation of antioxidant defense depends on the oxidizing agent used. The application of the phytocomplex extracts seem to be more effective than the application of the single phytocomplex components.

Keywords: phytocomplex; spinach extracts; colon cancer cell line; phytochemicals; antioxidants

check for
updates

Citation: Arru, L.; Mussi, F.; Forti, L.; Buschini, A. Biological Effect of Different Spinach Extracts in Comparison with the Individual Components of the Phytocomplex. *Foods* **2021**, *10*, 382. https://doi.org/10.3390/foods10020382

Academic Editor: Francisca Rodrigues

Received: 15 January 2021
Accepted: 4 February 2021
Published: 9 February 2021

Publisher's Note: MDPI stays neutral with regard to jurisdictional claims in published maps and institutional affiliations.

1. Introduction

The lifestyle of the most industrialized countries brings many benefits but can induce potential risks that can worsen the quality of life. Sedentary lifestyle, improper nutrition, unbalanced diet, chaotic pace of today's life, just to name a few, can have negative effects on human health; especially a fat-rich diet leads to oxidative stress which in turn can contribute to the onset of degenerative diseases [1].

There is increasing evidence of a close correlation between diet and risk of cancer, both in positive (prevention) and negative (development of the disease) sense [2]. The introduction of flavonoids, carotenoids, omega-3 fatty acids, vitamins, minerals, antioxidants through fruit and vegetables seems to have positive effects in reducing some types of cancer and chronic diseases, thanks to the ability of these molecules to reduce the damage caused by reactive oxygen species (ROS) [2–4].

Polyphenols, and antioxidants in general, act as free radical scavengers and metal chelators, helping the physiological cell response in counteracting the damage induced by ROS.

Intracellular ROS are normally generated during the cellular biochemical processes, and several cellular signaling pathways are regulated by ROS [5]. However, when their level happens to be increased by external agents (i.e., ionizing radiation, pollutants with chlorinated compounds or metal ions that may directly or indirectly generate ROS) they can damage proteins, lipids, and DNA, leading to impaired physiological functions with a decreased proliferative response, defective host defense and cell death [6].

Oxidative stress occurs when there is an imbalance between the intracellular levels of ROS and the cell defense systems; this can help the insurgence of diseases such as cardiovascular diseases, neurodegenerative diseases, and cancer [7]. Cells need to maintain the physiological homeostasis by balancing the ROS levels and the antioxidant defenses [6].

Not all antioxidant molecules have the same protective effect, but it is also known that the effect of a phytocomplex, a set of active ingredients contained in vegetable food, is often synergistic and greater than the effect that the single components can have. This can be in part explained because there is not a singular molecular target for a disease, but often disease is a result of a multi-factorial causality [8].

Furthermore, it should also be considered that the relative percentages of the constituents of the phytocomplex could play a decisive role in determining its effectiveness. Often, scientific attention focuses on a single active molecule, or on a few known constituents. However, the synergistic effect of the phytocomplex can be lost when testing single molecules or when, in the effort of extracting the phytocomplex, part of its minor components is lost [8–11].

Spinach (*Spinacia oleracea* L.) belongs to the family of Chenopodiaceae and it is a proven source of essential nutrients such as carotene (a precursor of vitamin A), ascorbic acid, and several types of minerals. According to the Agricultural Research Service of the U.S. Department of Agriculture, 100 g of fresh spinach provides at least 20% or more of the recommended dietary intake of β-carotene (provitamin A), lutein, folate (vitamin B9), α-tocopherol (vitamin E) and ascorbic acid (vitamin C). Moreover, spinach leaves contain flavonoids [12] and phenolic acids such as ferulic acid, ortho-coumaric and para-coumaric acids [13]. In 2009, Hait-Darshan and colleagues isolated from spinach leaves a mixture of antioxidants defined NAO (natural antioxidant) that contains aromatic polyphenols, including the phenolic acids and the derivatives of the glucuronic acid [14]. NAO can effectively counteract free radicals [15,16] resulting in an antiproliferative and anti-inflammatory potential, in vivo and in vitro [17].

In a previous study [18], we have already shown the ability of spinach leaf juice to inhibit the proliferation of the human HT29 colon cancer cell line in a time and dose-dependent manner. The juice significantly also reduced the damage induced by a known oxidant agent up to 80%.

In this study, we evaluated the biological effects of different spinach extracts (hydrophilic, liquid nitrogen, and water extraction) and of some of the main components, alone or mixed together, on the human colorectal adenocarcinoma HT29, which is a cell line representative of the gastrointestinal tract and a recognized good model for the study of the correlation between diet and carcinogenesis [19]. We chose this approach as a step forward our previous research [18]: the spinach juice from fresh leaves proved to have both anti-proliferative and antioxidant effect. This time we aimed (1) to investigate if different extraction methods could positively influence this outcome; (2) to test the effect of the main components of the phytocomplex previously identified, when submitted alone or mixed in a sort of artificial simplified phytocomplex.

2. Materials and Methods

2.1. Spinacia oleracea Extracts

In this study, three extraction methods have been used to obtain a phytocomplex rich in different polyphenols and other relevant biological active molecules fractions. To obtain an extract rich in hydrophilic compounds (hydrophilic extract, HE), water was added to spinach leaves in a 1:1 (*w/v*) ratio [15]. The leaves were ground in a mortar, filtered with sterile gauze, the liquid transferred into a micro-tube, and centrifuged twice for 12 min at 15,000× *g*. The collected supernatant was concentrated (Speed Vacuum Concentrator, Eppendorf 5301, Eppendorf, AG, Hamburg, Germany) to 25%, then added a 9:1 volume of acetone in water, vortexed for 5 min and centrifuged for 12 min at 15,000× *g*. The supernatant was carefully transferred to a fresh tube and completely dried. The dry extract was chilled on ice and stored at $-20\,°C$ until use. Water extract (WE) was obtained by simply adding sterile distilled water in a 1:1 ratio in the first step of the above-described procedure. Liquid nitrogen extract (NE) was obtained by grinding leaf samples in a mortar with liquid nitrogen, mashed material weighed, and transferred in a 10 mL syringe to be filtered with a 60 μm nylon filter. The extract was centrifuged for 10 min at $8000 \times g$ ($4°C$) and the supernatant was filtered (0.2 μm filter).

2.2. HT29 Cell Line

Cells were thawed and grown in tissue culture flasks as a monolayer in DMEM (Dulbecco's Modified Eagle Medium), supplemented with 1% L-glutamine (2 mM), 1% penicillin (5000 U/mL)/streptomycin (5000 μg/mL) and 10% fetal bovine serum (FBS) at $37\,°C$ in a humidified CO_2 (5%) incubator. The cultured cells were trypsinized with trypsin/EDTA for a maximum of 8 min and seeded with a subcultivation ratio of 1:3–1:8. Determination of cell numbers and viabilities was performed with the trypan blue exclusion test.

2.3. Modulation of the Proliferation

2.3.1. MTS Assay

To determine cell viability, in the exponential phase of the growth cells were seeded at 5×10^4/mL in 96-well plates in medium supplemented with 1% glutamine, 0.5% penicillin/streptomycin, and 5% fetal bovine serum. After seeding (24 h), cells were treated, in quadruplicate, with increasing concentrations of phytochemicals (1–500 μM) and incubated for 24 and 48 h. Ascorbic acid, 20-hydroxyecdysone, ferulic acid, 2-hydroxycinnamic acid, *p*-coumaric acid, β-carotene, and lutein were from Sigma-Aldrich Company Ltd. (Milan, Italy) and resuspended in dimethylsulfoxide. The cytotoxicity assay was performed by adding a small amount of the CellTiter 96R AQueous One Solution Cell Proliferation Assay (Promega Corporation, Madison, WI, USA) directly to culture wells, incubating for 4 h and then recording the absorbance at 450 nm with a 96-well plate reader (MULTISKAN EX, Thermo Electron Corporation, Vantaa, Finland). The percentage of cell growth is calculated as:

$$\text{growth \%} = 100 - [1 - (\text{OD450 treated}/\text{OD450 untreated})] \times 100 \tag{1}$$

2.3.2. Trypan Blue Exclusion Method

Different concentrations of spinach extracts (1%, 5%, 10%, 50%) were added to the cells medium. Cells were seeded in 6-well plates (2 mL/well) at the density of 2×10^5 cell/well. After 24 or 48 h of treatment, cells were trypsinized and resuspended in DMEM; a 1:1 dilution of the cell suspension was obtained using a 0.4% trypan blue solution (BioWhittake®, Lonza, Walkersville, MD, USA). The dilution was loaded on a counting chamber of a hemocytometer: since the dye freely passes only through the permeabilized membranes of dead cells, the percentage of viable cells can be evaluated. For each sample, 100 cells were scored.

2.3.3. Comet Assay

Cells were seeded at 1×10^5/mL in 6-well plates in DMEM supplemented with 1% glutamine, 0.5% penicillin/streptomycin, and 10% fetal bovine serum. After seeding (24 h), HT29 cells were treated with single phytochemicals, a mixture of them at the lower concentration, and spinach extracts (1%, 5%, 10%). The phytochemical concentration was chosen according to the amount that can be found in 100 g of fresh spinach. A concentration tenfold higher was also tested to evidence possible activity variation directly related to the concentration (Table 1). To assess possible synergic or antagonist effects, the activity of a mixture of the phytochemicals at the lower dosage was also evaluated. After 24 h of incubation at $37\,°C$, the cells were trypsinized and resuspended in DMEM at a concentration of 5×10^4 cell/mL; centrifuged (1 min, $800 \times g$) and the cell pellet resuspended in 90 μL Low Melting Agarose 0.7% (LMA), before being transferred onto degreased microscope slides previously dipped in 1% normal melting agarose (NMA) for the first layer. The agarose was allowed to set for 15 min at 4 °C before the addition of a final layer of LMA. Cell lysis was carried out at 4 °C overnight in lysis buffer (2.5 M NaCl, 100 mM Na2EDTA, 8 mM Tris-HCl, 1% Triton X-100, and 10% DMSO, pH 10). The electrophoretic migration was performed in alkaline buffer (1 mM Na_2EDTA, 300 mM NaOH, 0 °C) at pH > 13 (DNA unwinding: 20 min; electrophoresis: 20 min, 0.78 Vcm^{-1}, 300 mA). DNA was stained with 75 μL ethidium bromide (10 μg/mL) before the examination at 400× *g* magnification under a Leica DMLS fluorescence microscope (excitation filter BP 515–560 nm, barrier filter LP 580 nm), using an automatic image analysis system (Comet Assay III Perceptive Instruments Ltd., Bury St Edmunds, UK). Total fluorescence % in tail (TI, tail intensity) provided representative data on genotoxic effects. For each sample, coded and evaluated blind, at least three independent experiments were performed, 100 cells were analyzed, and the median value of TI was calculated.

Table 1. Concentration of pure phytochemicals tested in comet assay.

Samples	Content (mg/100 g)	Assayed Concentrations	
Ascorbic acid	4.0	3 μM	30 μM
20-Hydroxyecdysone	5.0	1.5 μM	15 μM
Lutein	7.2	2 μM	20 μM
β-Carotene	5.1	1.5 μM	15 μM
Ferulic acid	1.0	1 μM	10 μM
p-Coumaric acid	0.1	0.1 μM	1 μM
2-Hydroxycinnamic acid	2.8	2 μM	20 μM

2.3.4. Comet Assay—Antioxidant Activity

Cells were seeded and incubated with phytochemicals/extracts as described above. After incubation and trypsinization, cells were resuspended in DMEM (supplemented with 1% glutamine, 0.5% penicillin/streptomycin and 10% fetal bovine serum) at a concentration 1×10^5 cell/mL for further treatment in suspension before to perform the Comet assay, with H_2O_2 (100 μM) on ice for 5 min. The suspensions were then centrifuged twice (1 min, 800× *g*) to wash and recover the cells. The slides were prepared and analyzed as reported above.

2.4. Measurement of Reactive Oxygen Species (ROS) Production

Cells were seeded at 1×10^5/mL in 24-well plates in DMEM supplemented as described above. After seeding (24 h), cells were treated for 24 h with ascorbic acid 3 and 30 μM, 20-hydroxyecdysone 1.5 and 15 μM, ferulic acid 1 and 10 μM, 2-hydroxycinnamic acid 2 and 20 μM, *p*-coumaric acid 0.1 and 1 μM, β-carotene 1.5 and 15 μM, lutein 2 and 20 μM, with the phytochemicals mixture (at their lowest concentration) and with spinach extracts at 1–5–10%. After 24 h of treatment, cells were washed with PBS and pre-incubated for 30 min (37 °C) in the dark with DCFH-DA 10 μM diluted in PBS (pH 7.4). Cells were washed with PBS to remove extracellular DCFH-DA, resuspended in DMEM, and treated 30 min at 37 °C with menadione 100 μM, a known oxidant agent [20]. The medium was removed and a lysis solution (Tris-HCL 50 mM, 0.5% TritonX pH 7.4; cell dissociation solution, Sigma Aldrich, St. Louis, MO, USA) was added for 10 min. Cell lysates were scraped from the dishes and the extracts were centrifuged. The supernatant was collected, and the fluorescence was read with a fluorescence spectrophotometer (Spectra Fluor Plus, Tecan Group Ltd., Männedorf, Switzerland) looking at the fluorescence peak between 510 and 550 nm. Each experiment was performed in triplicate.

2.5. Statistical Analysis

Data were analyzed by univariate analysis of variance (ANOVA) with the Bonferroni multiple comparison post-hoc test through the SPSS 18.0 software (SPSS Inc., Chicago,

IL, USA). For each experiment, performed in triplicate, the significance was accepted for $p < 0.05$.

3. Results

3.1. Modulation of the Proliferation

3.1.1. MTS Assay

This colorimetric assay allows to evaluate if and to what extent the single phytochemicals tested can affect the proliferation of the HT29 cells, quantifying the number of cells in active proliferation. After 24 h and 48 h of treatment with increasing concentrations (1–500 μM) of ascorbic acid and hydroxycinnamic acids (ferulic-, *p*-coumaric- and 2-hydroxycinnamic acid) no variations have been found in the number of viable cells comparing treated and untreated samples (Figure 1). However, a significant decrease of cell proliferation has been recorded at the higher concentration tested (500 μM) after treatment with β-carotene, 20-hydroxyecdysone and lutein (Figure 1).

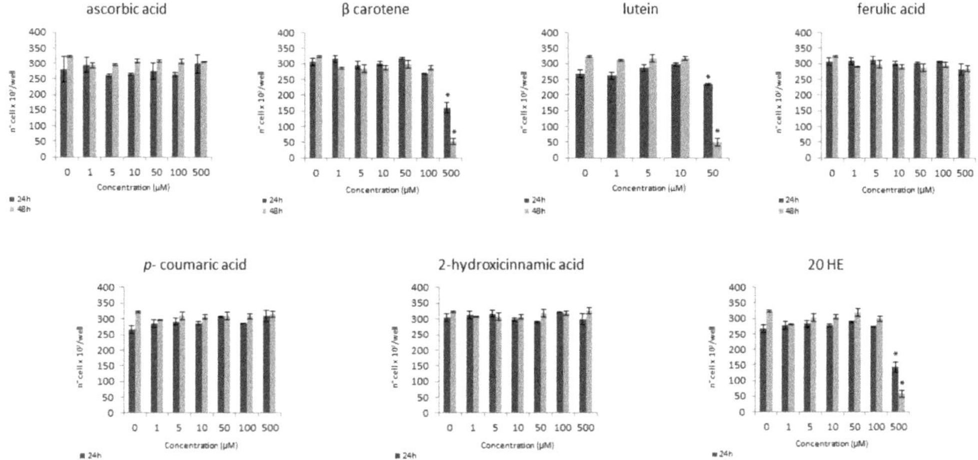

Figure 1. Modulation of the proliferation of HT29 cell line: number of cells/well on concentration after 24 and 48 h of treatment with increasing phytochemical concentration (* $p < 0.05$).

3.1.2. Trypan Blue Exclusion Method

The MTS assay is not recommended for testing cell viability to spinach extract, since the presence of fibers and debris can interfere with the assay. In this case, the trypan blue exclusion method has been chosen to evaluate the modulation of cell proliferation.

After 24 and 48 h of treatment with increasing concentration (1%, 5%, 10%, 50%) of water (WE) and liquid nitrogen (NE) extracts, data indicate an antiproliferative activity related to highest concentrations.

The hydrophilic extract (HE), at 48 h of treatment, shows a dose-dependent inhibition of proliferation and the highest concentration tested (50%) not only induces a reduction in cell division but also a strong cytotoxic effect. (Figure 2).

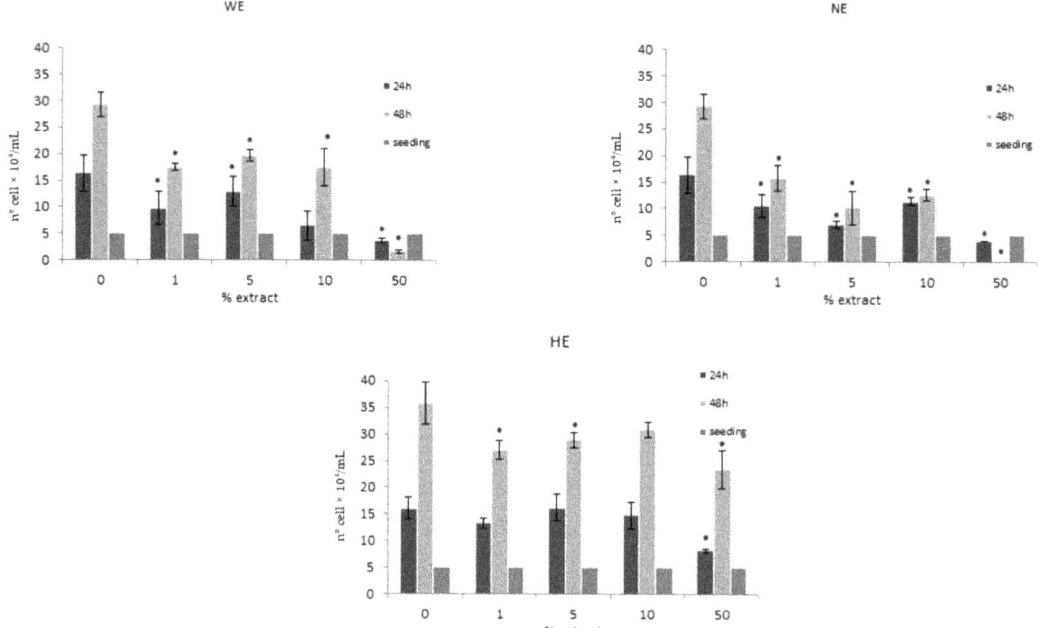

Figure 2. Modulation of the proliferation of HT29 cell line: number of cells/mL after 24 (dark grey) and 48 (light grey) hours of treatment with increasing concentrations of spinach extract. WE = water extract, NE = liquid nitrogen extract, HE = hydrophilic extract (* $p < 0.05$ taking into account the growth at 0 concentration). Seeding is the number of cells at time 0.

3.2. Genotoxic Activity

A Comet assay has been carried out to evaluate if the single phytochemicals, their mixture, or the spinach extracts exert a genotoxic effect on the HT29 cell line. This assay considers the onset of possible DNA damage by evaluating the presence, after electrophoresis, of fragmented DNA outside the core of the cell nucleus. Each phytochemical was tested considering the quantity that can be found in 100 g of fresh spinach as approximate mean in standard growth conditions (considering that cultivar, production method, and growing season can all impact on the nutrient composition), and at a concentration tenfold higher (Table 1); the mixture was prepared considering the lower concentration; the extracts were tested in the concentrations of 1%, 5%, and 10%.

After 24 h of treatment with the single phytochemicals, no genotoxic effect was observed except for lutein 20 µM that showed a significant increase in tail intensity (TI, Table 2). This could explain the antiproliferative activity observed at this concentration with the MTS assay (Figure 1), related to DNA damage somehow induced by lutein. The treatment with the mixture of phytochemicals does not lead to any observed genotoxic effect as well (Table 2). Apart WE, the 24 h treatment with NE and HE led to genotoxic effect when tested at higher concentration (Table 2). This behavior could be partially responsible of the results reported in the Comet assay after oxidative injury (Figure 3).

Table 2. Genotoxic effects of the different phytochemicals and extracts.

Samples	Concentration	TI% [1]	Sd
NT [2]		0.29	0.22
Lutein	2 μM	0.33	0.01
	20 μM	52.72	0.41
β-Carotene	1.5 μM	0.22	0.01
	15 μM	0.25	0.29
20HE	1.5 μM	0.18	0.01
	15 μM	0.59	0.23
Ascorbic acid	3 μM	0.32	0.35
	30 μM	0.25	0.29
p-Coumaric acid	0.1 μM	0.31	0.58
	1 μM	0.51	0.45
2-Hydroxicinnamic acid	2 μM	0.20	0.34
	20 μM	0.39	0.31
Ferulic acid	1 μM	0.21	0.14
	10 μM	0.20	0.07
Mix		0.23	0.11
WE	1%	0.69	0.17
	5%	1.03	0.01
	10%	0.77	0.06
	50%	4.33	0.70
NE	1%	0.86	0.08
	5%	0.38	0.08
	10%	1.33	0.46
	50%	toxic	
HE	1%	0.87	0.17
	5%	1.85	0.86
	10%	1.90	0.6
	50%	toxic	

[1] Total fluorescence % in tail. [2] Not treated.

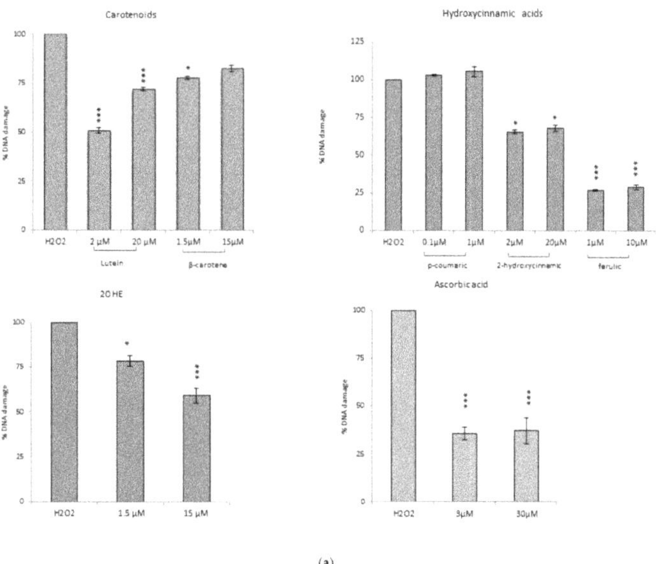

Figure 3. *Cont.*

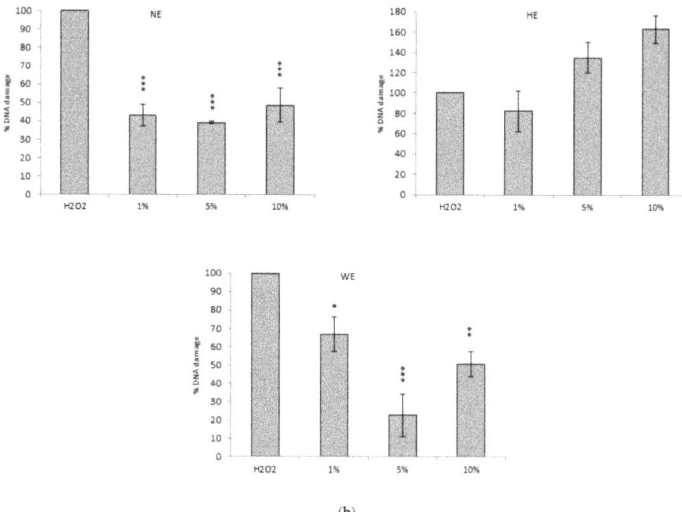

Figure 3. Percentage of reduction of DNA damage induced by H_2O_2 after a 24 h pre-treatment with different concentration (* $p < 0.05$; ** $p < 0.01$; *** $p < 0.001$): (**a**) pure phytochemicals; (**b**) extracts.

3.3. Antioxidant Activity

3.3.1. Comet Assay

The antioxidant activity was evaluated as the ability of the samples to increase the endocellular defenses against an external oxidative stress. The Comet assay was carried out after 24 h of treatment to measure the protection against oxidative DNA damage induced by H_2O_2 (100 μM).

Comet Assay with Phytochemicals

The chemicals that showed non-antiproliferative activity also exhibit a significant DNA damage reduction after H_2O_2 oxidative stress: up to 60% for ascorbic acid (3 and 30 μM) and 75% for ferulic acid (1 and 10 μM) without variations between concentrations tested. 2-hydroxycinnamic acid gives a DNA damage reduction of about 25% (2 and 20 μM). Only *p*-coumaric acid does not show any antioxidant activity; this might be related to the very low concentrations tested (0.1 and 1 μM) or to the fact that the molecule, acting as a scavenger, may have another main target (i.e., superoxide anion) (Figure 3).

Among the antiproliferative phytochemicals, DNA damage reduction up to 40% was recorded with the highest dose of 20-hydroxyecdysone (1.5 and 15 μM) and, among the carotenoids, up to 50% with lutein 2 μM. β-carotene (1.5 and 15 μM) reduces DNA damage of about 15–20%. Despite its genotoxicity, also lutein 20 μM seems to induce a certain level of protection (~30%); the same level (~32%) observed after pre-treatment with the mixture of phytochemicals (Figure 3).

Comet Assay with Spinach Extracts

All the extracts have been tested at 1%, 5%, and 10%. The WE shows a significant dose-dependent reduction of DNA damage: the 5% concentration seems to be the most active, with a DNA damage reduction of about 78%. The NE significantly reduces DNA damage up to 60%, without differences of activity among the concentrations tested. HE does not show the ability to counteract the damage induced by hydrogen peroxide: on the contrary, at the highest dosage, it shows a pro-oxidant activity (Figure 3).

3.3.2. Measurement of Variation in Reactive Oxygen Species (ROS) Concentration

The samples' ability to counteract the increase of ROS induced by an oxidizing agent has been investigated. For this purpose, menadione (vitamin K3) was used, a synthetic derivative of the natural vitamins K1 and K2 with a degree of toxicity against a wide variety of cancer cells. It can act directly, through the formation of reactive oxygen species, or indirectly through the depletion of the most important endogenous antioxidant, glutathione (GSH). The level of ROS was measured by a fluorescence assay with 2′,7′-dichlorofluorescein-diacetate (DCFH-DA), a non-fluorescent compound that crosses cell membranes. Once in the cytoplasm, esterases remove the acetates to produce 2′,7′-dichlorofluorescein (DCFH), which is not cell permeable anymore. DCFH is easily oxidized to 2′,7′-dichlorofluorescein (DCF), a highly fluorescent compound.

Only ascorbic acid, ferulic acid 10 μM, β-carotene at the highest dose and lutein can significantly inhibit the production of ROS, while 15 μM of 20-hydroxyecdysone weakly counteracts the increase of ROS levels induced by menadione (Figure 4). Interestingly, the synthetic mixture of the major components of the spinach complex presents no antioxidant activity, showing the presence of antagonistic effects (Figure 5). Among the extracts, only the lowest concentration of the liquid nitrogen extract (1%) and the highest concentration of the hydrophilic extract (10%) show the ability to significantly counteract the oxidative stress induced by this oxidizing agent (Figure 5).

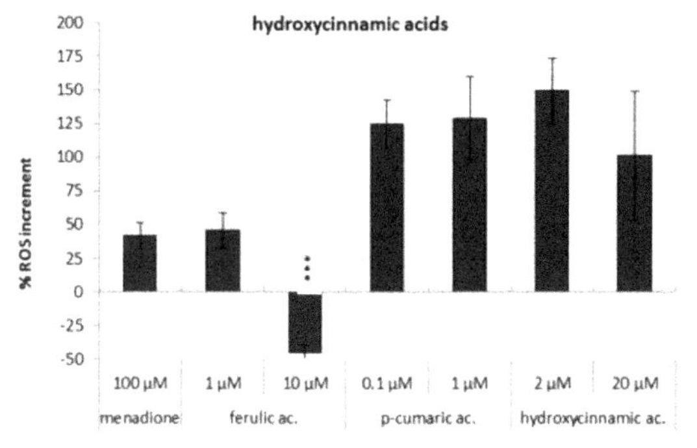
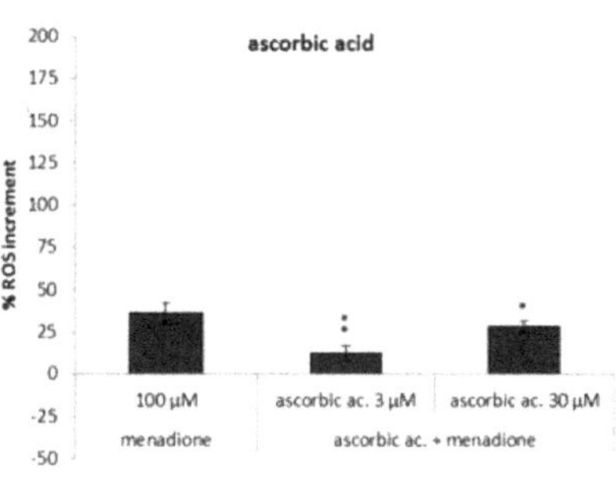
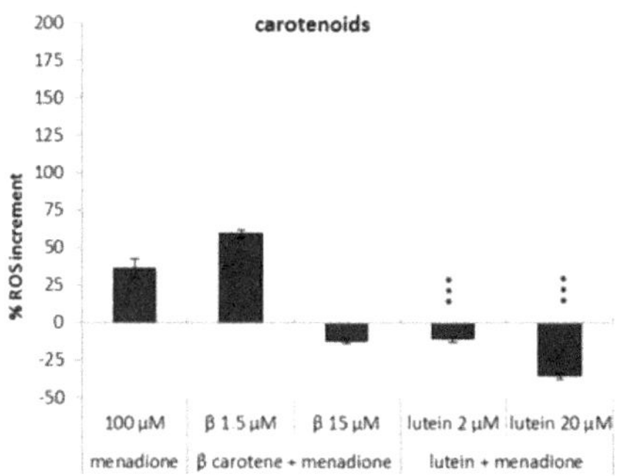
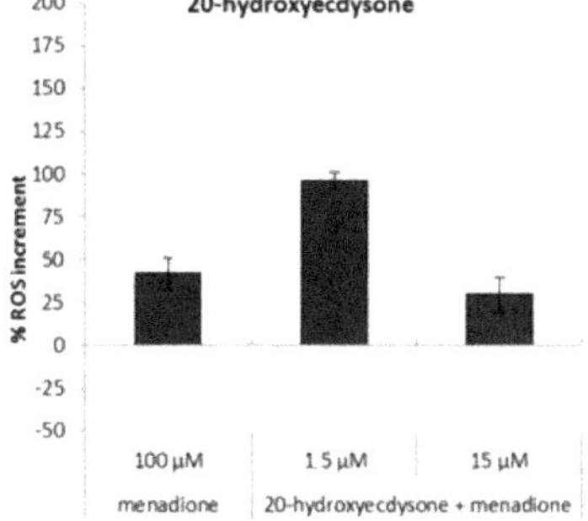

Figure 4. Evaluation of antioxidant ability of the phytochemicals to counteract the intracellular ROS levels caused by 24 h treatment with menadione oxidative insult (100 μM) on HT29 cell line, expressed as ROS increment % (* $p < 0.05$; ** $p < 0.01$; *** $p < 0.001$).

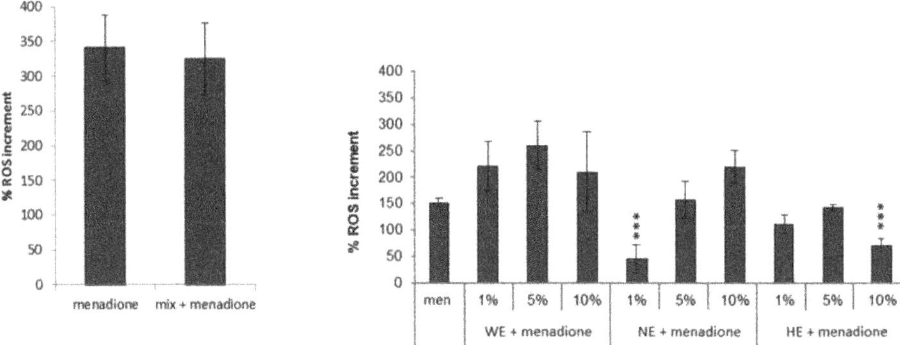

Figure 5. Evaluation of different spinach extract (left) and of single component mix (right) antioxidant ability to counteract menadione (100 µM) oxidative insult after 24 h of treatment expressed as ROS increment % (*** $p < 0.001$).

4. Discussion

Spinacia oleracea has a good antioxidant activity related to the presence of a pool of known phytochemicals such as ascorbic acid, carotenoids (β-carotene and lutein), hydroxycinnamic acids (ferulic acid, *p*-coumaric acid, 2-hydroxycinnamic acid), and 20-hydroxyecdysone [18]. The mechanism of action of these phytochemicals is not yet fully understood, even if it seems to be related both to their concentrations and the different types of cell lines involved. In this work, the ability has been evaluated of different extracts to modulate the proliferation of the human adenocarcinoma cell line (HT29). It has also tested the effect of the main phytochemicals belonging to the *Spinacia oleracea* phytocomplex as previously identified [18], submitted alone or merged together, in a molarity calculated as an approximate mean of concentration present in a standard condition of 100 g of fresh spinach (Table 1), considering that cultivar, production method, and growing season can all impact on nutrient composition. The activity of the phytochemicals seems to be related to the concentration applied, and on how the oxidative stress has been induced on the cells. In addition, in the case of the spinach extracts, the biological activity also reflects the method of extraction used, suggesting a greater level of complexity.

Considering the single antioxidants tested, hydroxycinnamic acids seem not to have antiproliferative activity on this cell line (Figure 1), independently from the concentration (1–500 µM) or the length time of the treatment (24 h or 48 h), in agreement also with Martini et al. [21]. The same considerations can be made for ascorbic acid, which does not induce any change in HT29 cells viability in the tested concentrations range (Figure 1), according to what observed by Fernandes et al. in 2017 on bone cancer cells cell line [22].

On the other hand, treating the cells with β-carotene (500 µM) leads to a severe inhibition of the proliferation (Figure 1). Similar results were obtained by Park et al. [23] on gastric cancer cells. Upadhyaya [24] reported a reduction of proliferation after 12 h of treatment with β-carotene starting from a concentration of 20 µM on the leukemic cell lines U937 and HL60. Compared to them, the HT29 cell line seems to be less sensitive to the effect of β-carotene, suggesting different effects of the same molecule on various cell lines. Different effects can also be found observing the response to lutein: according to the results presented in this paper, the HT29 cell line seems to be less sensitive (Figure 1) than the human breast cancer cell lines MCF7 and MDA-MB-157, as it undergoes a significant concentration-dependent reduction of viability after 24 h of treatment with lutein 5–120 µM [25].

On the other hand, all the spinach extracts demonstrate an antiproliferative activity on the HT29 cell line regardless of the type of extraction (Figure 2). While the HE shows dose-dependent inhibition of the proliferation, WE and NE show both time and dose-dependent effect, with a cytotoxic activity only at the highest concentrations tested. These results agree with what was observed by [26] on the colon cancer cell line Caco2 treated with *Amaranthus gangeticus* L. (red spinach) aqueous extracts and by Fornaciari et al. [18] on the HT29 cell line treated with *Spinacia oleracea* extracts.

Besides the antiproliferative activity, also the possible genotoxic effect of the single phytochemicals, of their mixture, and of the different spinach extracts have been evaluated by Comet assay, a useful approach to study the effect of nutrients and micronutrients. The data obtained with the hydroxycinnamic acids integrate what was already been reported by Ferguson et al. [27], namely the absence of genotoxicity at concentrations (0.5 and 1 mM) and time of treatment (7 days) higher than those assessed in the present article. Regarding the dose-dependent genotoxicity of lutein, similar behavior has been reported by Kalariya and colleagues [28]: after 9 h of treatment with lutein metabolites starting from the concentration 10 µM, a significant increment in tail intensity was observed on the retinal pigmented epithelium (ARPE-19). The presence of a genotoxic effect at 20 µM may suggest the involvement of a direct action of lutein on DNA, resulting in the reduction of the proliferation previously observed at concentrations higher than 10 µM.

The protective effect of foods rich in antioxidants carried out as decreased sensitivity to the damage induced by known oxidizing agents has been widely demonstrated [2,29]. In this work, we compared the antioxidant activity of extracts and chemicals using a modified protocol of the Comet assay that allows to verify the ability to reduce the extent of the DNA damage induced by hydrogen peroxide.

The phytochemicals mixture shows a significant DNA damage reduction, by 32% (Figure 3). However, it seems that there is not an additive antioxidant effect when the single phytochemicals are mixed, since the single molecules reached higher percentages of DNA damage reduction. Comparing the mixture to the natural phytocomplex, WE and NE show a significantly higher DNA damage reduction up to 75% (Figure 3). A similar result was observed by Ko and colleagues in 2014 on HepG2 cells and human leukocytes treated with *Spinacia oleracea* water extracts [1]. The activity of NE seems to be independent by its concentration, with a persistent reduction near 60%. WE shows a dose-dependent activity at the lowest doses with a DNA damage reduction up to 75%, while it loses effectiveness at the highest concentration. HE does not show activity at the lowest dose, while it proves a pro-oxidant effect at the highest ones. These observations suggest that the method of extraction strongly influences the molecular content of the phytocomplex and therefore the biological activity of the resulting extract. The WE activity recalls what has been observed for ascorbic acid (and other antioxidant molecules), high concentrations minimize the antioxidant effect or induce a pro-oxidant effect [30]. The antioxidant activity was further investigated by measuring the ability of the samples to modulate the HT29 physiological levels of ROS. Among the phytochemicals, only the highest concentration of ferulic acid, lutein, and 20-hydroxyecdysone significantly increases the intracellular ROS levels disturbing the cell oxidative balance (data not shown). A similar pro-oxidant effect was reported for ferulic acid in two cervical cancer cell lines (HeLa and ME-180) by Karthikeyan et al. in 2011 [31]. The induced imbalance does not seem to alter cell proliferation except for the treatment with lutein that, at the higher concentration, can increase the intracellular ROS levels, to inhibit cell proliferation and to induce significant DNA damage. The phytochemicals mixture does not induce ROS level variations, according to the behavior of the single molecules (Figure 5). The spinach extracts show an alteration of the intracellular ROS levels only at the higher tested concentration (10%), with a significant increase of ROS (Figure 5). Only NE shows a similar pro-oxidant effect already at a lower assayed concentration (5%).

Thereafter, also, the ability of the samples to counteract the increase of ROS induced by menadione has been evaluated. Among the single phytochemicals, β-carotene 15 µM, lutein 2 µM, and ascorbic acid 3 and 30 µM (which do not induce significant variations of intracellular ROS levels in physiological conditions, data not shown) significantly counteract the activity of menadione (Figure 4). The 10 µM ferulic acid and the 20 µM lutein induce an increase of ROS levels in the absence of stress and strongly reduce them in the presence of stress (Figure 4). The mixture of phytochemicals does not induce any variation of ROS levels (Figure 5); it seems that the simultaneous presence of all the phytochemicals can interfere with their activity. The different extracts action, able to exhibit pro-oxidant effect both in the absence and in the presence of oxidative stress and to significantly counteract the increase due to menadione, points out again conflicting results depending on the method of detection and on the oxidant agent used.

The *p*-coumaric acid alone does not reduce the oxidative damage induced either by hydrogen peroxide or by menadione. The *p*-coumaric acid normally acts as a scavenger of superoxide anions; this could explain its inability to counteract hydrogen peroxide. Moreover, since the *p*-coumaric acid is unable to counteract the effect of menadione, which can act reducing the levels of glutathione, it can be supposed that the antioxidant activity of the *p*-coumaric acid is closely associated with that of glutathione.

The 2-hydroxycinnamic acid does not alter the intracellular levels of ROS in physiological conditions, while, in the case of oxidative stress, its activity seems to be related to the type of oxidant agent involved: a pro-oxidant effect was observed in presence of menadione, suggesting an activity comparable to that of the *p*-coumaric acid; an antioxidant effect was observed in the presence of hydrogen peroxide, suggesting a scavenger activity towards a wider range of ROS.

In physiological conditions, the ferulic acid shows a dose-dependent activity, with a pro-oxidant effect at the highest dose. In case of stress, the ferulic acid has proved a dose-dependent antioxidant activity towards menadione and a greater dose-independent one towards the hydrogen peroxide. These observations suggest that the ferulic acid could act only directly in the presence of menadione, since it may interfere with the antioxidant endogenous systems. Moreover, the increased antioxidant activity showed towards the hydrogen peroxide, suggests that the ferulic acid can counteract this oxidative damage through a combination of a direct and indirect antioxidant activity.

In the absence of oxidative stress, 20 HE shows a dose-dependent activity, with a pro-oxidant effect at the highest concentration. In presence of stress, it shows a good antioxidant activity towards hydrogen peroxide and a dose-dependent activity towards menadione, counteracting it only at the highest dose. The β-carotene does not alter the intracellular levels of ROS in physiological conditions, while in the presence of stress it has shown a dose-independent antioxidant activity towards hydrogen peroxide and a dose-dependent activity towards menadione. Lutein counteracts both oxidant agents, even though in absence of stress it showed a pro-oxidant effect. The ascorbic acid does not induce alterations in the intracellular ROS levels in physiological conditions and it has shown significant antioxidant activity with both hydrogen peroxide and menadione.

In physiological conditions, the spinach extracts show a dose-dependent activity, increasing the intracellular ROS levels only at the highest doses. In the presence of oxidative stress, their activity is strongly related to the oxidant agent involved and to the method of extraction used. It is interesting to compare the biological activity of the single phytochemicals at the concentration that we can find in 100 g of fresh spinach, the "artificial" mixture of them, and the different spinach extracts at the lowest dose (1%), which corresponds to the available concentration coming from 100 g of fresh spinach. In physiological conditions, neither the single phytochemicals, nor the mixture, nor the spinach extracts induce alterations of the oxidative balance. In the presence of oxidative stress, a different behavior has been observed that is related to the kind of oxidative agent used. Almost all the single phytochemicals can reduce the oxidative damage induced by hydrogen peroxide, but when they are mixed there is not an additive antioxidant effect, since the single molecules reached higher percentages of DNA damage reduction. Considering the activity of the spinach extracts, they showed different activity towards the hydrogen peroxide depending on the method of extraction used and, consequently, on the bioactive molecules extracted. While the WE provides the same DNA damage reduction of the "artificial" mixture of phytochemicals, the NE shows improved effectiveness suggesting a synergic effect of the mixture of bioactive constituents contained, as often reported in the literature [8,9,32].

Against the oxidative damage induced by menadione, there is a more variable response. Among the single phytochemicals, only lutein and ascorbic acid are able to counteract the ROS levels increase, but when mixed with the other molecules they lose their ability. The spinach extracts also in this case show a different behavior depending on the method of extraction used: the HE is unable to counteract menadione, the WE shows a pro-oxidant effect and only NE reduces the intracellular ROS levels. Once again, it can be assumed that the biological effects of a plant extract are strongly related to a synergic effect of the bioactive molecules contained. A schematic overview of the activities described above is summarized in the following tables (Tables 3 and 4):

Table 3. Schematic overview of the antioxidant activity of the single phytochemicals and of their mixture ($\leftrightarrow$ = no activity; $\uparrow$ = increase of oxidative stress; $\downarrow$ = reduction of oxidative stress).

Phytochemical (Physiological Dose)	Antioxidant Activity		
	No Stress	Menadione	H_2O_2
p-Coumaric acid	$\leftrightarrow$	$\uparrow$	$\leftrightarrow$
2-Hydroxycinnamic acid	$\leftrightarrow$	$\uparrow$	$\downarrow$ (40%)
Ferulic acid	$\leftrightarrow$	$\uparrow$	$\downarrow$ (75%)
20 HE	$\leftrightarrow$	$\uparrow$	$\downarrow$ (20%)
Ascorbic acid	$\leftrightarrow$	$\downarrow$	$\downarrow$ (60%)
Lutein	$\leftrightarrow$	$\downarrow$	$\downarrow$ (50%)
β-Carotene	$\leftrightarrow$	$\leftrightarrow$	$\downarrow$ (20%)
Mix	$\leftrightarrow$	$\leftrightarrow$	$\downarrow$ (32%)

Table 4. Schematic overview of the antioxidant activity of the phytochemical mixture and of the spinach extracts ($\leftrightarrow$ = no activity; $\uparrow$ = increase of oxidative stress; $\downarrow$ = reduction of oxidative stress).

Sample	Antioxidant Activity		
	No Stress	Menadione	H_2O_2
Phytochemical mix	$\leftrightarrow$	$\leftrightarrow$	$\downarrow$ (32%)
HE 1%	$\leftrightarrow$	$\leftrightarrow$	$\leftrightarrow$
WE 1%	$\leftrightarrow$	$\uparrow$	$\downarrow$ (35%)
NE 1%	$\leftrightarrow$	$\downarrow$	$\downarrow$ (60%)

Another interesting aspect that comes out from this work is the complex relationship between inhibition of the proliferation and the antioxidant activity. Considering the oxidative balance, any variation in the amount of intracellular reactive oxygen species can induce impaired physiological functions and a ROS reduction seems to be related to a decreased proliferative response, but the relation is not so direct. With almost all the compounds at low concentration an antioxidant activity can be noted but no inhibition of the proliferation. This latter activity is observed just for 20-hydroxyecdysone and β-carotene but at high concentrations (100 μM). Similar behavior is observed with lutein but the antiproliferative concentration, in this case, is lower than the biologically active one of the other compounds.

5. Conclusions

In this work, the biological activity of several phytochemicals applied to a human cell line at different concentrations was evaluated. The results lead to a confirmation of their beneficial properties, acting on ROS and lowering the oxidative damage. This highlights, among other possibilities, also that of considering the development of supplements including compounds derived from spinach extract, in order to defend health and to trigger possible anticancer effects.

Author Contributions: Conceptualization, A.B. and L.A.; methodology, F.M.; validation, A.B., and L.F.; formal analysis, A.B.; investigation, F.M.; writing—original draft preparation, F.M.; writing—review and editing, L.A. and L.F.; supervision, A.B. All authors have read and agreed to the published version of the manuscript.

Funding: This research received no external funding.

Institutional Review Board Statement: Not applicable.

Informed Consent Statement: Not applicable.

Data Availability Statement: The remaining data are available on request from the corresponding author.

Conflicts of Interest: The authors declare no conflict of interest.

References

1. Ko, S.H.; Park, J.H.; Kim, S.Y.; Lee, S.W.; Chun, S.S.; Park, E. Antioxidant effects of spinach (*Spinacia oleracea* L.) supplementation in hyperlipidemic rats. *Prev. Nutr. Food Sci.* **2014**, *19*, 19–26. [CrossRef]
2. Parohan, M.; Anjom-Shoae, J.; Nasiri, M.; Khodadost, M.; Reza Khatibi, S.; Sadeghi, O. Dietary total antioxidant capacity and mortality from all causes, cardiovascular disease and cancer: A systematic review and dose–response meta-analysis of prospective cohort studies. *Eur. J. Nutr.* **2019**, *58*, 2175–2189. [CrossRef]
3. Rajoria, A.; Kumar, J.; Chauhan, A.K. Anti-oxidative and anti-carcinoginic role of lycopene in human health-A review. *J. Dairy Foods Home Sci.* **2010**, *29*, 3–4.
4. Singh, R.B.; Hristova, K.; Fedacko, J.; Singhal, S.; Khan, S.; Wilson, D.W.; Takahashi, T.; Sharma, Z. Antioxidant vitamins and oxidative stress in chronic heart failure. *World Heart J.* **2015**, *7*, 257–264.
5. Ray, P.D.; Huang, B.W.; Tsuji, Y. Reactive oxygen species (ROS) homeostasis and redox regulation in cellular signaling. *Cell Signal.* **2012**, *24*, 981–990. [CrossRef]
6. Valko, M.; Rhodes, C.J.; Moncol, J.; Izakovic, M.M.; Mazur, M. Free radicals, metals and antioxidants in oxidative stress-induced cancer. *Chem. Biol. Interact.* **2006**, *160*, 1–40. [CrossRef] [PubMed]
7. Waris, G.; Ahsan, H. Reactive oxygen species: Role in the development of cancer and various chronic conditions. *J. Carcinog.* **2006**, *5*, 14. [CrossRef] [PubMed]
8. Caesar, L.K.; Cech, N.B. Synergy and antagonism in natural product extracts: When 1 + 1 does not equal 2. *Nat. Prod. Rep.* **2019**, *36*, 869–888. [CrossRef] [PubMed]
9. Wagner, H.; Ulrich-Merzenich, G. Synergy research: Approaching a new generation of phytopharmaceuticals. *Phytomedicine* **2009**, *16*, 97–110. [CrossRef] [PubMed]
10. Ulrich-Merzenich, G.; Panek, D.; Zeitler, H.; Vetter, H.; Wagner, H. Drug development from natural products: Exploiting synergistic effects. *Indian J. Exp. Biol.* **2010**, *48*, 208–219. [PubMed]
11. Junio, H.A.; Sy-Cordero, A.A.; Ettefagh, K.A.; Burns, J.T.; Micko, K.T.; Graf, T.N.; Richter, S.J.; Cannon, R.E.; Oberlies, N.H.; Cech, N.B.J. Synergy-Directed Fractionation of Botanical Medicines: A Case Study with Goldenseal (*Hydrastis canadensis*). *Nat. Prod.* **2011**, *74*, 1621–1629. [CrossRef] [PubMed]
12. Gil, M.I.; Ferreres, F.; Tomas-Barberan, F. Effect of postharvest storage and processing on the antioxidant constituents (flavonoids and vitamin C) of fresh-cut spinach. *J. Agric. Food Chem.* **1999**, *47*, 2213–2217. [CrossRef]
13. Bunea, A.; Andjelkovic, M.; Socaciu, C.; Bobis, O.; Neacsu, M.; Verhé, R.; Van Camp, J. Total and individual carotenoids and phenolic acids content in fresh, refrigerated and processed spinach (*Spinacia oleracea* L.). *Food Chem.* **2008**, *108*, 649–656. [CrossRef] [PubMed]
14. Hait-Darshan, R.; Grossman, S.; Bergman, M.; Deutsch, M.; Zurgil, N. Synergistic activity between a spinach-derived natural antioxidant (NAO) and commercial antioxidants in a variety of oxidation systems. *Int. Food Res.* **2009**, *42*, 246–253. [CrossRef]
15. Bergman, M.; Varshavsky, L.; Gottlieb, H.E.; Grossman, S. The antioxidant activity of aqueous spinach extract: Chemical identification of active fractions. *Phytochemistry* **2001**, *58*, 143–152. [CrossRef]
16. Lomnitski, L.; Bergman, M.; Nyska, A.; Ben-Shaul, V.; Grossman, S. Composition, efficacy and safety of spinach extracts. *Nutr. Cancer* **2003**, *46*, 222–231. [CrossRef]
17. Lomnitski, L.; Carbonatto, M.; Ben-Shaul, V.; Peano, S.; Conz, A.; Corradin, L.; Maronpot, R.R.; Grossman, S.; Nyska, A. The prophylactic effects of natural water-soluble antioxidant from spinach and apocynin in a rabbit model of lipopolysaccharide-induced endotoxemia. *Toxicol. Pathol.* **2000**, *28*, 588–600. [CrossRef]
18. Fornaciari, S.; Milano, F.; Mussi, F.; Pinto-Sanchez, L.; Forti, L.; Buschini, A.; Arru, L. Assessment of antioxidant and antiproliferative properties of spinach plants grown under low oxygen availability. *J. Sci. Food Agric.* **2015**, *95*, 490–496. [CrossRef]
19. Grajek, W.; Olejnik, A. Epithelial cell cultures in vitro as a model to study functional properties of food. *Pol. J. Food Nutr. Sci.* **2004**, *13*, 5–24.
20. Maioli, E.; Greci, L.; Soucek, K.; Hyzdalova, M.; Pecorelli, A.; Fortino, V.; Valacchi, G. Rottlerin inhibits ROS formation and prevents NFkappaB activation in MCF-7 and HT-29 cells. *J. Biomed. Biotechnol.* **2009**, *2009*, 742936. [CrossRef]

21. Martini, S.; Conte, A.; Tagliazucchi, D. Antiproliferative Activity and Cell Metabolism of Hydroxycinnamic Acids in Human Colon Adenocarcinoma Cell Lines. *J. Agric. Food Chem.* **2019**, *67*, 3919–3931. [CrossRef]
22. Fernandes, G.; Barone, A.W.; Dziak, R. The effect of ascorbic acid on bone cancer cells *in vitro*. *Cogent Biol.* **2017**, *3*, 1288335. [CrossRef]
23. Park, Y.; Choi, J.; Lim, J.W.; Kim, H. β-Carotene-induced apoptosis is mediated with loss of Ku proteins in gastric cancer AGS cells. *Genes Nutr.* **2015**, *10*, 467. [CrossRef] [PubMed]
24. Upadhyaya, K.R.; Radha, K.S.; Madhyastha, H.K. Cell cycle regulation and induction of apoptosis by beta-carotene in U937 and HL-60 leukemia cells. *J. Biochem. Mol. Biol.* **2007**, *40*, 1009–1015. [PubMed]
25. Li, Y.; Zhang, Y.; Liu, X.; Wang, M.; Wang, P.; Yang, J.; Zhang, S. Lutein inhibits proliferation, invasion and migration of hypoxic breast cancer cells via downregulation of HES1. *Int. J. Oncol.* **2018**, *52*, 2119–2129. [CrossRef] [PubMed]
26. Sani, H.A.; Rahmat, A.; Ismail, M.; Rosli, R.; Endrini, S. Potential anticancer effect of red spinach (*Amaranthus gangeticus*) extract. *Asia Pac. J. Clin. Nutr.* **2004**, *13*, 396–400.
27. Ferguson, L.R.; Zhu, S.T.; Harris, P.J. Antioxidant and antigenotoxic effects of plant cell wall hydroxycinnamic acids in cultured HT-29 cells. *Mol. Nutr. Food Res.* **2005**, *49*, 585–593. [CrossRef]
28. Kalariya, N.M.; Ramana, K.V.; Srivastava, S.K.; Van Kuijk, F.J. Carotenoid derived aldehydes-induced oxidative stress causes apoptotic cell death in human retinal pigment epithelial cells. *Exp. Eye Res.* **2008**, *86*, 70–80. [CrossRef]
29. Griffiths, K.; Aggarwal, B.; Singh, R.; Buttar, H.; Wilson, D.; De Meester, F. Food Antioxidants and Their Anti-Inflammatory Properties: A Potential Role in Cardiovascular Diseases and Cancer Prevention. *Diseases* **2016**, *4*, 28. [CrossRef]
30. Milano, F.; Mussi, F.; Fornaciari, S.; Altunoz, M.; Forti, L.; Arru, L.; Buschini, A. Oxygen Availability during Growth Modulates the Phytochemical Profile and the Chemo-Protective Properties of Spinach Juice. *Biomolecules* **2019**, *9*, 53. [CrossRef]
31. Karthikeyan, S.; Kanimozhi, G.; Prasad, N.R.; Mahalakshmi, R. Radiosensitizing effect of ferulic acid on human cervical carcinoma cells in vitro. *Toxicol. In Vitro* **2011**, *25*, 1366–1375. [CrossRef] [PubMed]
32. Van Vuuren, S.; Viljoen, A. Plant-based antimicrobial studies—Methods and approaches to study the interaction between natural products. *Planta Med.* **2011**, *77*, 1168–1182. [CrossRef] [PubMed]

Article

Mayten Tree Seed Oil: Nutritional Value Evaluation According to Antioxidant Capacity and Bioactive Properties

Rosanna Ginocchio [1,2], Eduardo Muñoz-Carvajal [3], Patricia Velásquez [3,4], Ady Giordano [3,*], Gloria Montenegro [4], Germán Colque-Perez [5] and César Sáez-Navarrete [5,6]

1 Departamento de Ecosistemas y Medio Ambiente, Facultad de Agronomía e Ingeniería Forestal, Pontificia Universidad Católica de Chile, Av. Vicuña Mackenna, Macul 4860, Chile; rginocch@uc.cl
2 Center of Applied Ecology and Sustainability (CAPES), Pontificia Universidad Católica de Chile, Av. Vicuña Mackenna, Macul 4860, Chile
3 Departamento de Química Inorgánica, Facultad de Química y de Farmacia, Pontificia Universidad Católica de Chile, Av. Vicuña Mackenna, Macul 4860, Chile; eamunoz1@uc.cl (E.M.-C.); pdvelasquez@uc.cl (P.V.)
4 Departamento de Ciencias Vegetales, Facultad de Agronomía e Ingeniería Forestal, Pontificia Universidad Católica de Chile, Av. Vicuña Mackenna, Macul 4860, Chile; gmonten@uc.cl
5 Departamento de Ingeniería Química y Bioprocesos, Facultad de Ingeniería, Pontificia Universidad Católica de Chile, Av. Vicuña Mackenna, Macul 4860, Chile; gjcolque@uc.cl (G.C.-P.); csaez@ing.puc.cl (C.S.-N.)
6 Centro de Energía UC, Pontificia Universidad Católica de Chile, Av. Vicuña Mackenna, Macul 4860, Chile
* Correspondence: agiordano@uc.cl

Citation: Ginocchio, R.; Muñoz-Carvajal, E.; Velásquez, P.; Giordano, A.; Montenegro, G.; Colque-Perez, G.; Sáez-Navarrete, C. Mayten Tree Seed Oil: Nutritional Value Evaluation According to Antioxidant Capacity and Bioactive Properties. *Foods* **2021**, *10*, 729. https://doi.org/10.3390/foods10040729

Academic Editors: Francisca Rodrigues and Cristina Delerue-Matos

Received: 26 February 2021
Accepted: 26 March 2021
Published: 30 March 2021

Publisher's Note: MDPI stays neutral with regard to jurisdictional claims in published maps and institutional affiliations.

Abstract: The Mayten tree (*Maytenus boaria* Mol.), a native plant of Chile that grows under environmentally limiting conditions, was historically harvested to extract an edible oil, and may represent an opportunity to expand current vegetable oil production. Seeds were collected from Mayten trees in north-central Chile, and seed oil was extracted by solvent extraction. The seed oil showed a reddish coloration, with quality parameters similar to those of other vegetable oils. The fatty acid composition revealed high levels of monounsaturated and polyunsaturated fatty acids. Oleic and linoleic acids, which are relevant to the human diet, were well represented in the extracted Mayten tree seed oil. The oil displayed an antioxidant capacity due to the high contents of antioxidant compounds (polyphenols and carotenoids) and may have potential health benefits for diseases associated with oxidative stress.

Keywords: antioxidant capacity; DPPH; solvent extraction; nutrition; *Maytenus boaria*; ABTS

1. Introduction

In recent decades, the oil crop sector has been one of the most dynamic agricultural segments worldwide, with a 4.3% per annum (p.a.) growth rate compared with an average of 2.1% p.a. for all agriculture [1]. Worldwide production, consumption, and trade in this sector have increasingly become dominated by a small number of crops [2,3], as palm, soybean, and rapeseed oils have represented 82% of the total global oil crop production since 1990 (according to oil equivalent measurements). Secondary oil crops represent major elements of the food supply and food security in several countries, including sesame oil (e.g., in Sudan, Uganda, Ethiopia, and Myanmar), groundnut oil (e.g., in Sudan, Ghana, Myanmar, Vietnam, Senegal, the United Republic of Tanzania, and Benin), coconut oil (e.g., in the Philippines, Sri Lanka, Vietnam, and Mexico), olive oil (Mediterranean countries), and cottonseed oil (e.g., in Central Asia, the Sahel, Pakistan, Turkey, and the Syrian Arabic Republic) [1].

Vegetable oils are derived from the seeds and fruits of plants. Among the vegetable oils that are derived from seeds (seed oils), most are currently obtained from only a few commercially significant species (i.e., soybeans, sunflowers, rapeseed, flax, oil palm nuts, castor beans, groundnuts, cottonseed, and shea nuts) [4]. The most common energy-rich compounds contained in the endosperm or cotyledons of seeds are carbohydrates (starches) and fatty acids (oils) [4,5]. However, several plant species that grow under environmentally limiting conditions (i.e., arid and semiarid climates or nutrient-poor soils) worldwide that are not currently used as oil crops are known to feature oil-bearing seeds [6]. These crops may constitute an opportunity for expanding vegetable oil production to regions where crop production is not feasible, either currently or in the future, due to global climate change scenarios. Therefore, many indigenous tree species, which may be more resistant than current agricultural crops to limiting environmental factors (i.e., heat, water stress, salinity, frosts, and pests) but are not yet grown commercially, are becoming increasingly recognized as potentially valuable sources of vegetable oils [7], such as the Mayten tree (*Maytenus boaria* Mol.).

The Mayten tree is a medium-sized evergreen tree (up to 15 m in height) that is native to Chile, Argentina, Peru, and Brazil [8]. In Chile, the species has a wide geographic distribution (28° to 45° Southern latitude and 15 to 1800 msl) [9,10], presenting great adaptability to different site conditions, such as precipitation levels (mean annual values from 355 to 2351 mm), soil pH (neutral to acidic), and soil water availability (dry to saturated) [8]. These trees have been described as having high seed oil contents. Scattered historical information has suggested that this seed oil was extracted from seeds collected from wild trees and used for human consumption during Colonial times (from 1600 to 1810) in central Chile [11]. Large-scale seed oil extraction and exportation to France has been described during the 19th century [12], likely for lamp oil use [13]. However, scarce information is available regarding the amounts produced, the extraction methods used, and the specific physicochemical characteristics of this seed oil. According to the work of Vicente Bustillos, from 1846 [13], Mayten tree seeds are easy to grind and press (cold or heat press), and large amounts of oil can be obtained (approx. 25% of total weight); the oil is transparent, reddish-yellowish in color, featuring a bitter taste and a density similar to that of olive oil (specific weight of 0.92), and it begins to freeze at 4–5 °C. Recent literature has indicated high oil contents in the seeds of the Mayten tree (40%) [14], which may be used for cooking [15] and industrial purposes (i.e., as a substitute for linseed oil) [16]. However, none of these applications are currently in use.

The Mayten tree belongs to the *Celastraceae* family, a plant group known to be rich in secondary metabolites (i.e., sesquiterpenes and agafurans), some of which have been reported to demonstrate interesting pharmacological behaviors [17], antiseptic properties [18] or pesticidal activities [19]. Other chemical compounds (i.e., agafurans) that have been isolated from the Mayten leaves feature biopesticidal or natural insecticidal properties [20]. Most of the studies that have been performed to explore secondary metabolites and bioactive properties have examined vegetative aerial tissues (leaves and stems); however, no studies have examined seeds or seed oil [8,14,21,22]. Therefore, the primary objective of the present study was to perform a physicochemical characterization of Mayten tree seed oil, including the evaluation of antioxidant capacity and bioactive properties to determine the nutritional value.

2. Materials and Methods

2.1. Seed Materials and Seed Oil Extraction

Seeds from Mayten trees (Figure S1) were collected in north-central Chile (Elqui Valley, Santiago Metropolitan Park, and Quirihue) from February 2016 to April 2016. The seeds were hand-cleaned, allowed to air dry, and stored at 5 °C. Seeds were ground using an electric coffee grinder, and the oil was extracted with solvents (2:4:2 *w:v:v* ratio of seeds:methanol:chloroform), according to the method described by Bligh and Dyer [23]. The mixture of ground seeds and solvents was agitated at 200 rpm for 3 h. The method described by Bligh and Dyer was modified by adding water only in the second stage of the extraction. Distilled water and chloroform were then added (1.8:2.0:2.0, *v:v:v*, mixture:water:chloroform), by forming a ternary system, and the mixture was vacuum filtered through a 7 µm pore filter. Two phases were collected, the phase of methanol–water composition (top), and the chloroform–seed oil (bottom). The chloroform was evaporated using a rotary evaporator (40 °C for 30 min), and the seed oil was stored at 4 °C for 30 min in the dark (Figure S1).

2.2. Physicochemical Analysis

Determination of the seed oil color was performed according to the method described by Popa and Doran [24] using a colorimeter (WR10, DANSTLEE). Each extract was placed in a quartz cuvette for measurement. Coordinate values were obtained, where L represents lightness and varies between 0 (black) and 100 (white), a expresses red (+) or green (−), and b indicates yellow (+) or blue (−). Chroma (C°) and hue (H°) values were obtained from the following Equations:

$$C° = \sqrt{a^2 + b^2} \tag{1}$$

$$H° = \arctan\frac{a}{b} \tag{2}$$

The seed oil density was determined using pre-calibrated volumetric flasks [25]. The peroxide index, which measures the initial oxidation state, is expressed in milliequivalent of oxygen per kilogram of oil. The iodine value indicates the degree of unsaturation, expressed in gram of iodine absorbed by 100 g of oil. Free acidity, is the percentage of free acid present in the oil, expressed in oleic acid percentage. Rancimet 743 measures the stability oxidation of the oil, corresponds to the induction period, expressed in hours, and thiobarbituric acid reactive substances (TBARS) measure malondialdehyde (MDA) present in the sample. All analyses were determined according to the American Oil Chemists' Society (AOCS) standard method [26], as described by Petropoulus et al. (2020) [27].

2.3. Fatty Acid Profile

Gas chromatography (GC) coupled with a flame ionization detector (FID) was used according to the AOCS standard method [26]. The following standard measurements were used: butanoic acid (C4:0); caproic acid (C6:0), caprylic acid (C8:0); capric acid (C10:0); undecylic acid (C11:0); lauric acid (C12:0); tridecylic acid (C13:0); myristic acid (C14:0); pentadecylic acid (C15:0); palmitic acid (C16:0); margaric acid (C17:0); stearic acid (C18:0); arachidic acid (C20:0); japonic acid (C21:0); behenic acid (C22:0); tetrasanoic acid (C24:0); tetradecene acid (C14:1); pentadecylic acid (C15:1); palmitoleic acid (C16:1); heptadecene acid (C17:1); oleic acid (C18:1); timnodonic acid (C20:1n9); erucic acid (C22:1n9); tetrasaenoic acid (C24:1); linoleic acid (C18:2n6); γ-linoleic acid (C18:3n6); α-linoleic acid (C18:3n3); gondoic acid (C20:2n6); di-homo-γ–linoleic acid (C20:3n6); dihomolinolenic acid (C20:3n3); eicosatetranoic acid (C20:4n6); eicosapentaenoic acid (EPA) (C20:5n3); docosadienenoic acid (C22:2); and docosahexaenoic acid (DHA) (C22:6n3).

2.4. Lipidic Indices

The qualities of the lipids in the seed oil were determined by evaluating the ratio of polyunsaturated fatty acids (PUFAs) to saturated fatty acids (SFAs) [28]. Two indices of coronary heart disease risk—the atherogenic index (AI), used to obtain the standard cardiac risk estimation; and the thrombogenic index (TI), which indicates the tendency to form clots in the blood vessels—were calculated as functions of the composition of monounsaturated fatty acids (MUFAs), PUFAs, SFAs, and specific fatty acids [29], as follows:

$$AI = \frac{(C12:0 + 4 \times C14:0 + C16:0)}{(MUFA + PUFA)} \tag{3}$$

$$TI = \frac{(C14:0 + C16:0 + C18:0)}{\left(0.5 \times MUFA + 0.5 \times n-6PUFA + 3 \times n-3PUFA \times \left(\frac{n-3PUFA}{n-6PUFA}\right)\right)} \tag{4}$$

2.5. Extraction of Antioxidant Compounds

Seed oil was mixed with methanol and hexane (1:5:1, *v:v:v*, oil:methanol:hexane) and maintained in an ultrasonic bath at 20 Hz for 20 min at a room temperature bath [30]. The solution was then centrifuged at 4000 rpm for 20 min at room temperature. The mixture was stored at 4 °C for 1 h, and the supernatant was then filtered through a 0.22 μm membrane with a syringe filter for the antioxidant extraction.

2.6. Quantification of Phenols and Flavonoids, and Determination of Antioxidant Capacity

Total polyphenolic contents (TPC) were estimated using the Folin–Ciocalteu (FC) method based on the antioxidant extract from oil, as described by Velásquez et al. (2019) [31]. Gallic acid (GA) was used as the standard; therefore, the resulting values are expressed in mg GAE $(100\ \text{g})^{-1}$ of seed oil.

Total flavonoid contents (TFC) were estimated using the aluminum chloride method based on the antioxidant extract from oil, as described by Velásquez et al. (2019) [31]. Quercetin (Q) was used as the standard, and results are expressed in mg QE $(100\ \text{g})^{-1}$ of seed oil. Some specific phenolic acids and flavonoids, such as 4-hydroxy benzoic acid, apigenin, caffeic acid, catechin, chlorogenic acid, cinnamic acid, chrysin, coumaric acid, epigenin, ferulic acid, gallic acid, kaempferol, luteolin, pinocembrin, quercetin, resveratrol, rutin, sinapic acid, syringic acid, vanillic acid, and phytohormone abscisic acid were quantified through using ultra-performance liquid chromatography (UPLC)-tandem mass spectrometry (MS/MS), according to the method described by Giordano et al. (2019) [32], by interpolation from standard calibration curves.

Antioxidant capacity was determined using the ferric reducing antioxidant power (FRAP) and 2,2′-azino-bis-(3-ethylbenzothiazoline-6-sulfonic acid (ABTS) radical-stabilization methods, according to the methods described by Diniyah et al. (2020) [33]. The antioxidant extract was measured using a $FeSO_4$ standard and is expressed in mg $FeSO_4/100$ g of seed oil, whereas Trolox (T) was used as the standard for radical stabilization, which is expressed in T equivalents $(\text{TE}\ 100\ \text{g})^{-1}$ of seed oil. The 2,2-diphenyl-1-picrylhydrazyl (DPPH) radical was also assessed as described by Diniyah et al. (2020) [33], with some modifications, using several antioxidant extract dilutions. A curve was generated comparing the inhibition percentage against the tested dilutions (logarithmic relationship), interpolating the half-maximal inhibitory concentration (IC_{50}), which is expressed in g of seed oil per mL^{-1} of methanolic extract. The Agilent 8453 UV-visible spectrophotometer (Palo Alto, CA, USA) was used for these analyses.

2.7. Extraction and Quantification of Carotenoids and β-Carotene

A carotene extract was obtained through seed oil saponification, as described by Varzakas and Kiokias (2016) [34]. The quantification of total carotenes was performed according to Bihler et al. (2010) [35], using β-carotene as the standard and expressed as mg β-carotene/100 g of seed oil. Absorbance was measured using the Agilent 8453 UV-visible spectrophotometer (Palo Alto, CA, USA).

The carotene profile and β-carotene levels were obtained using a high-performance liquid chromatography (HPLC)-diode-array detector (DAD). A reverse phase C18 column was used (150 mm × 4 mm × 5 μm), with a mobile phase flow of 1.5 mL min^{-1}. Methanol (A), water (B), and n-butanol (C) were used as solvents in the following concentration gradient: 0 min 3% A, 92% B, and 5% C; 4 min 0% A, 92% B, and 8% C; 18.1 min 3% A, 92% B, and 5% C, until 23 min. The carotene profile was measured at 440 nm. The β-carotene level was interpolated from the calibration curve of a 10–100 mg/L β-carotene standard.

2.8. Statistical Analysis

All analyses were realized in triplicate ($n = 3$), and the data were expressed as the mean ± standard deviation using the statistical software Minitab 19.

3. Results and Discussion

3.1. Seed Oil Extraction

A yield of 61.77 ± 8.24% w/w seed oil was extracted from Mayten tree seeds, which is a yield larger than those for seed oils extracted from sunflower (51.00% w/w), sesame (48,00% w/w), and pumpkin seeds (46.00% w/w) [36].

3.2. Physicochemical Characteristics

The seed oil extracted from Mayten tree seeds had a reddish coloration, according to the tone angle (H: 59.40°), which is categorized among the purple–red colors [37]. The oil had a lower yellow and a higher red color composition (Table 1) compared with palm oil (b* = 56.87–74.50; a* = 2.21–13.0) [37]. According to the clarity analysis, Mayten tree seed oil is closer to white, based on a 1 to 100 scale, and is slightly darker than palm oil [37].

Table 1. Physicochemical characterization of Mayten tree seed oil.

Analysis		Value
Color		
	L	53.94 ± 7.66 *
	a	24.92 ± 8.20 *
	b	41.02 ± 5.15 *
	C°	48.16 ± 8.34 *
	H°	59.40 ± 6.23 *
Density		1.06 ± 0.07 (g/mL)
Peroxide value		5.10 ± 0.18 (meq O_2/kg oil)
Free Acidity		3.89 ± 0.19 (% oleic acid)
TBARS		5.74 ± 0.21(nmol/g lipids)
Rancimat		52.15 ± 2.15 (h)
Iodine value		57.63 ± 2.00 *

* adimentional value. All values reported in this table correspond to mean ± standard deviation.

The Blight and Dyer solvent extraction method allows for the extraction of high molecular weight lipophilic compounds that exist in a solid state at 25 °C. This method resulted in oil dispersion [37,38], with a density greater than 1 (1.06 ± 0.07 g mL^{-1}); as reported in Table 1, this is a value similar to the density of pine oil (1.042–1.224 g mL^{-1}) extracted using organic solvents [25]. The seed oil density value was higher than that reported for palm oil (1.06 g mL^{-1} versus 0.892–0.899 g mL^{-1}); such a value could indicate the presence of high molecular weight compounds, such as carotenoids and fatty acid [38].

The peroxide value identified for the Mayten tree seed oil was 5.10 ± 0.18 meq O_2/kg oil (Table 1), a value similar to that reported for olive oil [23], and was within the range reported for palm oil (1.0–10.0 mg eq O_2 g^{-1} oil) [38]. In addition, this peroxide value is below that allowed by the Chilean Sanitary Regulations of Food for the year 2015 (10.0 mg eq O_2 g^{-1} oil) and also lower than that obtained from the seeds of the maqui berry [23].

The iodine value of 57.63 identified for Mayten tree seed oil was similar to the value for palm oil (46.0–56.0), but higher than that for coconut oil (6.0–11.0) [38,39], showing a similar degree of unsaturation with these oils. The free acidity of Mayten tree seed oil was within the range identified for palm oil (3.7–5%, Table 1) and within the range allowed by the Chilean Sanitary Regulations of Food for the year 2015 [23].

The TBARS index of the Mayten tree seed oil had a lower value than those for avocado and olive seed oils [40]. When examining the oxidative stability under acceleration conditions (Rancimet analysis), the Mayten tree seed oil demonstrated a longer induction period (52.15 ± 2.15 h) than soybean oil (11.2 h) or extra virgin olive oil (24–26 h) [41]. According to the obtained physicochemical properties, Mayten tree see oil has values similar to other commercial oils indicating that it could be used for human consumption.

3.3. Fatty Acid Profile and Nutritional Quality

The fatty acid profile and nutritional qualities of Mayten tree seed oil were as shown in Figure 1 and Table 2, respectively. Mayten tree seed oil had a high unsaturated fatty acid content (Figure 1). The MUFA concentration of this seed oil (25.70 g/100 g oil) (Table 2) was lower than that for avocado seed oil (49–75.96%) [40,42,43], palm oil (37.1–39.2%) [38], olive oil (80.53%) [40], and extra virgin olive oil (71.32–79.62%) [41]. However, the PUFA composition of 20.98 g/100 g oil (20.98%) (Table 2) was greater than that for palm oil (8.1–10.5%) [38] and olive oil (5.43%) [40] and was within the range of that for avocado seed oil (11.75–37.13%) [40,42,43]. Based on these results, Mayten tree seed oils would likely provide health benefits, as unsaturated fatty acids favor specific enzymatic reactions such as cyclooxygenases, lipoxygenases and cytochrome P450 enzymes that resolve different processes of inflammation and protect against brain or renal dysfunctions [43,44], and PUFAs can provide the essential fatty acids that must be obtained through the diet.

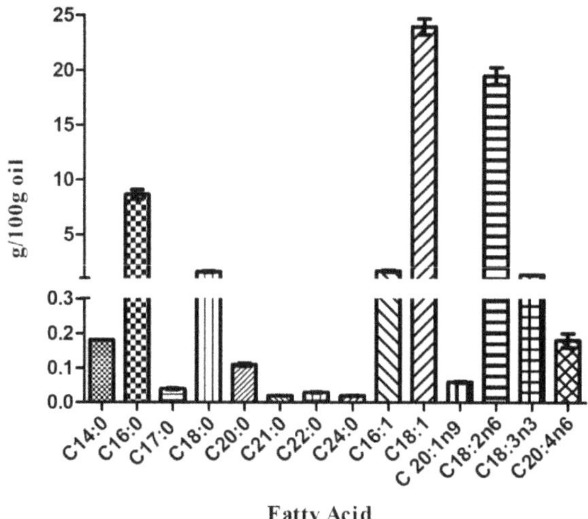

Figure 1. Profile of fatty acids of Mayten tree seed oil.

Table 2. Composition of the relevant groups of fatty acids and the nutritional qualities of Mayten tree seed oil.

	Value
SFA	10.66 (g/100 g oil)
MUFA	25.70 (g/100 g oil)
PUFA	20.98 (g/100 g oil)
n-6	19.66 (g/100 g oil)
n-3	1.32 (g/100 g oil)
PUFA/SFA	1.97
n-6/n-3	12.62
AI	0.20
TI	0.07

The ingestion of foods with PUFA/SFA ratios in the range of 1.25 to 2.4 has been described to confer beneficial effects for the prevention of cardiovascular diseases (CVDs) [45]. As Mayten tree seed oil was found to have a PUFA/SFA value of 1.97 (Table 2), this oil would provide a comparative advantage over other commonly consumed seed oils, such as olive oil and menhaden oil [28].

The oleic acid (C18:1), linoleic acid (C18:2n6), and palmitic acid (C16:0) contents were high (in decreasing order) in Mayten tree seed oil (Figure 1). The oleic acid contents of Mayten tree seed oil were higher than in sunflower oil, similar to contents in soybean oil [46,47], and lower than in palm oil [38]. Oleic acid (C18:1) is an omega-9 MUFA with anti-thrombosis and other bioactive properties, which has been used in cosmetic and pharmacological applications [48].

The linoleic acid (C18:2n6) content in Mayten tree seed oil was higher than that in olive oil and palm oil [38]. Linoleic acid is often described as a precursor for other lipid mediators with anti-inflammatory properties [43]. Linoleic acid (C18:2n6) is an essential omega-6 fatty acid that has been shown to be involved in the maintenance of the skin's permeable barrier against water, and the topical application of linoleic acid (C18:2n6) has been shown to improve dermatitis. Furthermore, linoleic acid is metabolized to arachidonic acid, which is a precursor of the eicosanoid compounds that regulate a range of physiological processes [49].

Palmitoleic acid (C16:1) and α-linoleic acid (C18:3n3) were found among the unsaturated fatty acid composition of Mayten tree seed oil (Figure 1), at higher levels than are found in sesame, sunflower, and rice brain [50]. Canola seed oil contains 3.4 g of palmitic acid per 100 g of oil [51], whereas the palmitic acid contents of Mayten tree seed oil were two-fold higher that in canola seed oil.

Two indices of coronary heart disease risk, AI and TI [29], were calculated for Mayten tree seed oil and other commonly consumed seed oils, as shown in Figure 2. The Mayten tree seed oil's AI value was lower than for rice bran oil (Figure 2) and menhaden oil [28] but higher than for sunflower oil, canola oil, and olive oil (Figure 2). By contrast, the TI value for Mayten tree seed oil was lower than values found in other commonly used seed oils, such as olive, sunflower [28], rice bran, and canola oils [50] (Figure 2), likely due to the Mayten tree seed oil's higher contents of oleic acid, a compound that reduces the risks of thrombosis [52]. These results indicate that Mayten tree seed oils could potentially reduce the risks of generating thrombi in blood vessels compared with other commonly consumed seed oils.

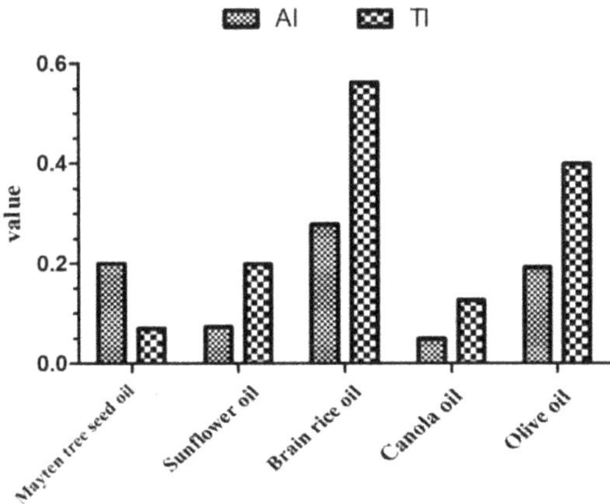

Figure 2. Atherogenic index (AI) and thrombogenic index (TI) values among Mayten tree seed oil and other commonly consumed seed oils (adapted from ref. [50]).

3.4. Polyphenol and Flavonoid Compounds

Polyphenol compounds (Table 3) were co-extracted with fatty acids during the seed oil extraction process from Mayten tree seeds, similar to other seed oils [23]. The TPC value (16.1 mg GAE/100 g oil) in Mayten tree seed oil represents 0.016% of the total oil contents, which is a larger proportion than is found in canola (11.26 mg GAE/100 g oil) [53] and olive oils (2–13 mg GAE/100 g oil) [54]. Flavonoids are a type of polyphenol component that dissolves in oil, and Mayten tree seed oil was found to contain a TFC of 15.5 mg EQ/100 g seed oil, which is similar to that reported for grape, rice bran, and chia seed oils but higher than those reported for sunflower, canola, soybean, and cottonseed oils [55].

Table 3. Phenolic and carotenoid compound contents in Mayten tree seed oil.

Parameter	Value
Total Phenolic content	16.1 ± 4.3 (mg GAE/100 g oil)
Total Flavonoids content	15.5 ± 2.3 (mg QE/100 g oil)
Total Carotenoids content	153.2 ± 6.5 (mg β-carotene E/100 g oil)
β-carotene content	106.8 ± 40.2 (mg β-carotene/100 g oil)
Coumaric acid	5.47 ± 0.03 (µg/100 g oil)
Quercetin	3.21 ± 0.05 (µg/100 g oil)
Myricetin	7.74 ± 0.04 (µg/100 g oil)
Ferulic acid	49.64 ± 0.03 (µg/100 g oil)

All values reported in this table correspond to mean ± standard deviation.

Flavonoids, such as quercetin and myricetin; phenolic acids, such as ferulic, coumaric, and syringic acids; and the sesquiterpene abscisic acid (Table 3) were found in Mayten tree seed oil. Ferulic acid, the most abundant compound in mayten tree seed oil, has been demonstrated to display various pharmacological properties, including anti-inflammatory activities [56,57]. Ferulic acid has been shown to reduce cholesterol synthesis in the liver, followed by an increase in sterol acid secretion, and has been shown to act as a chemoprotective agent against coronary heart disease, preventing thrombi and sclerosis production [57,58]. Ferulic acid also inhibits the population growth of influenza, syncytial, and human immunodeficiency viruses [56] and has been shown to exert anticancer activity against colon and rectal cancer [57]. Similar to ferulic acid, the other polyphenolic compounds identified in this seed oil are known to provide health benefits, such as antioxidant properties against free radical formation and the prevention of diseases and infections due to antimicrobial activities [59].

3.5. Total Carotenoids and β-Carotene

The reddish coloration observed for Mayten tree seed oil was likely derived from its high carotene content (Table 3), which is present in the seed aril (Figure S1) [42,43]. This relationship between seed oil color and presence of carotenoids has also been described for palm and cucumber seed oils [60,61]. Mayten tree seed oil is darker in color because it has a total carotene content (TCC) that is three times higher than that of palm seed oil [38].

Approximately 70% of the total carotene content in Mayten tree seed oil is represented by β-carotene (Table 3). As a precursor of vitamin A, β-carotene plays an important role in the prevention of cataracts and macular hatching, in addition to improving night blindness and dry eyes [62]. The difference between the TCC and β-carotene contents in Mayten tree seed oil is expected to indicate the presence of other carotenoid compounds that were not specifically identified in the present study. The high carotene contents, together with the presence of polyphenols, may give this seed oil resistance to fatty acid peroxidation, which would prevent rancidity [63].

3.6. Antioxidant Capacity

The resistance against factors that cause rancidity in seeds oils is a function of the antioxidant capacity of seed oils, which can provide nutritional value without requiring the incorporation of synthetic antioxidants, such as those used by the food industry [64]. The antioxidant capacity of Mayten tree seed oil against the ABTS radical appeared to be lower than the capacities found in other seed oils (Table 4). However, this characteristic is typically determined based on the presence of polyphenolic compounds (as in the present study) and carotenoids because these molecules also react to the ABTS radical [65].

Table 4. Antioxidants capacity in Mayten tree seed oil.

Properties	Value
Antioxidant Capacity-ABTS	1.25 ± 0.45 (mg TE/100 g of oil)
IC_{50}	1.06 ± 0.03 (g oil)
FRAP	149 ± 7.45 (mg $FeSO_4$/100 g of oil)

All values reported in this table correspond to mean ± standard deviation.

The FRAP value is a measure of the antioxidant capability based on the evaluation of electron donation that occurs due to the activity of antioxidant compounds (Table 4) [66]. For Mayten tree seed oil, the FRAP value was three-fold higher that of butylhydroxytoluene (BHT), an artificial antioxidant commonly used by the food industry [64], indicating that Mayten tree seed oil displays a high antioxidant capacity. According to the IC_{50} value, a low amount of Mayten tree seed oil dissolved in methanol was necessary to inhibit 50% of the DPPH radical [64], indicating a better response against the proton donation mechanism of oxidated compounds. Based on the FRAP results, the antioxidant capacity of Mayten tree seed oil was found to be three-fold that of rapeseed oil and 15-fold those for sunflower, rice brain, and olive oils [50,67]. The percentage of DPPH radical inhibition of a decreasing methanol dilution range of Mayten tree seed oil is shown in Figure 3b. Initially, the percentage of inhibition increased as the contents of the seed oil increases, followed by a stabilization when 60–70% inhibition of the DPPH radical was achieved. Therefore, the examined Mayten tree seed oil appears to have a higher antioxidant capacity than the BHT supplement commonly used by the food industry (Figure 3).

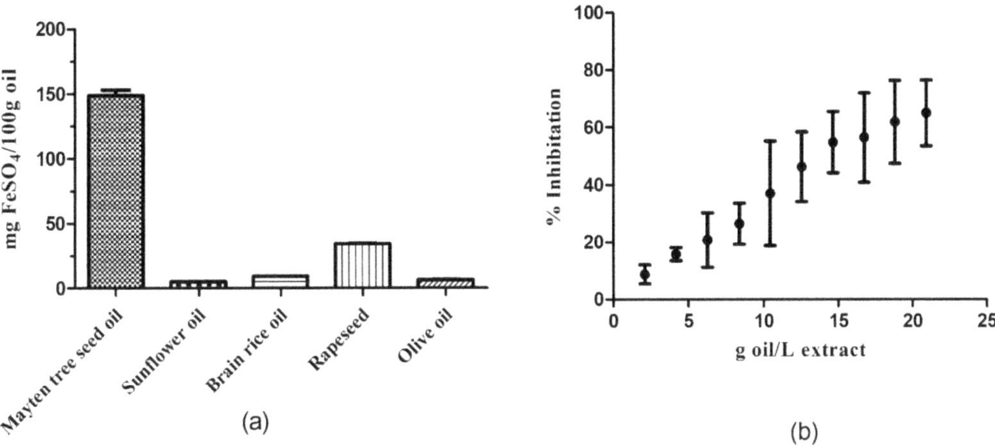

Figure 3. (**a**) Ferric reducing antioxidant power (FRAP) values for Mayten tree seed oil, sunflower oil, brain rice oil, canola oil, and olive oil. (Adapted with permission from Szydłowska-Czerniak, A. et al. [67], Copyright 2008 Elsevier); (**b**) Inhibition of the 2,2-diphenyl-1-picrylhydrazyl (DPPH) radical by Mayten tree seed oil. Values are represent as mean ± standard deviation.

4. Conclusions

Mayten tree seed oil was obtained through the application of the Blight and Dyer solvent (methanol–chloroform–water) extraction method, resulting in a 61.77% yield, which may be further improved using other extraction procedures. This seed oil was found to have high omega-6 and omega-9 fatty acid contents, with a higher PUFA content than that found in most major commercial vegetable oils.

The analyzed quality parameters indicated that Mayten tree seed oil has a high resistance against rancidity, with values similar to those of other commercial seed oils and within the requirements of Chilean Health Regulations. The high TPC and TFC values for this seed oil provide protection against lipid peroxidation and result in high ABTS, DPPH, and FRAP antioxidant capacities, which may be beneficial for human health and pharmacological uses.

The reddish coloration of Mayten tree seed oil is associated with a high content of carotenoids, which may constitute a comparative advantage relative to other vegetable oils currently used for human consumption. β-carotene is the dominant carotenoid in this seed oil, potentially providing elements that can contribute to the functional properties of this oil.

The results of the present study indicate that Mayten tree seed oil has nutritional value, based on the antioxidant capacity and bioactive properties identified for this oil. This seed oil would be an interesting alternative to other vegetable oils intended for human consumption, particularly since it could be produced in areas affected by global climate change with higher yields than other traditional oils. It could be considered as a functional food, a carotenoids supplement or an antioxidants additive ingredient for the food industry. However, studies exploring the domestication of this tree species are necessary.

Supplementary Materials: The following are available online at https://www.mdpi.com/article/10.3390/foods10040729/s1. Figure S1: (a) Fruits and seeds (orange-reddish color) of Mayten tree, (b) seed oil of Mayten tree extracted with solvents (methanol and chloroform).

Author Contributions: Conceptualization, R.G. and A.G.; methodology, E.M.-C., G.C.-P., A.G. and G.M.; validation, C.S.-N.; formal analysis, E.M.-C., P.V. and A.G.; investigation, E.M.-C.; writing—original draft preparation, E.M.-C., P.V. and A.G.; writing—review and editing, R.G. and C.S.-N.; supervision, R.G., A.G., G.M. and C.S.-N.; funding acquisition, R.G. and C.S.-N. All authors have read and agreed to the published version of the manuscript.

Funding: This research was funded by Proyecto Interdisciplina VRI-UC N°6/2015 and FONDEQUIP under Grants EQM 130032 and EQM160042.

Acknowledgments: The authors acknowledgment the analytical support of the Instituto de Nutrición y Tecnología de Alimentos (INTA) for determination of fatty acids and quality analysis. Authors thanks to Víctor Ahumada for his contribution to the antimicrobial capacity analysis.

References

1. Alexandros, N.; Bruinsma, J. *World Agriculture towards 2030/2050: The 2012 Revision ESA Working Papers*; FAO Agricultural Development Economics Division: Rome, Italy, 2012; Volume 54. [CrossRef]
2. Fine, F.; Lucas, J.L.; Chardigny, J.M.; Redlingshöfer, B.; Renard, M. Food losses and waste in the French oilcrops sector. *OCL* **2015**, *22*, A302. [CrossRef]
3. Varnham, A. *Seed Oil: Biological Properties, Health Benefits and Commercial Applications*; Nova Science Publishers: New York, NY, USA, 2014; ISBN 9781634630955.
4. McDonald, M.B.; Copeland, L.O. *Seed Production*; Springer: Boston, MA, USA, 1997; ISBN 978-1-4613-6825-0.
5. Huang, A.H.C. Oleosins and oil bodies in seeds and other organs. *Plant Physiol.* **1996**, *110*, 1055–1061. [CrossRef] [PubMed]
6. Kapoor, S.; Gandhi, N.; Tyagi, S.K.; Kaur, A.; Mahajan, B.V.C. Extraction and characterization of guava seed oil: A novel industrial byproduct. *LWT* **2020**, *132*, 109882. [CrossRef]
7. Naderi, M.; Torbati, M.; Azadmard-Damirchi, S.; Asnaashari, S.; Savage, G.P. Common ash (*Fraxinus excelsior* L.) seeds as a new vegetable oil source. *LWT* **2020**, *131*, 109811. [CrossRef]
8. Marcora, P.I.; Tecco, P.A.; Zeballos, S.R.; Hensen, I. Influence of altitude on local adaptation in upland tree species from central Argentina. *Plant Biol.* **2017**, *19*, 123–131. [CrossRef] [PubMed]
9. Rodríguez, R.R.; Matthei, S.; Quezada, M. *Flora Arbórea De Chile*; Biblioteca de Recursos Renovables y no Renovables de Chile; Editorial de la Universidad de Concepción: Concepción, Chile, 1983.
10. Wendler, J.; Donoso, C. Antecedentes morfológicos y genecológicos de *Maytenus boaria*. *Bosque* **1985**, *6*, 93–99.
11. Mackenna, B.V. Historia crítica y social de la ciudad de Santiago desde su fundación hasta nuestros días (1541–1868). *Impr. Mercur.* **1869**, *2*, 1541–1868.
12. Boletín de las leyes y de las órdenes y decretos del gobierno. *Imprenta Mercur.* **1854**, *1853*, 132–147.
13. Memorias, Anales de la Universidad de Chile. Available online: https://anales.uchile.cl/index.php/ANUC/article/view/1815 (accessed on 1 November 2020).

14. Tinto, J. Utilización de los recursos forestales argentinos, Instituto Forestal Nacional. Buenos Aires. *Folleto Técnico For.* **1979**, *41*, 97.
15. Facciola, S. *Cornucopia II. A Source Book of Edible Plants*; Kompong Publications: Vista, CA, USA, 1998.
16. Montenegro, G.; Timmermann, B. *Chile, Nuestra Flora Útil: Guía de Plantas de Uso Apícola, en Medicina Folklórica, Artesanal y Ornamental*; Ediciones Pontificia Universidad Católica de Chile: Santiago, Chile, 2000; ISBN 9562884945.
17. Spivey, A.C.; Weston, M.; Woodhead, S. *Celastraceae* sesquiterpenoids: Biological activity and synthesis. *Chem. Soc. Rev.* **2002**, *31*, 43–59. [CrossRef]
18. Gullo, F.P.; Sardi, J.C.O.; Santos, V.A.F.F.M.; Sangalli-Leite, F.; Pitangui, N.S.; Rossi, S.A.; de Paula Silva, A.C.A.; Soares, L.A.; Silva, J.F.; Oliveira, H.C.; et al. Antifungal Activity of Maytenin and Pristimerin. *Evid.-Based Complement. Altern. Med.* **2012**, *2012*. [CrossRef]
19. Céspedes, C.; Alarcón, J. Biopesticidas de origen botánico, fitoquímicos y extractos de *Celastraceae, Rhamnaceae y Scrophulariaceae*. *Lat. Am. Caribb. Bull. Med. Aromat. Plants* **2011**, *10*, 175–181.
20. Hidalgo, D. *Actividad Nematicida Sobre Meloidogynehapla de Extractos Acuosos de Especies Arbóreas y Arbustivas de la Zona sur de Chile*; Universidad Austral de Chile: Valdivia, Chile, 2008.
21. Zavala, H.A.; Hormazabal, U.E.; Montenegro, R.G.; Rosalez, V.M.; Quiroz, C.A.; Paz, R.C.; Rebolledo, R.R. Effects of extracts from *Maytenus* on *Aegorhinus superciliosus* (coleoptera: *Curculionidae*) and *Hippodamia convergens* (coleoptera: *Coccinellidae*). *Rev. Colomb. Entomol.* **2017**, *43*, 233–244. [CrossRef]
22. Cabello, A.; Camelio, M. Germinación de semillas de maitén (*Maytenus boaria*) y producción de plantas en vivero. Universidad de Chile. *Rev. Ciencias For.* **1996**, *11*, 3–17.
23. Bastías-Montes, J.M.; Monterrosa, K.; Muñoz-Fariña, O.; García, O.; Acuña-Nelson, S.M.; Vidal-San Martín, C.; Quevedo-Leon, R.; Kubo, I.; Avila-Acevedo, J.G.; Domiguez-Lopez, M.; et al. Chemoprotective and antiobesity effects of tocols from seed oil of Maqui-berry: Their antioxidative and digestive enzyme inhibition potential. *Food Chem. Toxicol.* **2020**, *136*, 111036. [CrossRef] [PubMed]
24. Popa, S.; Boran, S. CIELAB and thermal properties of sesame food oil under antocyanin and UV influence. *Rev. Chim.* **2017**, *68*, 1499–1503. [CrossRef]
25. Islam, M.N.; Sabur, A.; Ahmmed, R.; Hoque, M.E. Oil extraction from pine seed (*Polyalthia longifolia*) by solvent extraction method and its property analysis. *Procedia Eng.* **2015**, *105*, 613–618. [CrossRef]
26. *AOCS Official Methods and Recommended Practices of the AOCS*; American Oil Chemist's Society: Urbana, IL, USA, 2009; ISBN 978-1-893997-74-5.
27. Petropoulos, S.A.; Fernandes, Â.; Arampatzis, D.A.; Tsiropoulos, N.G.; Petrović, J.; Soković, M.; Barros, L.; Ferreira, I.C.F.R. Seed oil and seed oil byproducts of common purslane (*Portulaca oleracea* L.): A new insight to plant-based sources rich in omega-3 fatty acids. *LWT* **2020**, *123*. [CrossRef]
28. Otero, P.; Gutierrez-Docio, A.; Navarro del Hierro, J.; Reglero, G.; Martin, D. Extracts from the edible insects *Acheta domesticus* and *Tenebrio molitor* with improved fatty acid profile due to ultrasound assisted or pressurized liquid extraction. *Food Chem.* **2020**, *314*, 126200. [CrossRef]
29. Basdagianni, Z.; Papaloukas, L.; Kyriakou, G.; Karaiskou, C.; Parissi, Z.; Sinapis, E.; Kasapidou, E. A comparative study of the fatty acid and terpene profiles of ovine and caprine milk from Greek mountain sheep breeds and a local goat breed raised under a semi-extensive production system. *Food Chem.* **2019**, *278*, 625–629. [CrossRef]
30. Chanioti, S.; Tzia, C. Optimization of ultrasound-assisted extraction of oil from olive pomace using response surface technology: Oil recovery, unsaponifiable matter, total phenol content and antioxidant activity. *LWT Food Sci. Technol.* **2017**, *79*, 178–189. [CrossRef]
31. Velásquez, P.; Montenegro, G.; Giordano, A.; Retamal, M.; Valenzuela, L.M.; Montenegro, G.; Giordano, A.; Retamal, M.; Valenzuela, L.M. Bioactivities of phenolic blend extracts from Chilean honey and bee pollen. *CyTA J. Food* **2019**, *17*, 754–762. [CrossRef]
32. Giordano, A.; Fuentes-Barros, G.; Castro-Saavedra, S.; González-Cooper, A.; Suárez-Rozas, C.; Salas-Norambuena, J.; Acevedo-Fuentes, W.; Leyton, F.; Tirapegui, C.; Echeverría, J.; et al. Variation of Secondary Metabolites in the Aerial Biomass of *Cryptocarya alba*. *Nat. Prod. Commun.* **2019**, *14*, 1–11. [CrossRef]
33. Diniyah, N.; Alam, M.B.; Lee, S.-H. Antioxidant potential of non-oil seed legumes of Indonesian's ethnobotanical extracts. *Arab. J. Chem.* **2020**. [CrossRef]
34. Varzakas, T.; Kiokias, S. HPLC Analysis and Determination of Carotenoid Pigments in Commercially Available Plant Extracts. *Curr. Res. Nutr. Food Sci.* **2016**, *4*, 1–14. [CrossRef]
35. Biehler, E.; Mayer, F.; Hoffmann, L.; Krause, E.; Bohn, T. Comparison of 3 Spectrophotometric Methods for Carotenoid Determination in Frequently Consumed Fruits and Vegetables. *J. Food Sci.* **2010**, *75*, C55–C61. [CrossRef] [PubMed]
36. Marcus, J.B. Lipids Basics: Fats and Oils in Foods and Health. In *Culinary Nutrition*; Elsevier: Amsterdam, The Netherland, 2013; pp. 231–277. [CrossRef]
37. de Almeida, D.T.; Curvelo, F.M.; Costa, M.M.; Viana, T.V.; de Lima, P.C. Oxidative stability of crude palm oil after deep frying akara (Fried bean paste). *Food Sci. Technol.* **2018**, *38*, 142–147. [CrossRef]
38. Mba, O.I.; Dumont, M.J.; Ngadi, M. Palm oil: Processing, characterization and utilization in the food industry—A review. *Food Biosci.* **2015**, *10*, 26–41. [CrossRef]

39. Tiefenbacher, K.F.; Tiefenbacher, K.F. Technology of Main Ingredients—Sweeteners and Lipids. *Wafer Waffle* **2017**, 123–225. [CrossRef]
40. Berasategi, I.; Barriuso, B.; Ansorena, D.; Astiasarán, I. Stability of avocado oil during heating: Comparative study to olive oil. *Food Chem.* **2012**, *132*, 439–446. [CrossRef] [PubMed]
41. Souza, P.T.; Ansolin, M.; Batista, E.A.C.; Meirelles, A.J.A.; Tubino, M. Identification of extra virgin olive oils modified by the addition of soybean oil, using ion chromatography. *J. Braz. Chem. Soc.* **2019**, *30*, 1055–1062. [CrossRef]
42. Méndez-Zúñiga, S.M.; Corrales-García, J.E.; Gutiérrez-Grijalva, E.P.; García-Mateos, R.; Pérez-Rubio, V.; Heredia, J.B. Fatty Acid Profile, Total Carotenoids, and Free Radical-Scavenging from the Lipophilic Fractions of 12 Native Mexican Avocado Accessions. *Plant Foods Hum. Nutr.* **2019**, 501–507. [CrossRef]
43. Corrales-García, J.E.; del Rosario García-Mateos, M.; Martínez-López, E.; Barrientos-Priego, A.F.; Ybarra-Moncada, M.C.; Ibarra-Estrada, E.; Méndez-Zúñiga, S.M.; Becerra-Morales, D. Anthocyanin and Oil Contents, Fatty Acids Profiles and Antioxidant Activity of Mexican Landrace Avocado Fruits. *Plant Foods Hum. Nutr.* **2019**, *74*, 210–215. [CrossRef]
44. Tao, L. Oxidation of Polyunsaturated Fatty Acids and its Impact on Food Quality and Human Health Protective potential of dietary soy View project Enrichment of omega-3 fatty acids by microalgae View project. *Adv. Food Technol. Nutr. Sci. Open J.* **2015**, *1*, 135–142. [CrossRef]
45. Siri-Tarino, P.W.; Chiu, S.; Bergeron, N.; Krauss, R.M. Saturated Fats Versus Polyunsaturated Fats Versus Carbohydrates for Cardiovascular Disease Prevention and Treatment. *Annu. Rev. Nutr.* **2015**, *35*, 517–543. [CrossRef]
46. Choi, S.-G.; Won, S.-R.; Rhee, H.-I. Oleic Acid and Inhibition of Glucosyltransferase. *Olives Olive Oil Health Dis. Prev.* **2010**, 1375–1383. [CrossRef]
47. Zamani, S.; Naderi, M.R.; Soleymani, A.; Nasiri, B.M. Sunflower (*Helianthus annuus* L.) biochemical properties and seed components affected by potassium fertilization under drought conditions. *Ecotoxicol. Environ. Saf.* **2020**, *190*, 110017. [CrossRef]
48. Zanqui, A.B.; da Silva, C.M.; Ressutte, J.B.; de Morais, D.R.; Santos, J.M.; Eberlin, M.N.; Cardozo-Filho, L.; da Silva, E.A.; Gomes, S.T.M.; Matsushita, M. Extraction and assessment of oil and bioactive compounds from cashew nut (*Anacardium occidentale*) using pressurized n-propane and ethanol as cosolvent. *J. Supercrit. Fluids* **2020**, *157*, 104686. [CrossRef]
49. Sanders, T.A.B. *Introduction: The Role of Fats in Human Diet*. The Role of Fats in Human Diet; Elsevier Inc.: Amsterdam, The Netherland, 2016. [CrossRef]
50. Orsavova, J.; Misurcova, L.; Vavra Ambrozova, J.; Vicha, R.; Mlcek, J. Fatty acids composition of vegetable oils and its contribution to dietary energy intake and dependence of cardiovascular mortality on dietary intake of fatty acids. *Int. J. Mol. Sci.* **2015**, *16*, 12871–12890. [CrossRef] [PubMed]
51. Confortin, T.C.; Todero, I.; Luft, L.; Ugalde, G.A.; Mazutti, M.A.; Oliveira, Z.B.; Bottega, E.L.; Knies, A.E.; Zabot, G.L.; Tres, M.V. Oil yields, protein contents, and cost of manufacturing of oil obtained from different hybrids and sowing dates of canola. *J. Environ. Chem. Eng.* **2019**, *7*, 102972. [CrossRef]
52. Ghaeni, M.; Ghahfarokhi, K.N. Fatty Acids Profile, Atherogenic (IA) and Thrombogenic (IT) Health Lipid Indices in *Leiognathus-bindus* and *Upeneussulphureus*. *J. Mar. Sci. Res. Dev.* **2013**, *3*, 3–5. [CrossRef]
53. Ghazani, S.M.; García-Llatas, G.; Marangoni, A.G. Micronutrient content of cold-pressed, hot-pressed, solvent extracted and RBD canola oil: Implications for nutrition and quality. *Eur. J. Lipid Sci. Technol.* **2014**, *116*, 380–387. [CrossRef]
54. Pérez, M.M.; Yebra, A.; Melgosa, M.; Bououd, N.; Asselman, A.; Boucetta, A. Caracterización colorimétrica y clasificación del aceite de oliva virgen de la cuenca mediterránea hispano-marroquí. *Grasas Aceites* **2003**, *54*, 392–396.
55. Xuan, T.D.; Ngoc, Q.T.; Khanh, T.D. An Overview of Chemical Profiles, Antioxidant and Edible Oils Marketed in Japan. *Foods* **2018**, *7*, 21. [CrossRef]
56. Kumar, N.; Pruthi, V. Potential applications of ferulic acid from natural sources. *Biotechnol. Rep.* **2014**, *4*, 86–93. [CrossRef]
57. Ou, S.; Kwok, K.C. Ferulic acid: Pharmaceutical functions, preparation and applications in foods. *J. Sci. Food Agric.* **2004**, *84*, 1261–1269. [CrossRef]
58. Srinivasan, M.; Sudheer, A.R.; Menon, V.P. Ferulic Acid: Therapeutic Potential through Its Antioxidant Property. *J. Clin. Biochem. Nutr.* **2007**, *40*, 92–100. [CrossRef] [PubMed]
59. Maalik, A.; Khan, F.A.; Mumtaz, A.; Mehmood, A.; Azhar, S.; Atif, M.; Karim, S.; Altaf, Y.; Tariq, I. Pharmacological applications of quercetin and its derivatives: A short review. *Trop. J. Pharm. Res.* **2014**, *13*, 1561–1566. [CrossRef]
60. Bardaa, S.; Halima, N.B.; Aloui, F.; Mansour, R.B.; Jabeur, H.; Bouaziz, M.; Sahnoun, Z. Oil from pumpkin (*Cucurbita pepo* L.) seeds: Evaluation of its functional properties on wound healing in rats. *Lipids Health Dis.* **2016**, *15*, 1–12. [CrossRef] [PubMed]
61. Othman, N.; Manan, Z.A.; Alwi, S.W.; Sarmidi, M. A review of extraction technology of carotenoids. *J. Appl. Sci.* **2010**, *10*, 1187–1191. [CrossRef]
62. Neha, K.; Haider, M.R.; Pathak, A.; Yar, M.S. Medicinal prospects of antioxidants: A review. *Eur. J. Med. Chem.* **2019**, *178*, 687–704. [CrossRef]
63. Sandmann, G. Antioxidant protection from UV-and light-stress related to carotenoid structures. *Antioxidants* **2019**, *8*, 219. [CrossRef]
64. Hasheminya, S.-M.; Dehghannya, J. Composition, phenolic content, antioxidant and antimicrobial activity of *Pistacia atlantica* subsp. kurdica hulls' essential oil. *Food Biosci.* **2020**, *34*, 100510. [CrossRef]
65. Mueller, L.; Boehm, V. Antioxidant Activity of β-Carotene Compounds in Different In Vitro Assays. *Molecules* **2011**, *16*, 1055–1069. [CrossRef] [PubMed]

66. Gupta, S.; Caraballo, M.; Agarwal, A. Total Antioxidant Capacity Measurement by Colorimetric Assay. In *Oxidants, Antioxidants and Impact of the Oxidative Status in Male Reproduction*; Elsevier: Amsterdam, The Netherland, 2019; pp. 207–215. [CrossRef]
67. Szydłowska-Czerniak, A.; Dianoczki, C.; Recseg, K.; Karlovits, G.; Szłyk, E. Determination of antioxidant capacities of vegetable oils by ferric-ion spectrophotometric methods. *Talanta* **2008**, *76*, 899–905. [CrossRef] [PubMed]

Article

Bioactive Compounds in Wild Nettle (*Urtica dioica* L.) Leaves and Stalks: Polyphenols and Pigments upon Seasonal and Habitat Variations

Maja Repajić [1], Ena Cegledi [1], Zoran Zorić [1], Sandra Pedisić [1], Ivona Elez Garofulić [1,*], Sanja Radman [2], Igor Palčić [3] and Verica Dragović-Uzelac [1]

[1] Faculty of Food Technology and Biotechnology, University of Zagreb Pierottijeva 6, 10000 Zagreb, Croatia; maja.repajic@pbf.unizg.hr (M.R.); ecegledi@pbf.hr (E.C.); zzoric@pbf.hr (Z.Z.); sandra.pedisic@pbf.unizg.hr (S.P.); vdragov@pbf.hr (V.D.-U.)

[2] Faculty of Agriculture, University of Zagreb, Svetošimunska cesta 25, 10000 Zagreb, Croatia; sradman@agr.hr

[3] Institute of Agriculture and Tourism, Karla Huguesa 8, 52440 Poreč, Croatia; palcic@iptpo.hr

* Correspondence: ivona.elez@pbf.unizg.hr

Abstract: This study evaluated the presence of bioactives in wild nettle leaves and stalks during the phenological stage and in the context of natural habitat diversity. Thus, wild nettle samples collected before flowering, during flowering and after flowering from 14 habitats situated in three different regions (continental, mountain and seaside) were analyzed for low molecular weight polyphenols, carotenoids and chlorophylls using UPLC-MS/MS and HPLC analysis, while the ORAC method was performed for the antioxidant capacity measurement. Statistical analysis showed that, when compared to the stalks, nettle leaves contained significantly higher amounts of analyzed compounds which accumulated in the highest yields before flowering (polyphenols) and at the flowering stage (pigments). Moreover, nettle habitat variations greatly influenced the amounts of analyzed bioactives, where samples from the continental area contained higher levels of polyphenols, while seaside region samples were more abundant with pigments. The levels of ORAC followed the same pattern, being higher in leaves samples collected before and during flowering from the continental habitats. Hence, in order to provide the product's maximum value for consumers' benefit, a multidisciplinary approach is important for the selection of a plant part as well as its phenological stage with the highest accumulation of bioactive compounds.

Keywords: nettle leaves and stalks; phenological stage; location; accelerated solvent extraction; UPLC-MS/MS; polyphenols; chlorophylls; carotenoids; antioxidant capacity; ORAC

Citation: Repajić, M.; Cegledi, E.; Zorić, Z.; Pedisić, S.; Elez Garofulić, I.; Radman, S.; Palčić, I.; Dragović-Uzelac, V. Bioactive Compounds in Wild Nettle (*Urtica dioica* L.) Leaves and Stalks: Polyphenols and Pigments upon Seasonal and Habitat Variations. *Foods* **2021**, *10*, 190. https://doi.org/10.3390/foods10010190

Received: 23 November 2020
Accepted: 14 January 2021
Published: 18 January 2021

Publisher's Note: MDPI stays neutral with regard to jurisdictional claims in published maps and institutional affiliations.

1. Introduction

Nettle (*Urtica dioica* L.) is a perennial wild plant of the Urticaceae family, genus Urtica, which is widespread in Europe, Asia, America and part of Africa, and has been adapted to different climatic conditions [1,2]. Nettle has long been used in the food, cosmetic and pharmaceutical industries due to its nutritional and health potential, as all parts of nettle (leaves, stalks and roots) show a rich composition of bioactive compounds with high antioxidant capacity [2,3] Previous studies have shown that nettle leaves and stalks are a rich source of vitamins A, B and C, minerals (iron, potassium, calcium, magnesium), polyphenols such as phenolic acids and flavonoids as well as pigments, especially chlorophyll and carotenoids [4–11]. In accordance with the above, aerial parts of nettle have anti-inflammatory and therapeutic effects; these nettle parts are used in the treatment of arthritis, anemia, allergies, joint pain and urinary tract infections, have a diuretic effect and are used to strengthen hair [3,12]. Besides aerial parts, nettle root also presents a rich source of various compounds such as protein lectin, sterols, polysaccharides, lignans and phenols [5,7,13,14] and is mostly used in the treatment of benign prostatic hyperplasia [15]. Apart from medicinal use, other applications of nettle include food preparation, where it is consumed in the form of tea, soup, stew or salad [3], or for commercial extraction of chlorophyll, which is used as a green coloring agent (E140) [16].

For medicinal purpose and medicinal preparations, nettle is mostly often used in the form of liquid or dry extract; thus, it is important to apply extraction method that will give a highly stabile extract with the greatest possible content of bioactive ingredients. Therefore, new extraction methods are increasingly being used and one of them is accelerated solvent extraction (ASE). In addition to being an efficient method, it uses less solvent, shortens the extraction time and more effectively isolates the target components [17].

Aside from extraction method, extract quality and richness in bioactives also depends on used plant material, either wild or cultivated, where its chemical composition and consequently antioxidant capacity are influenced by environmental, genotypic and phenotypic factors.

Different parts of plant may contain different amounts of particular compounds, e.g., nettle leaves accumulate higher amounts of polyphenols and chlorophylls in comparison with stalks [6,7,18]. In general, leaves are the richest part of a nettle in bioactive compounds, therefore they are mostly used in processing. However, changes in chemical composition and compounds' distribution occur with plant's maturity, where bioactive compounds are present in different proportions during different phenological stages. For example, the content of polyphenols decreases with growth and maturity of the plant [19]. Bioactive compounds are produced in response to different forms of (a)biotic stresses, as well as to fulfil important physiological tasks (attracting pollinators, establishing symbiosis, providing structural components to lignified cell walls of vascular tissues, etc.) [20]. These processes are often connected to specific phenological stages. Hence, harvest time depends on the type of final product. Although opinion on nettle optimal harvest time differs among various authors [3], Moore (1993) [21] stated that for juices and other fresh preparations, nettle leaves are best picked in spring or early summer (before flowering), and according to Upton (2013) [3] for dried preparations, it is best to harvest from mid-spring to late summer. If nettle is used for food purposes, the recommended harvest should be at the pre-flowering and flowering phases, certainly before the appearance of the seeds when it contains the least bioactive ingredients [3].

Nevertheless, nettle herb is mostly wild-harvested [3]. Concerning the natural habitat and climate, nettle is a quite adaptable plant. It grows in areas characterized by mild to temperate climates and prefers open or partly shady habitats with plenty of moisture such as forests, by rivers or streams and on roadsides [2]. Still, accumulation of polyphenols and pigments varies upon climate and habitat diversity. Plants grown in cold climates often show greater antioxidant properties, as a result of oxidative stress defense [22], while pigments synthesis is enhanced due to exposure to higher temperatures and more sunlight [23,24].

Although mentioned scientific literature provides data regarding nettle chemical composition, to our best knowledge there are no comprehensive studies on polyphenols and pigments constituents and their accumulation in wild nettle leaves and stalks during different vegetation periods of growing across diverse regions. These cognitions could be beneficial input data in a production of liquid and dry extracts. Therefore, the current study aimed to examine the presence and profile of low molecular weight polyphenols, carotenoids and chlorophylls as well as to determine antioxidant capacity in wild nettle leaves and stalks collected during three phenological stages (before flowering, during flowering and after flowering) from 14 different natural habitats situated in three regions in Croatia.

2. Materials and Methods

2.1. Chemicals

HPLC grade acetonitrile was procured from J.T. Baker Chemicals (Deventer, Netherlands). Purified water was obtained by a Milli-Q water purification system (Millipore, Bedford, MA, USA). Ethanol (96%) was purchased from Gram–mol d.o.o. (Zagreb, Croatia) and formic acid (98–100%) from T.T.T. d.o.o. (Sveta Nedjelja, Croatia). Commercial standards of quercetin-3-glucoside, kaempferol-3-rutinoside, myricetin, caffeic acid, gallic acid, ferulic acid, sinapic acid, quinic acid, chlorogenic acid, *p*-coumaric acid, esculetin, scopoletin, α-carotene, β-carotene, chlorophyll *a* and chlorophyll *b* were purchased from Sigma–Aldrich (St. Louis, MO, USA). Epicatechin, catechin, epigallocatechin gallate, epicatechin gallate, apigenin, luteolin and naringenin were obtained from Extrasynthese (Genay, France), while quercetin-3-rutinoside was procured from Acros Organics (Thermo Fisher Scientific, Geel, Belgium). Apigenin was dissolved in ethanol with 0.5% (*v/v*) dimethyl sulfoxide, standards of carotenoids and chlorophylls in *n*-hexane. Other standards were prepared as a stock solution in methanol, and working standard solutions were prepared by diluting the stock solutions to yield five concentrations.

2.2. Plant Material

Samples of wild nettle (*Urtica dioica* L.) were collected at three phenological stages [(I) before flowering, (II) during flowering and (III) after flowering] during 2019 from different habitats in Croatia belonging to three regions (continental, mountain and seaside) (Table 1). Plant material was identified by using usual keys and iconographies with support of Department of Vegetable Crops, Faculty of Agriculture, University of Zagreb (Croatia). Immediately after harvesting, leaves were separated from stalks and samples were stored at −18 °C, freeze-dried (Alpha 1-4 LSCPlus, Martin Christ Gefriertrocknungsanlagen GmbH, Osterode am Harz, Germany) and afterwards grinded into fine powder using a commercial grinder (GT11, Tefal, Rumilly, France). Obtained powders were immediately analyzed for total solids by drying at 103 ± 2 °C to constant mass [25] and further used for the extraction. Content of dry matter in samples was >95%.

Table 1. Location and weather characteristics of wild nettle (*Urtica dioica* L.) habitats.

Region	Location	Altitude/ Latitude/Longitude	Weather Parameters	Phenological Stage					
				I Before Flowering		II During Flowering		III After Flowering	
				April	May	June	July	September	October
Continental	Sela Žakanjska	244 m 45°36′27.80″N 15°20′38.21″E	a.d. T (°C)	11.0	13.4	22.6	22.0	16.3	12.7
			T min (°C)	−0.1	0.4	11.4	9.2	2.9	0.7
			T max (°C)	27.7	26.5	34.7	35.6	30.5	26.7
			a.p. (mm)	143.4	170.1	73.8	85.4	101.8	55.6
	Sopčić Vrh	177 m 45°34′14.88″N 15°20′24.98″E	a.d. T (°C)	11.0	13.4	22.6	22.0	16.3	12.7
			T min (°C)	−0.1	0.4	11.4	9.2	2.9	0.7
			T max (°C)	27.7	26.5	34.7	35.6	30.5	26.7
			a.p. (mm)	143.4	170.1	73.8	85.4	101.8	55.6
	Žakanje	178 m 45°36′34.38″N 15°20′14.96″E	a.d. T (°C)	11.0	13.4	22.6	22.0	16.3	12.7
			T min (°C)	−0.1	0.4	11.4	9.2	2.9	0.7
			T max (°C)	27.7	26.5	34.7	35.6	30.5	26.7
			a.p. (mm)	143.4	170.1	73.8	85.4	101.8	55.6
	Zagreb I (Gračani)	119 m 45°51′31.10″N 15°58′19.34″E	a.d. T (°C)	12.4	13.7	23.8	22.9	17.2	13.2
			T min (°C)	1.9	2.1	13.3	10.4	4.6	2.2
			T max (°C)	27.1	26.1	34.6	35.9	33.1	25.9
			a.p. (mm)	81.1	147.7	70.8	76.8	150.1	42.3
	Zagreb II (Vrapče)	119 m 45°49′8.69″N 15°52′49.84″E	a.d. T (°C)	13.6	14.3	24.8	24.1	18.4	14.8
			T min (°C)	5.4	5.6	15.1	14.0	8.5	5.2
			T max (°C)	22.9	27.1	27.3	35.5	32.7	24.6
			a.p. (mm)	85.2	123.1	83.9	65.8	131.6	39.5
	Koretići	410 m 45°48′47.23″N 15°33′36.18″E	a.d. T (°C)	9.0	9.9	20.4	19.6	14.5	13.3
			T min (°C)	−3.0	−1.8	9.8	7.2	1.5	3.5
			T max (°C)	22.4	21.5	31.2	30.0	28.2	22.8
			a.p. (mm)	135.7	283.7	81.4	184.8	120.2	59.5

Table 1. *Cont.*

Region	Location	Altitude/ Latitude/Longitude	Weather Parameters	Phenological Stage					
				I Before Flowering		II During Flowering		III After Flowering	
				April	May	June	July	September	October
Mountain	Ogulin	320 m 45°15′47.84″N 15°13′42.36″E	a.d. T (°C)	10.8	12.4	21.6	21.4	15.6	13.0
			T min (°C)	0.5	0.5	11.8	8.3	2.8	3.0
			T max (°C)	25.5	25.1	33.4	33.0	29.3	25.9
			a.p. (mm)	167.4	319.2	139.5	109.4	143.6	64.2
	Čovići I	456 m 44°49′44.07″N 15°17′57.29″E	a.d. T (°C)	9.4	11.1	20.1	19.7	14.0	10.6
			T min (°C)	−2.1	−1.3	7.6	5.5	−2.0	1.2
			T max (°C)	24.8	25.0	34.1	34.5	29.6	25.1
			a.p. (mm)	138.6	189.3	25.1	106.2	106.9	31.8
	Čovići II	456 m 44°49′50.05″N 15°17′57.18″E	a.d. T (°C)	9.4	11.1	20.1	19.7	14.0	10.6
			T min (°C)	−2.1	−1.3	7.6	5.5	−2.0	1.2
			T max (°C)	24.8	25.0	34.1	34.5	29.6	25.1
			a.p. (mm)	138.6	189.3	25.1	106.2	106.9	31.8
Seaside	Poreč	0.34 m 45°13′37.03″N 13°35′39.64″E	a.d. T (°C)	13.0	14.5	24.3	24.9	19.4	15.7
			T min (°C)	3.9	6.0	13.2	13.4	7.3	6.3
			T max (°C)	23.5	22.7	33.6	33.6	30.9	25.7
			a.p. (mm)	116.1	210.0	7.3	58.7	143.2	38.6
	Limski zaljev	17 m 45°7′56.45″N 13°39′13.78″E	a.d. T (°C)	13.0	14.5	24.3	24.9	19.4	15.7
			T min (°C)	3.9	6.0	13.2	13.4	7.3	6.3
			T max (°C)	23.5	22.7	33.6	33.6	30.9	25.7
			a.p. (mm)	116.1	210.0	7.3	58.7	143.2	38.6
	Bale	129 m 45°2′25.93″N 13°47′8.88″E	a.d. T (°C)	13.4	14.4	23.9	24.5	19.8	15.5
			T min (°C)	4.9	4.5	13.8	13.6	7.5	5.8
			T max (°C)	23.7	24.5	34.0	34.3	33.0	25.5
			a.p. (mm)	129.5	264.7	37.4	71.5	91.1	42.0
	Vodnjan	141 m 44°57′28.79″N 13°51′6.10″E	a.d. T (°C)	13.4	14.4	23.9	24.5	19.8	15.5
			T min (°C)	4.9	4.5	13.8	13.6	7.5	5.8
			T max (°C)	23.7	24.5	34.0	34.3	33.0	25.5
			a.p. (mm)	129.5	264.7	37.4	71.5	91.1	42.0
	Muntrilj	342 m 45°14′30.84″N 13°48′38.44″E	a.d. T (°C)	11.1	12.5	22.2	22.3	16.4	13.1
			T min (°C)	0.5	1.1	11.2	9.5	2.3	2.5
			T max (°C)	23.3	23.5	35.8	36.1	31.7	25.2
			a.p. (mm)	135.1	295.1	26.0	72.6	90.5	26.4

a.d. T = average day temperature, T min = minimal day temperature, T max = maximal day temperature, a.p. = accumulated precipitation.

2.3. Extraction Conditions

Extraction of polyphenols and pigments from dry nettle leaves and stalks was carried out by ASE. Extraction conditions and procedure were adopted from the study of Repajić et al. (2020) [11]: extraction was performed in Dionex™ ASE™ 350 Accelerated Solvent Extractor (Thermo Fisher Scientific Inc., Sunnyvale, CA, USA) using ethanol (96%) as the extraction solvent. Extraction was accomplished in 34 mL stainless steel cells fitted with 2 cellulose filters (Dionex™ 350/150 Extraction Cell Filters, Thermo Fisher Scientific Inc., Sunnyvale, CA, USA), within which 1 g of sample was mixed with 2 g of diatomaceous earth, placed in cell and filled up with diatomaceous earth to the full cell volume. Extraction parameters differed for leaves and stalks: leaves were extracted under 110 °C with 10 min of static extraction time and 4 cycles, while stalk extracts were obtained at 80 °C, 5 min of static extraction time and 4 cycles (parameters previously optimized). Other extraction parameters remained fixed for the extraction of both plant parts, namely pressure 10.34 MPa, 30 s of purge with nitrogen and 50% of flushing. Obtained extracts were collected in 250 mL glass vessel with Teflon septa, transferred into 50 mL volume flask and made up to volume with the extraction solvent. All extracts were filtered through a 0.45 μm membrane filter (Macherey-Nagel GmbH, Düren, Germany) prior to further analysis. All extracts have been prepared in a duplicate (n = 2).

2.4. UPLC-MS/MS Conditions

Identification and quantification of phenolics were performed on UPLC–MS/MS in both ionization modes on a 6430 QQQ mass spectrometer Agilent Technologies (Agilent, Santa Clara, CA, USA). Analytes were ionized using ESI ion source with nitrogen as desolvation and collision gas (temperature 300 °C, flow 11 L h^{-1}), capillary voltage, +4 −3.5 kV^{-1} and the pressure of nebulizer was set at 40 psi. The mass spectrometer was linked to UPLC system (Agilent series 1290 RRLC instrument) consisted of binary pump, autosampler and a column compartment thermostat. Reversed phase separation was performed on a Zorbax Eclipse Plus C18 column 100 × 2.1 mm with 1.8 μm particle size (Agilent). Column temperature was set at 35 °C and the injection volume was 2.5 μL. The solvent compositions and the gradient conditions used were as described previously by Elez Garofulić et al. (2018) [26]. For instrument control and data processing, Agilent MassHunter Workstation Software (ver. B.04.01) was used. Quantitative determination was carried out using the calibration curves of the standards, where protocatechuic acid, gentisic acid, syringic acid and *p*-hydroxybenzoic acid were calculated as gallic acid equivalents and cinnamic acid according to *p*-coumaric acid. Isorhamnetin rutinoside, quercetin rhamnoside, quercetin, isorhamnetin, quercetin pentoside, quercetin acetylhexoside, quercetin acetylrutinoside and quercetin pentosylhexoside were calculated according to quercetin-3-glucoside, kaempferol hexoside, kaempferol pentoside, kaempferol rhamnoside, kaempferol pentosylhexoside and kaempferol according to kaempferol-3-rutinoside, apigenin hexoside and genistein according to apigenin, while umbelliferone was expressed as scopoletin equivalents. All analyses have been performed in a duplicate and concentrations of analyzed compounds are expressed as mg 100 g^{-1} of dry matter (dm) (N = 4).

2.5. HPLC-UV-VIS/PDA Conditions

The carotenoids and chlorophylls identification and quantification were performed using Agilent Infinity 1260 system equipped with Agilent 1260 photodiode array detector (PDA; Agilent, Santa Clara, CA, USA) with an automatic injector and Chemstation software (ver. C.01.03).

The separation of carotenoids and chlorophylls was performed using Develosil RP-Aqueus C 30 column (250 × 4.6 mm i.d. 3 μm, Phenomenex, Torrance, CA, USA). The solvent composition and the used gradient conditions were described previously by Castro–Puyana et al. (2017) [27]. The mixture of MeOH:MTBE:water (90:7:3, *v/v/v*) (A) and MeOH:MTBE (10:90, *v/v*) (B) formed the mobile phase. The injection volume was 10 μL and the flow rate was kept at 0.8 mL min^{-1}. The chromatogram was monitored by scanning from 240 to 770 nm and the signal intensities detected at 450 nm and 660 nm were used for carotenoid and chlorophyll quantitation. Identification was carried out by comparing retention times and spectral data with those of the authentic standards (α- and β-carotene, chlorophyll *a* and *b*) or in case of unavailability of standards by comparing the absorption spectra reported in the literature [28,29]. Quantifications were made by the external standard calculation, using calibration curves of the standards α-carotene, β-carotene, chlorophyll *a* and chlorophyll *b*. The quantification of individual carotenoid compounds (neoxantine, violaxantine, lutein and its derivatives, derivative of zeaxantine and lycopene) was calculated as β-carotene equivalents and derivatives of chlorophylls as chlorophyll *a* and *b* equivalents using the equation based on the calibration curves, respectively. All determinations have been performed in a duplicate and results are expressed as mg 100 g^{-1} dm (N = 4).

2.6. ORAC Determination

The procedure was based on a previously reported method [30,31] with slight modifications. Briefly, a 96 wells black microplate was prepared containing 150 μL of fluorescein solution (70.30 nM) and 25 μL of blank (75 μM phosphate buffer, pH 7.4), Trolox standard (3.24–130.88 μM) or sample (appropriate diluted) were added. The plate was incubated for 30 min at 37 °C. After the first three cycles (representing the baseline signal), AAPH

(240 mM) was injected into each well to initiate the peroxyl radical generation. Fluorescence intensity (excitation at 485 nm and emission at 528 nm) was monitored every 90 sec over a total measurement period of 120 min using an automated plate reader (BMG LABTECH, Offenburg, Germany) and data were analyzed by MARS 2.0 software. The results were expressed as mmol Trolox equivalent (TE) 100 g^{-1} of dm. Determinations were carried out in duplicate (N = 4).

2.7. Statistical Analysis

Statistica ver. 10.0 software (Statsoft Inc., Tulsa, OK, USA) was applied for the statistical analysis. Full factorial randomized design was designated for the experimental part and descriptive statistic was used for the basic data evaluation. Continuous variables (polyphenols, pigments and antioxidant capacity) were analyzed by multifactorial analysis of variance (MANOVA) and marginal mean values were compared with Tukey's HSD test. Relationships between determined compounds and antioxidant capacity were examined by calculated Pearson's correlation coefficients, while possible grouping of the samples according to the examined sources of variations was tested using Principal Component Analysis (PCA). Significance level $p \leq 0.05$ was assigned for all tests.

3. Results and Discussion

This study examined the influence of plant part (leaves and stalks), phenological stage (before flowering, during flowering and after flowering) and habitat (Table 1) on the concentrations of polyphenols and pigments in wild nettle grown in Croatia. A total of 84 nettle samples were analyzed, where target compounds (polyphenols and pigments) were extracted using ASE and their identification/quantification was assessed by UPLC-MS/MS (polyphenols) and HPLC-UV-VIS/PDA (pigments). Moreover, obtained extracts were characterized for their antioxidant capacity by the ORAC method.

3.1. Influence of Phenological Stage and Habitat on Polyphenols in Nettle Leaves and Stalks

Table 2 shows detailed polyphenolic profile and mass spectrometric data obtained by UPLC-MS/MS analysis of nettle leaves and stalks. A total of 41 polyphenolic compounds were identified, belonging to the classes of benzoic, cinnamic and other phenolic acids, flavonols, flavan-3-ols, flavones, isoflavones, flavanones and coumarins (Supplementary files 1 & 2). Among the benzoic acids, compound **35** was identified as gallic acid by comparison of its retention time and mass spectra data with those of an authentic standard. Other benzoic acids were tentatively identified according to their mass fragmentation patterns. Compounds **2** and **14** showed same fragmentation pattern with molecular ion at m/z 153 and fragment ion at m/z 109, corresponding to the loss of carbon dioxide moiety and implicating the structure of dihydroxybenzoic acids and were therefore according to their polarity tentatively identified as protocatehuic (3,4-dihydroxybenzoic acid) and gentisic acid (1,3-dihydroxybenzoic acid), respectively [32]. Compound **31** showed precursor ion at m/z 197 and fragmentation loss of -15 amu corresponding to the loss of methyl radical characteristic for methoxylated phenolic acids and was tentatively identified as syringic acid. Compound **34** showed precursor ion at m/z 137 and characteristic fragmentation pattern for deprotonated phenolic acid with loos of -44 amu due to decarboxylation [33] and was assigned as *p*-hydroxybenzoic acid. The composition of benzoic acids in nettle leaves and stalks is in accordance with previous reports [9,14]. Among the cinnamic acids, compounds **12, 15, 19, 25** and **32** were identified using authentic standards as caffeic, chlorogenic, *p*-coumaric, ferulic and sinapic acid, respectively. Compound **21** was presented with precursor ion at m/z 147, and fragment ion at m/z 103 as a result of decarboxylation and was due to its mass spectra data assigned as cinnamic acid [32]. Compound **16** was identified as quinic acid comparing its spectral data and retention time with those of an authentic standard. The composition of cinnamic acids is in accordance with previous reports by Orčić et al. (2014) [14] and Francišković et al. (2017) [34] with the exception of cinnamic acid which was not detected in their research, but was reported previously in composition of nettle leaves by Zeković et al. (2017) [35]. The most numerous class of flavonoid polyphenols identified in nettle were flavonols and their glycosides. Compounds **4, 8, 17** and **18** were identified by the authentic standard comparison as kaempferol-3-rutinoside, myricetin, quercetin-3-glucoside and quercetin-3-rutinoside, respectively. Other compounds were tentatively identified according to their mass spectra and characteristic fragmentation patterns reported previously. Among the aglycones, compounds **10, 24** and **41** were assigned as quercetin, isorhamnetin and kaempferol due to characteristic molecular ion at m/z 301, m/z 315 and m/z 285 [36]. The presence of this aglycones in nettle aerial parts was confirmed previously by Bucar et al. (2006) [37]. Flavonol glycosides lacking authentic standards were tentatively identified according to the characteristic loss of sugar moiety and formation of aglycon fragment ion. Therefore, because of fragment ion at m/z 317, compound **3** was assigned as isorhamnetin glycoside. Precursor ion at m/z 625 implicated glycosylation with rhamnose (+146 amu) and glucose (+162 amu), so it was assigned as isorhametin rutinoside.

Table 2. Mass spectrometric data and identification of polyphenols.

Compound	Rt (min)	Cone Voltage (V)	Collision Energy (V)	Ionization Mode	Precursor Ion (m/z)	Fragment Ions (m/z)	Tentative Identification
				Benzoic acids			
2	0.828	105	9	-	153	109	Protocatechuic acid (3,4-dihydoxybenzoic acid)
14	0.992	100	9	-	153	109	Gentisic acid (2,5-dihydroxybenzoic acid)
31	8.837	90	7	-	197	182	Syringic acid
34	11.358	80	10	-	137	93	*p*-hydroxybenzoic acid
35	11.375	100	10	-	169	125	Gallic acid *
				Cinnamic acids			
12	0.975	80	10	-	179	135	Caffeic acid *
15	1.254	80	10	-	353	191	Chlorogenic acid *
19	3.332	80	10	-	163	119	*p*-coumaric acid *
21	4.490	100	5	-	147	103	Cinnamic acid
25	6.158	80	5	-	193	178	Ferulic acid *
32	11.012	100	17	-	223	193	Sinapic acid *
				Other phenolic acids			
16	1.620	150	20	-	191	85	Quinic acid *
				Flavonols			
3	0.842	120	15	+	625	317	Isorhamnetin rutinoside
4	0.856	120	15	+	595	287	Kaempferol-3-rutinoside *
5	0.880	100	5	+	449	303	Quercetin rhamnoside
6	0.880	30	5	+	449	287	Kaempferol hexoside
8	0.907	140	25	+	319	273	Myricetin *
10	0.938	130	15	-	301	151	Quercetin
17	1.855	100	5	+	465	303	Quercetin-3-glucoside *
18	2.461	120	5	+	611	303	Quercetin-3-rutinoside *
24	5.963	160	21	-	315	300	Isorhamnetin
27	7.106	100	5	+	419	287	Kaempferol pentoside
28	7.256	100	5	+	435	303	Quercetin pentoside
29	7.930	100	5	+	433	287	Kaempferol rhamnoside
30	8.242	100	10	+	507	303	Quercetin acetylhexoside
33	11.232	100	15	+	653	303	Quercetin acetylrutinoside
36	11.391	100	15	+	597	303	Quercetin pentosylhexoside
39	11.758	120	15	+	581	287	Kaempferol pentosylhexoside
41	11.822	130	0	-	285	285	Kaempferol
				Flavan-3-ols			
23	4.728	100	5, 15	+	459	289, 139	Epigallocatechin gallate *
37	11.615	100	10	+	291	139	Epicatechin *
38	11.621	100	5	+	291	165	Catechin *
40	11.792	100	5	+	443	291	Epicatechin gallate *

Table 2. Cont.

Compound	Rt (min)	Cone Voltage (V)	Collision Energy (V)	Ionization Mode	Precursor Ion (*m/z*)	Fragment Ions (*m/z*)	Tentative Identification
				Flavones			
7	0.890	135	5	+	433	271	Apigenin hexoside
9	0.924	140	35	+	287	153	Luteolin *
22	4.615	80	30	+	271	153	Apigenin *
				Isoflavones			
20	4.468	145	32	-	269	133	Genistein
				Flavanones			
11	0.945	130	16	-	271	151	Naringenin *
				Coumarins			
1	0.821	120	19	-	161	133	Umbelliferone (7-hydroxycoumarin)
13	0.979	105	15	-	177	133	Esculetin *
26	6.333	80	8	-	191	176	Scopoletin *

* Identification confirmed using authentic standards.

Its presence in nettle leaves and stalks was reported previously by Pinelli et al. (2008) [6]. Compounds **5, 28, 30, 33** and **36** were identified as quercetin glycosides due to MS/MS ion at *m/z* 303 and were assigned as quercetin rhamnoside, quercetin pentoside, quercetin acetylhexoside, quercetin acetylrutinoside and quercetin pentosylhexoside due to fragmentation losses corresponding to rhamnose (−146 amu), pentose (−132 amu), hexose with acetyl residue (−162 and −42 amu), rutinose with acetyl residue (−308 and −42 amu) and pentose with hexose moiety (−132 and −162 amu) [38]. Previous reports on quercetin glycosides composition in nettle mostly included quercetin glucoside [6,14,34] and quercetin rutinoside [8,14,34], while not reporting the presence of acylated glycosides and diglycosides identified in this study. The latter provides the valuable contribution to detailed insight into nettle polyphenolic profile. Because of the characteristic fragment ion at *m/z* 287 corresponding to the kaempferol aglycon, compounds **6, 27, 29** and **39** were assigned as kaempferol hexoside, pentoside, rhamnoside and pentosylhexoside, respectively, due to fragment losses of corresponding sugar moieties. Similar to the previous literature reports on quercetin glycosides, the ones on kaempferol glycosides mostly only include kaempferol rutinoside [6,8] or glucoside [14,34,39], while not reporting the presence of kaempferol pentoside, rhamnoside and pentosylhexoside which are therefore being confirmed here for the first time. All compounds belonging to the class of flavan-3-ols (**23, 37, 38** and **40**), namely epigallocatechingallate, epicatechin, catechin and epicatechingallate were identified and confirmed according to the authentic standard. Orčić et al. (2014) [14] identified catechin in nettle stalks, epicatechin was reported by Proestos et al. (2006) [40] in leaves, while there are no available reports on previous identification of epicatechingallate and epigallocatechingallate. Compounds **9** and **22** were assigned as luteolin and apigenin due to molecular ions at *m/z* 287 and *m/z* 271 and confirmed by comparison with standards, while compound **7** was tentatively identified as apigenin hexoside based on fragment ion at *m/z* 271 and fragmentation loss of -162 amu specific for hexose residue. Nencu et al. (2012) [41] reported the polyphenolic composition of nettle leaves including aglycones luteolin and apigenin, which is in accordance with our findings, while literature reports on flavone aglycones are scarce. Compound **20** showed precursor ion at *m/z* 269 and fragment ion at *m/z* 133, corresponding to the previously reported fragmentation mechanism of genistein anion [42], confirmed in nettle leaves extract by Zeković et al. (2017) [35]. Compounds **11, 13** and **26** were identified by its corresponding authentic standards as naringenin, esculetin and scopoletin, while compound **1** was tentatively assigned as umbelliferone due to molecular ion at *m/z* 161 and fragment ion at *m/z* 133 formed after the loss of one carbon monoxide molecule [43]. The composition of flavanones and coumarines reported in our study is in accordance with previous literature data [14,34,35].

To examine the influence of phenological stage and habitat on the content of polyphenols in nettle leaves and stalks, identified polyphenols were arranged in corresponding classes, following which their individual concentrations accordingly summarized and subjected to statistical analysis, as shown in Table 3. Total polyphenols grand mean (GM) was 380.90 mg 100 g^{-1} dm, among which cinnamic acids were the most abundant group (GM 179.22 mg 100 g^{-1} dm), followed by flavonols (GM 134.60 mg 100 g^{-1} dm), flavones (GM 24.56 mg 100 g^{-1} dm), flavan-3-ols (GM 20.70 mg 100 g^{-1} dm) and benzoic acids (GM 10.20 mg 100 g^{-1} dm). Coumarins, isoflavones and other acids were present in lower concentrations: GM values 5.31, 3.09 and 2.88 mg 100 g^{-1} dm, respectively, while the least represented group of polyphenols were flavanones (GM 0.34 mg 100 g^{-1} dm). Moreover, obtained results are in accordance with the results of other authors [6,8,11,14], who reported quite similar phenolic profile in nettle extracts where cinnamic acids accounted for the most of presented total polyphenols.

As can be observed, the plant part, phenological stage and habitat had a significant influence ($p < 0.01$) on amounts of all polyphenols' groups. When comparing amounts of polyphenols between nettle leaves and stalks, it can be seen that leaves accumulated significantly higher concentrations of all polyphenols' groups (Table 3). Otles and Yalcin (2012) [7] also documented higher polyphenols content in wild nettle leaves extracts when compared to stalks extracts, as well as Pinelli et al. (2008) [6] who studied the content of polyphenols in cultivated and wild nettle and reported higher total polyphenols in leaves of both types of nettle (cultivated 7.364 mg g^{-1} fw, wild 2.58 mg g^{-1} fw) as opposed to nettle stalks (cultivated 3.670 mg g^{-1} fw, wild 0.750 mg g^{-1} fw).

Same authors documented the abundance of nettle stalks with fibers, consisting of several components of the lignin. However, in study of Orčić et al. (2014) [14], who examined nettle samples picked at three different locations, several identified polyphenols were recorded in higher levels in stalks, but the cinnamic acids presented in their study with chlorogenic acid were also more abundant in leaves.

Considering the phenological stage, it can be noticed that the 1st phenological stage (before flowering) resulted with higher concentrations of all polyphenols, except flavan-3-ols which were significantly higher during the 2nd phenological stage (flowering) (Table 3). Overall, total polyphenols decreased for almost 50% by the 3rd phenological stage. Similar to our results, in two studies of Nencu et al. (2012, 2013) [41,44], it was concluded that the optimal time for nettle leaves harvest was March, since the polyphenols content greatly decreased (over 80%) by June and September, respectively. Authors reported that the total polyphenols decrease is due to the decrease of non-tannin phenols (phenolcarboxylic acids and flavonoids), which are the most important compounds from nettle leaves. This was also confirmed by Roslon et al. (2003) [45] who reported a sudden drop of phenolcarboxylic acids in leaves harvested at the plant flowering stage. Furthermore, the results of Biesiada et al. (2009, 2010) [46,47] and Kőszegi et al. (2020) [19] also indicated that the beginning of the nettle vegetation period was optimal for harvesting, giving the highest yield of polyphenols, which then decreased by autumn for over 50%. Therefore, in order to obtain extracts with the highest polyphenols content, the optimal time to harvest the aerial parts of the nettle is spring (before the flowering of the plant). It can be assumed that the total polyphenols decrease starting at the flowering stage is a result of the physiological switch from the vegetative to the generative phase and the formation of flowers [48].

Table 3. The differences in polyphenols content (mg 100 g^{-1} dm) in wild nettle (*Urtica dioica* L.) due to the plant part, phenological stage and habitat.

Source of Variation		Benzoic Acids	Cinnamic Acids	Other Acids	Flavonols	Flavan-3-ols	Flavones	Isoflavones	Flavanones	Coumarins	Total Polyphenols
Plant part		$p < 0.01$ *	$p < 0.01$ *	$p < 0.01$ *	$p < 0.01$ *	$p < 0.01$ *	$p < 0.01$ *	$p < 0.01$ *	$p < 0.01$ *	$p < 0.01$ *	$p < 0.01$ *
	leaves	12.55 ± 0.04b	209.46 ± 0.26b	4.30 ± 0.03b	160.26 ± 0.14b	25.99 ± 0.04b	29.28 ± 0.05b	3.37 ± 0.02b	0.40 ± 0.01b	6.53 ± 0.01b	452.14 ± 0.39b
	stalks	7.86 ± 0.04a	148.98 ± 0.26a	1.45 ± 0.03a	108.94 ± 0.14a	15.42 ± 0.04a	19.84 ± 0.05a	2.81 ± 0.02a	0.29 ± 0.01a	4.09 ± 0.01a	309.67 ± 0.39a
Phenological stage		$p < 0.01$ *	$p < 0.01$ *	$p < 0.01$ *	$p < 0.01$ *	$p < 0.01$ *	$p < 0.01$ *	$p < 0.01$ *	$p < 0.01$ *	$p < 0.01$ *	$p < 0.01$ *
	1st	12.65 ± 0.05c	223.32 ± 0.32c	3.66 ± 0.03c	169.53 ± 0.17c	22.23 ± 0.05b	31.89 ± 0.06c	3.70 ± 0.02c	0.48 ± 0.01c	7.28 ± 0.02c	474.75 ± 0.48c
	2nd	11.55 ± 0.05b	202.70 ± 0.32b	3.18 ± 0.03b	141.72 ± 0.17b	23.88 ± 0.05c	22.28 ± 0.06b	2.99 ± 0.02b	0.33 ± 0.01b	5.11 ± 0.02b	413.75 ± 0.48b
	3rd	6.42 ± 0.05a	111.63 ± 0.32a	1.78 ± 0.03a	92.54 ± 0.17a	16.00 ± 0.05a	19.50 ± 0.06a	2.58 ± 0.02a	0.22 ± 0.01a	3.53 ± 0.02a	254.21 ± 0.48a
Region/Habitat		$p < 0.01$ *	$p < 0.01$ *	$p < 0.01$ *	$p < 0.01$ *	$p < 0.01$ *	$p < 0.01$ *	$p < 0.01$ *	$p < 0.01$ *	$p < 0.01$ *	$p < 0.01$ *
C	Sela Žakanjska	12.06 ± 0.10g	200.25 ± 0.70h	2.10 ± 0.07b	134.87 ± 0.38g	16.68 ± 0.11b	24.18 ± 0.12g	2.92 ± 0.05d	0.31 ± 0.02bcd	4.41 ± 0.04a	397.78 ± 1.03g
	Sopčić Vrh	9.11 ± 0.10cd	215.63 ± 0.70i	4.12 ± 0.07f	150.83 ± 0.38i	26.97 ± 0.11i	26.69 ± 0.12h	3.25 ± 0.05e	0.49 ± 0.02fg	5.20 ± 0.04d	442.29 ± 1.03j
	Žakanje	19.39 ± 0.10i	227.10 ± 0.70j	2.56 ± 0.07c	177.87 ± 0.38j	32.13 ± 0.11j	42.44 ± 0.12j	5.29 ± 0.05i	0.41 ± 0.02def	5.92 ± 0.04f	513.12 ± 1.03l
	Zagreb I	10.21 ± 0.10e	172.62 ± 0.70e	4.22 ± 0.07fg	130.21 ± 0.38f	15.63 ± 0.11a	23.33 ± 0.12f	1.95 ± 0.05a	0.33 ± 0.02cde	6.76 ± 0.04h	365.26 ± 1.03e
	Zagreb II	11.06 ± 0.10f	185.09 ± 0.70f	4.21 ± 0.07fg	125.81 ± 0.38d	18.80 ± 0.11cd	21.49 ± 0.12d	3.47 ± 0.05ef	0.31 ± 0.02bcd	6.51 ± 0.04g	376.74 ± 1.03f
	Koretići	10.87 ± 0.10f	195.03 ± 0.70g	4.53 ± 0.07gh	144.36 ± 0.38h	22.09 ± 0.11g	18.87 ± 0.12b	3.77 ± 0.05g	0.45 ± 0.02efg	5.06 ± 0.04d	405.02 ± 1.03h
M	Ogulin	13.18 ± 0.10h	212.80 ± 0.70i	4.62 ± 0.07h	182.65 ± 0.38k	20.88 ± 0.11f	36.09 ± 0.12i	3.51 ± 0.05f	0.56 ± 0.02g	6.77 ± 0.04h	481.06 ± 1.03k
	Čovići I	9.44 ± 0.10cd	203.51 ± 0.70h	3.73 ± 0.07e	152.62 ± 0.38i	19.69 ± 0.11e	23.92 ± 0.12fg	4.47 ± 0.05h	0.44 ± 0.02efg	5.63 ± 0.04e	423.46 ± 1.03i
	Čovići II	8.96 ± 0.10c	194.18 ± 0.70g	3.08 ± 0.07d	127.91 ± 0.38e	21.62 ± 0.11g	17.14 ± 0.12a	2.91 ± 0.05d	0.53 ± 0.02g	4.70 ± 0.04bc	381.01 ± 1.03f
S	Poreč	9.37 ± 0.10cd	130.43 ± 0.70a	1.31 ± 0.07a	107.51 ± 0.38a	16.43 ± 0.11b	20.64 ± 0.12c	1.87 ± 0.05a	0.16 ± 0.02a	4.40 ± 0.04a	292.12 ± 1.03a
	Limski zaljev	9.53 ± 0.10d	141.67 ± 0.70c	1.07 ± 0.07a	111.21 ± 0.38b	18.32 ± 0.11c	23.43 ± 0.12f	2.36 ± 0.05b	0.23 ± 0.02abc	4.80 ± 0.04c	312.61 ± 1.03c
	Bale	7.14 ± 0.10b	143.55 ± 0.70c	1.82 ± 0.07b	114.39 ± 0.38c	19.17 ± 0.11de	22.41 ± 0.12e	2.68 ± 0.05cd	0.16 ± 0.02a	4.49 ± 0.04a	315.81 ± 1.03cd
	Vodnjan	6.14 ± 0.10a	134.44 ± 0.70b	1.03 ± 0.07a	112.89 ± 0.38bc	22.74 ± 0.11h	22.13 ± 0.12e	2.35 ± 0.05b	0.19 ± 0.02ab	5.15 ± 0.04d	307.05 ± 1.03b
	Muntrilj	6.40 ± 0.10a	152.75 ± 0.70d	1.87 ± 0.07b	111.27 ± 0.38b	18.72 ± 0.11cd	21.06 ± 0.12cd	2.46 ± 0.05bc	0.22 ± 0.02abc	4.55 ± 0.04ab	319.29 ± 1.03d
Grand mean		10.20	179.22	2.88	134.60	20.70	24.56	3.09	0.34	5.31	380.90

C = continental, M = mountain, S = seaside. * Statistically significant variable at $p \leq 0.05$. Results are expressed as mean ± SE (N = 4). Values with different letters within column are statistically different at $p \leq 0.05$.

108

Habitats of wild nettle samples differed according to the climate conditions and could be grouped into three different regions: continental, mountain and seaside (Table 1). As presented in Table 3, habitats significantly ($p < 0.01$) differed regarding polyphenols content, with no uniform pattern regarding individual polyphenolic groups. Thus, Žakanje, belonging to the continental region, was characterized with the highest concentrations of total polyphenols (513.12 ± 1.03 mg 100 g^{-1} dm), benzoic (19.39 ± 0.10 mg 100 g^{-1} dm) and cinnamic acids (227.10 ± 0.70 mg 100 g^{-1} dm), flavan-3-ols (32.13 ± 0.11 mg 100 g^{-1} dm), flavones (42.44 ± 0.12 mg 100 g^{-1} dm) and isoflavones (5.29 ± 0.05 mg 100 g^{-1} dm). Contrarily, Ogulin, situated in mountain areas, was characterized with the highest amounts of other acids (4.62 ± 0.07 mg 100 g^{-1} dm), flavonols (182.65 ± 0.38 mg 100 g^{-1} dm), flavanones (0.56 ± 0.02 mg 100 g^{-1} dm) and coumarins (6.77 ± 0.04 mg 100 g^{-1} dm). Moreover, seaside habitats generally showed the lowest presence of all polyphenols. Still, based on total polyphenols content, a difference between seaside samples and ones from other two regions can be observed, where continental and mountain samples showed significantly higher levels of total polyphenols when compared to the samples from seaside zone. This could be explained as a plant's self-defense against oxidative stress caused by lower temperatures. According to Di Virgillo et al. (2015) [1] habitat greatly affects the accumulation of polyphenolic compounds in nettle. Just as in the current study, other authors also confirmed a diversity in nettle polyphenols content in growing areas [7,14].

3.2. Influence of Phenological Stage and Habitat on Pigments in Nettle Leaves and Stalks

The presence of nettle natural color carriers, carotenoids and chlorophylls was monitored by HPLC analysis, which has detected a total of 13 carotenoids and 9 chlorophylls in wild nettle leaves and stalks, namely neoxanthin and its two derivatives, violaxanthin and its two derivatives, 13′-*cis*-lutein, lutein 5,6-epoxide, lutein, zeaxanthin, 9′-*cis*-lutein, α-carotene, β-carotene, chlorophyll *a* and its six derivatives and chlorophyll *b* and its derivative (Figure 1, Supplementary file 1). A similar chlorophylls and carotenoids composition was previously reported [4,11]. For statistical purposes, identified pigments were grouped and analyzed as total carotenoids and total chlorophylls, as well as their sum (total pigments) (Table 4). Total pigments GM was 644.22 mg 100 g^{-1} dm, most of which were chlorophylls (GM 611.19 mg 100 g^{-1} dm), while carotenoids were less present (GM 33.03 mg 100 g^{-1} dm). Other authors also reported higher chlorophylls content in nettle leaves extracts in comparison with the content of carotenoids [9,11,47,49].

As presented in Table 4, all sources of variation significantly ($p < 0.01$) affected both groups of pigments as well as their sum. When comparing the pigments distribution in examined plant parts, abundance in pigments was expectedly higher in leaves since they are major photosynthesis organs [50]. Accordingly, Hojnik et al. (2007) [18] also reported a much higher concentration of chlorophylls in nettle leaves in comparison with stalks (147.1 vs. 16 mg g^{-1} extract). Furthermore, determined values for total chlorophylls in leaves were similar to previously reported results by Biesiada et al. (2010) [46], Zeipiņa et al. (2014) [49] and Repajić et al. (2020) [11], but were higher than in Đurović et al.'s (2017) [9] study. Also, the obtained total carotenoids content was in accordance with the values documented in Repajić et al.'s (2020) [11] study, but it showed dissimilarity in comparison with the data of other authors [4,9,46,49], probably due to environmental differences.

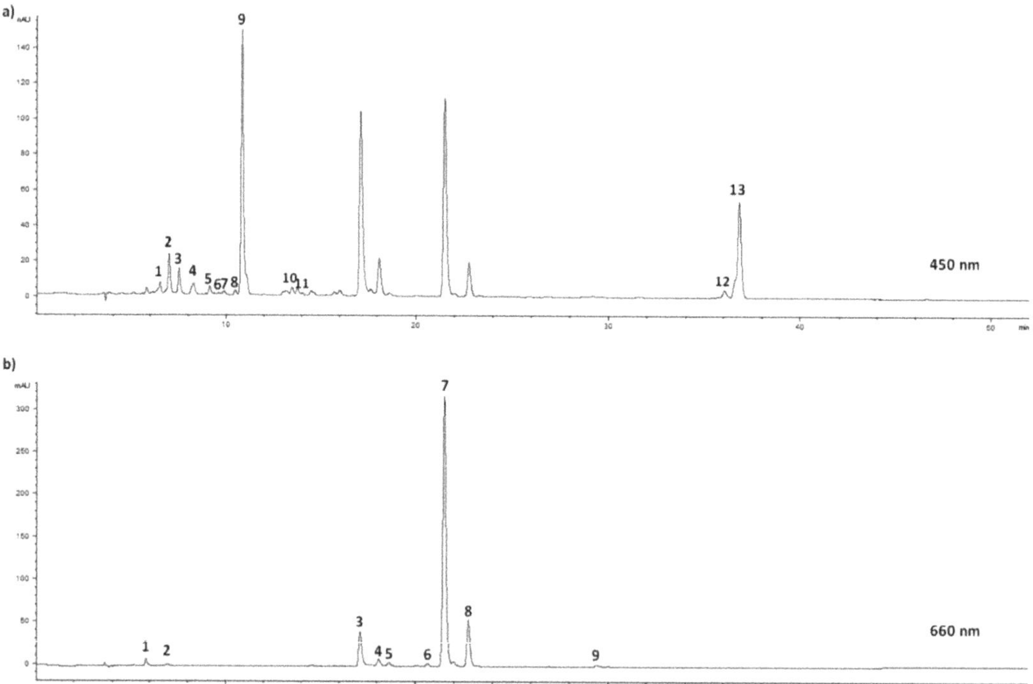

Figure 1. HPLC UV-VIS/PDA detection of pigments in wild nettle leaves (*Urtica dioica* L.) collected from Poreč before flowering: (**a**) at 450 nm (1 = violaxanthin derivative 1, 2 = neoxanthin derivative 1, 3 = neoxanthin, 4 = violaxanthin, 5 = violaxanthin derivative 2, 6 = 13′-*cis*-lutein, 7 = neoxanthin derivative 2, 8 = lutein 5,6-epoxide, 9 = lutein, 10 = zeaxanthin, 11 = 9′-*cis*-lutein, 12 = α-carotene, 13 = β-carotene); (**b**) at 660 nm (1 = chlorophyll *a* derivative 1, 2 = chlorophyll *a* derivative 2, 3 = chlorophyll *b*, 4 = chlorophyll *b* derivative 1, 5 = chlorophyll *a* derivative 3, 6 = chlorophyll *a* derivative 4, 7 = chlorophyll *a*, 8 = chlorophyll *a* derivative 5, 9 = chlorophyll *a* derivative 6).

Regarding the phenological stage, the highest amounts of all analyzed pigments were observed during the 2nd stage (flowering), where chlorophylls were the dominant pigments present in almost a 19-fold higher concentration (691.46 mg 100 g^{-1} dm) when compared to the amount of carotenoids (36.97 mg 100 g^{-1} dm). Similarly, Biesiada et al. (2009) [47] reported increased content of chlorophylls and carotenoids in nettle leaves when harvested in July in comparison with the harvest in May. Additionally, Marchetti et al. (2018) [10] observed that the highest lutein and β-carotene concentrations in nettle leaves occurred during the flowering stage (184 and 6.7 µg g^{-1} dm, respectively). Pajević et al. (1999) [51] also determined the maximum levels of chlorophylls and carotenoids in leaves of five alfalfa (*Medicago sativa* L.) genotypes just before and during the flowering stage. These similar patterns can be explained by enhanced production of secondary metabolites, such as plant pigments, during the flowering stage as a plant mechanism for fulfilling important physiological tasks like attracting pollinators [20].

When observing the differences in nettle pigments among the examined habitats, generally samples grown in seaside regions (particularly in the Limski zaljev and Bale habitats) had the highest pigments content. As this area was generally characterized by higher temperatures and lower accumulated precipitation (Table 1), these results are expected since the level of pigments in nettle is influenced by environmental factors, primarily the climate and growing location, where exposure to higher temperatures and more solar energy will result in a higher pigments content [24]. The results of Candido et al.'s (2015) [52] study, in which they examined carotenoid content in buriti palms pulp grown in two different regions (Amazon and Cerrado, Brazil), supported the aforementioned results. They concluded that a higher content of carotenoids was measured in samples from the Amazon area, characterized by higher temperatures and humidity which prevent photodegradation of fruit pigments.

Table 4. The differences in pigments content (mg 100 g^{-1} dm) and ORAC values (mmol TE 100 g^{-1} dm) in wild nettle (*Urtica dioica* L.) upon plant part, phenological stage and habitat.

Source of Variation		Carotenoids	Chlorophylls	Total Pigments	ORAC
Plant Part		$p < 0.01$ *	$p < 0.01$ *	$p < 0.01$ *	$p < 0.01$ *
leaves		61.46 ± 0.08b	1126.94 ± 0.66b	1188.40 ± 0.71b	11.96 ± 0.02b
stalks		4.60 ± 0.08a	95.45 ± 0.66a	100.05 ± 0.71a	7.37 ± 0.02a
Phenological Stage		$p < 0.01$ *	$p < 0.01$ *	$p < 0.01$ *	$p < 0.01$ *
1st		32.49 ± 0.10b	589.07 ± 0.81b	621.56 ± 0.86b	11.26 ± 0.04b
2nd		36.97 ± 0.10c	691.46 ± 0.81c	728.44 ± 0.86c	12.10 ± 0.04c
3rd		29.64 ± 0.10a	553.04 ± 0.81a	582.67 ± 0.86a	5.63 ± 0.04a
Region/Habitat		$p < 0.01$ *	$p < 0.01$ *	$p < 0.01$ *	$p < 0.01$ *
C	Sela Žakanjska	37.29 ± 0.21h	701.89 ± 1.76g	739.18 ± 1.87h	11.76 ± 0.06i
	Sopčić Vrh	31.95 ± 0.21d	558.11 ± 1.76c	590.06 ± 1.87d	11.78 ± 0.06j
	Žakanje	27.94 ± 0.21c	466.84 ± 1.76a	494.78 ± 1.87a	12.25 ± 0.06m
	Zagreb I	26.70 ± 0.21b	480.27 ± 1.76b	506.97 ± 1.87b	11.89 ± 0.06k
	Zagreb II	32.14 ± 0.21d	600.73 ± 1.76d	632.87 ± 1.87e	9.46 ± 0.06f
	Koretići	31.81 ± 0.21d	596.83 ± 1.76d	628.64 ± 1.87e	11.22 ± 0.06h
M	Ogulin	33.23 ± 0.21e	598.67 ± 1.76d	631.91 ± 1.87e	12.20 ± 0.06l
	Čovići I	25.65 ± 0.21a	472.09 ± 1.76ab	497.74 ± 1.87a	10.59 ± 0.06g
	Čovići II	34.29 ± 0.21f	650.39 ± 1.76e	684.68 ± 1.87f	9.46 ± 0.06f
S	Poreč	27.40 ± 0.21bc	552.88 ± 1.76c	580.28 ± 1.87c	6.28 ± 0.06b
	Limski zaljev	40.21 ± 0.21j	719.69 ± 1.76h	759.90 ± 1.87i	6.26 ± 0.06a
	Bale	38.35 ± 0.21i	760.95 ± 1.76i	799.30 ± 1.87j	6.58 ± 0.06c
	Vodnjan	35.93 ± 0.21g	678.67 ± 1.76f	714.60 ± 1.87g	8.08 ± 0.06e
	Muntrilj	39.55 ± 0.21j	718.67 ± 1.76h	758.23 ± 1.87i	7.52 ± 0.06d
Grand mean		**33.03**	**611.19**	**644.22**	**9.67**

C = continental, M = mountain, S = seaside. * Statistically significant variable at $p \leq 0.05$. Results are expressed as mean ± SE (N = 4). Values with different letters within column are statistically different at $p \leq 0.05$.

3.3. Influence of Phenological Stage and Habitat on Antioxidant Capacity in Nettle Leaves and Stalks

The results of nettle antioxidant capacity measured by the ORAC method are given in Table 4 and Supplementary file 1. ORAC GM was 9.67 mmol TE 100 g^{-1} dm. Moreover, the nettle antioxidant capacity was significantly influenced ($p < 0.01$) by all examined sources of variation. Nettle leaves showed higher antioxidant capacity in comparison with stalks (11.96 mmol TE 100 g^{-1} dm vs. 7.37 mmol TE 100 g^{-1} dm). Similar ORAC values in nettle leaves were recorded in study of Repajić et al. (2020) [11], while Česlova et al. (2016) [53] obtained the same results by measuring the antioxidant capacity of different nettle parts infusions, where nettle leaves gained higher DPPH levels when compared to stalks. In support, Kırca and Arslan (2008) [54] concluded that leaves and flowers of different examined plants had a higher antioxidant capacity when compared to stalks and seeds.

When observing the influence of phenological stage, the highest ORAC value was observed during flowering, after which it significantly decreased and was the lowest after flowering. Similar to the results of the current study, other authors [19,46] documented that the antioxidant capacity of nettle leaves was higher in the earliest periods (April/May and June/July), after which it decreased (September/October).

Nettle samples showed diversity in antioxidant capacity upon habitat variations. As can be observed, samples from the continental and mountain part were described with the highest ORAC levels as opposed to nettles grown in seaside areas, which were characterized with the lowest antioxidant capacity levels. These results are in accordance with previously discussed contents of polyphenols and pigments, where a certain grouping of the samples according to the presence of polyphenols and pigments by the growing area is evident. Moreover, calculated correlation coefficients supported this observation, since they showed a strong correlation between ORAC values and cinnamic acids, flavonols and total phenols (Table 5).

Table 5. Pearson's correlations between analyzed compounds (mg 100 g^{-1} dm) and ORAC values (mmol TE 100 g^{-1} dm).

Group of Compounds	ORAC Value
Benzoic acids	0.53 *
Cinnamic acids	0.71 *
Other acids	0.59 *
Flavonols	0.68 *
Flavan-3-ols	0.47 *
Flavones	0.36 *
Isoflavones	0.36 *
Flavanones	0.39 *
Coumarins	0.60 *
Total phenols	0.71 *
Carotenoids	0.46 *
Chlorophylls	0.44 *
Total pigments	0.44 *

* $p \leq 0.05$.

Obtained results clearly demonstrated the importance of the appropriate plant part selection as well as its phenological stage with the presence of the highest bioactive compounds accumulation in order to obtain the maximally enriched product, which will be beneficial for consumers.

3.4. PCA Analysis

Additionally, in order to examine a possible grouping of the nettle samples according to the applied sources of variations, PCA was carried out and obtained results are presented in Figure 2.

According to the preliminary PCA, a communality value of $\geq$0.5 described all 14 variables, thus they were all included in the test. The first two components (PC1 and PC2) explained 71.31% of total variance, where PC1 accounted for 53.47% of total variance, while PC2 attributed to 17.84% of total variance. Since PC1 strongly/very strongly negatively correlated ($-0.77 \leq r \leq -0.96$) with benzoic and cinnamic acids, flavonols, flavan-3-ols, flavones, ORAC values and total polyphenols, while PC2 had a strong/very strong correlation with carotenoids, chlorophylls and total pigments ($-0.79 \leq r \leq -0.81$), these variables could be considered as the most discriminating variables.

As can be seen in Figure 2a, separation of the samples clearly occurs based on the plant part. Most of the leaf samples were distributed at negative PC2 values, while all samples of stalks were situated at positive PC2 values. Regarding the phenological stage, a certain grouping appeared between samples from the 1st and 3rd phenological stage, where samples collected before flowering were mainly situated at negative PC1 values and almost all of the post-flowering samples were located at the positive PC1 values (Figure 2b). a partial grouping of nettle samples is visible in Figure 2c based on the growing region, where the most of separation can be seen to be present between continental and seaside samples, although this did not completely occur.

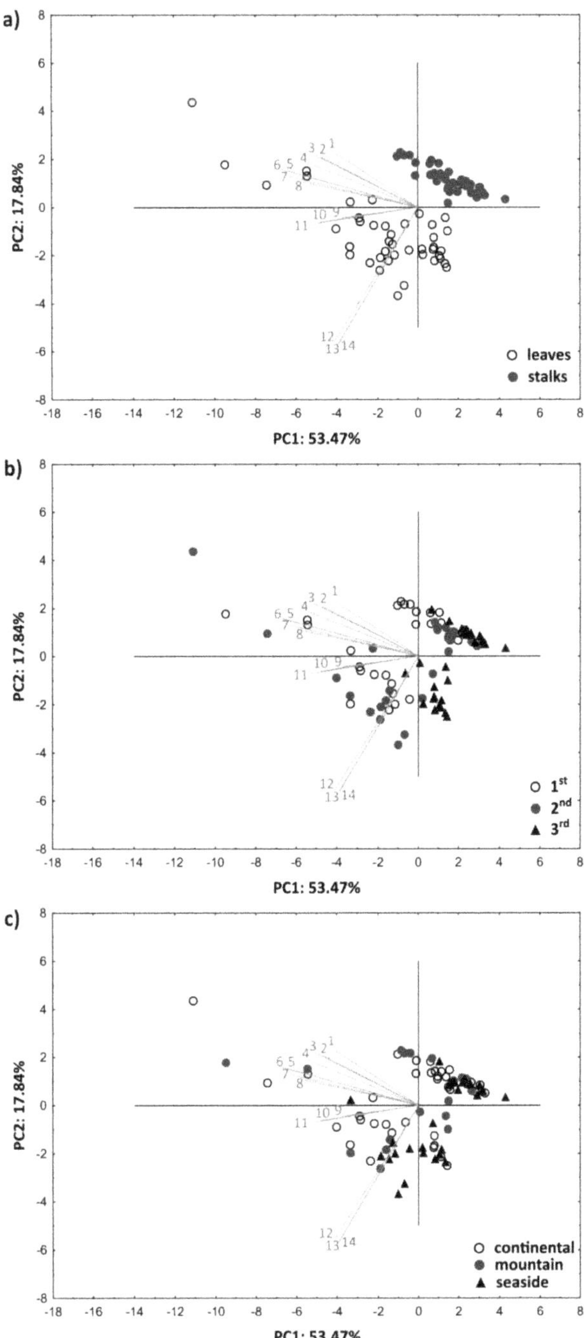

Figure 2. Distribution of wild nettle samples in two-dimensional coordinate system defined by the first two principal components (PC1 and PC2) according to the (**a**) plant part; (**b**) phenological stage; (**c**) growing region (1 = isoflavones, 2 = flavanones, 3 = flavones, 4 = benzoic acids, 5 = cinnamic acids, 6 = total polyphenols, 7 = flavonols, 8 = flavan-3-ols, 9 = other acids, 10 = coumarins, 11 = ORAC, 12 = carotenoids, 13 = chlorophylls, 14 = total pigments).

4. Conclusions

The current study confirmed the abundance of wild nettle with diverse bioactive molecules such as low molecular weight polyphenols and pigments, where 41 phenolic compounds, 13 carotenoids and 9 chlorophylls were documented. By using applied extraction conditions, cinnamic acids and flavonols were found to be the dominant classes of identified polyphenols (33.10–519.81 mg 100 g^{-1} dm and 57.44–383.25 mg 100 g^{-1} dm, respectively), while chlorophylls were the most abundant natural pigments (4.26–1934.38 mg 100 g^{-1} dm). Moreover, the ORAC values of obtained nettle extracts ranged from 3.05 to 19.83 mmol TE 100 g^{-1} dm. However, in order to obtain high valuable wild nettle extracts that are abundant in natural antioxidants, it is of the utmost importance to select appropriate plant parts as well as an appropriate harvest time. Obtained results evidenced that the highest levels of nettle bioactives accompanied by high antioxidant capacity were present in leaves, which should be collected during the early phenological period (before and at the flowering stage). Moreover, the amounts of wild nettle polyphenols and pigments greatly differed based on the natural habitat, as samples from the seaside region were characterized with elevated accumulation of pigments, while higher polyphenols amounts were present in habitats located in continental and mountain areas. This research will surely contribute to the selection of plant part and phenological stage for nettle optimal harvest, as well as to designate nettle natural habitats that have been shown to be a source of valuable plant material. These findings present the basis for the production of nettle seedlings with high bioactives content, which could further be used in the production of liquid and dry extracts. Furthermore, they showed the importance of a multidisciplinary approach for the selection of a plant part as well as its phenological stage in order to provide highly enriched products intended for the benefit of consumers.

In addition, besides low molecular weight polyphenols and pigments covered by this research, future studies could also include other beneficial compounds present in nettle such as oligomers and polymers as well as sterols, to provide a full insight into the nettle's bioactive potential.

Supplementary Materials: The following figures and tables are available online at https://www.mdpi.com/2304-8158/10/1/190/s1, (file 1) Figure S1: LC-MS/MS chromatogram in dMRM acquisition from the extract of wild nettle leaves (*Urtica dioica* L.) collected from Poreč before flowering, (file 2) Tables S1–S3: Concentrations of individual compounds and ORAC values in nettle (*Urtica dioica* L.) samples.

Author Contributions: Conceptualization, M.R.; Data curation, M.R., Z.Z., S.P. and I.E.G.; Formal analysis, M.R., E.C., Z.Z. and S.P.; Methodology, M.R. and I.E.G.; Project administration, V.D.-U.; Resources, S.R. and I.P.; Supervision, M.R. and V.D.-U.; Visualization, M.R.; Writing—original draft, M.R., E.C. and I.E.G.; Writing—review & editing, M.R., S.P., I.E.G., S.R., I.P. and V.D.-U. All authors have read and agreed to the published version of the manuscript.

Funding: This research was funded by the Croatian Science Foundation project, grant number IP-01-2018-4924.

Acknowledgments: Authors wish to thank Valentina Kruk from mag.ing.techn.aliment for her assistance during experimental work and the Meteorological and Hydrological Institute of Croatia for providing the meteorological data.

Conflicts of Interest: The authors declare no conflict of interest. The funders had no role in the design of the study; in the collection, analyses, or interpretation of data; in the writing of the manuscript, or in the decision to publish the results.

References

1. Di Virgilio, N.; Papazoglou, E.G.; Jankauskiene, Z.; Di Lonardo, S.; Praczyk, M.; Wielgusz, K. The potential of stinging nettle (*Urtica dioica* L.) as a crop with multiple uses. *Ind. Crop. Prod.* **2015**, *68*, 42–49. [CrossRef]
2. Kregiel, D.; Pawlikowska, E.; Antolak, H. Urtica spp.: Ordinary plants with extraordinary properties. *Molecules* **2018**, *23*, 1664. [CrossRef] [PubMed]
3. Upton, R. Stinging nettles leaf (*Urtica dioica* L.): Extraordinary vegetable medicine. *J. Herb. Med.* **2013**, *3*, 9–38. [CrossRef]

4. Guil-Guerrero, J.L.; Rebolloso-Fuentes, M.M.; Isasa, M.E.T. Fatty acids and carotenoids from Stinging Nettle (*Urtica dioica* L.). *J. Food Compos. Anal.* **2003**, *16*, 111–119. [CrossRef]

5. Sajfrtová, M.; Sovová, H.; Opletal, L.; Bártlová, M. Near-critical extraction of β-sitosterol and scopoletin from stinging nettle roots. *J. Supercrit. Fluids* **2005**, *35*, 111–118. [CrossRef]

6. Pinelli, P.; Ieri, F.; Vignolini, P.; Bacci, L.; Baronti, S.; Romani, A. Extraction and HPLC analysis of phenolic compounds in leaves, stalks, and textile fibers of *Urtica dioica* L. *J. Agric. Food Chem.* **2008**, *56*, 9127–9132. [CrossRef]

7. Otles, S.; Yalcin, B. Phenolic compounds analysis of root, stalk, and leaves of nettle. *Sci. World J.* **2012**, *2012*. [CrossRef]

8. Carvalho, A.R.; Costa, G.; Figueirinha, A.; Liberal, J.; Prior, J.A.V.; Lopes, M.C.; Cruz, M.T.; Batista, M.T. Urtica spp.: Phenolic composition, safety, antioxidant and anti-inflammatory activities. *Food Res. Int.* **2017**, *99*, 485–494. [CrossRef]

9. Đurović, S.; Pavlić, B.; Šorgić, S.; Popov, S.; Savić, S.; Pertonijević, M.; Radojković, M.; Cvetanović, A.; Zeković, Z. Chemical composition of stinging nettle leaves obtained by different analytical approaches. *J. Funct. Foods* **2017**, *32*, 18–26. [CrossRef]

10. Marchetti, N.; Bonetti, G.; Brandolini, V.; Cavazzini, A.; Maietti, A.; Meca, G.; Mañes, J. Stinging nettle (*Urtica dioica* L.) as a functional food additive in egg pasta: Enrichment and bioaccessibility of Lutein and β-carotene. *J. Funct. Foods* **2018**, *47*, 547–553. [CrossRef]

11. Repajić, M.; Cegledi, E.; Kruk, V.; Pedisić, S.; Çınar, F.; Bursać Kovačević, D.; Žutić, I.; Dragović-Uzelac, V. Accelerated Solvent Extraction as a Green Tool for the Recovery of Polyphenols and Pigments from Wild Nettle Leaves. *Processes* **2020**, *8*, 803. [CrossRef]

12. Dhouibi, R.; Affes, H.; Salem, M.B.; Hammami, S.; Sahnoun, Z.; Zeghal, K.M.; Ksouda, K. Screening of pharmacological uses of Urtica dioica and others benefits. *Prog. Biophys. Mol. Biol.* **2020**, *150*, 67–77. [CrossRef] [PubMed]

13. Wagner, H.; Willer, F.; Samtleben, R.; Boos, G. Search for the antiprostatic principle of stinging nettle (Urtica dioica) roots. *Phytomedicine* **1994**, *1*, 213–224. [CrossRef]

14. Orcic, D.; Franciškovic, M.; Bekvalac, K.; Svircev, E.; Beara, I.; Lesjak, M.; Mimica-Dukic, N. Quantitative determination of plant phenolics in Urtica dioica extracts by high-performance liquid chromatography coupled with tandem mass spectrometric detection. *Food Chem.* **2014**, *143*, 48–53. [CrossRef] [PubMed]

15. Chrubasik, J.E.; Roufogalis, B.D.; Wagner, H.; Chrubasik, S. A comprehensive review on the stinging nettle effect and efficacy profiles. Part II: Urticae radix. *Phytomedicine* **2007**, *14*, 568–579. [CrossRef] [PubMed]

16. Brown, D. Encyclopedia of herbs and their uses. In *Encyclopedia of Common Natural Ingredients Used in Food, Drugs, and Cosmetics*; Leung, Y.A., Foster, S., Eds.; Wiley: Hoboken, NJ, USA, 1995.

17. Mottaleb, M.A.; Sarker, S.D. Accelerated solvent extraction for natural products isolation. In *Natural Products Isolation*; Springer: Berlin/Heidelberg, Germany, 2012; pp. 75–87.

18. Hojnik, M.; Škerget, M.; Knez, Ž. Isolation of chlorophylls from stinging nettle (*Urtica dioica* L.). *Sep. Purif. Technol.* **2007**, *57*, 37–46. [CrossRef]

19. Kőszegi, K.; Békássy-Molnár, E.; Koczka, N.; Kerner, T.; Stefanovits-Bányai, É. Changes in Total Polyphenol Content and Antioxidant Capacity of Stinging Nettle (*Urtica dioica* L.) from Spring to Autumn. *Period. Polytech. Chem. Eng.* **2020**, *64*, 548–554. [CrossRef]

20. Ncube, B.; Van Staden, J. Tilting plant metabolism for improved metabolite biosynthesis and enhanced human benefit. *Molecules* **2015**, *20*, 12698–12731. [CrossRef]

21. Moore, M. *Medicinal Plants of the Mountain West*; Museum of New Mexico Press: Santa Fe, NM, USA, 2003; ISBN 0890134545.

22. Tolić, M.T.; Krbavčić, I.P.; Vujević, P.; Milinović, B.; Jurčević, I.L.; Vahčić, N. Effects of weather conditions on phenolic content and antioxidant capacity in juice of chokeberries (*Aronia melanocarpa* L.). *Pol. J. Food Nutr. Sci.* **2017**, *67*, 67–74. [CrossRef]

23. González-Talice, J.; Yuri, J.A.; del Pozo, A. Relations among pigments, color and phenolic concentrations in the peel of two Gala apple strains according to canopy position and light environment. *Sci. Hortic.* **2013**, *151*, 83–89. [CrossRef]

24. Saini, R.K.; Nile, S.H.; Park, S.W. Carotenoids from fruits and vegetables: Chemistry, analysis, occurrence, bioavailability and biological activities. *Food Res. Int.* **2015**, *76*, 735–750. [CrossRef] [PubMed]

25. AOAC. *Official Methods of Analysis: Changes in Official Methods of Analysis Made at the Annual Meeting. Supplement*; Association of Official Analytical Chemists: Rockville, MD, USA, 1990; Volume 15.

26. Elez Garofulić, I.; Zorić, Z.; Pedisić, S.; Brnčić, M.; Dragović-Uzelac, V. UPLC-MS2 Profiling of Blackthorn Flower Polyphenols Isolated by Ultrasound-Assisted Extraction. *J. Food Sci.* **2018**, *83*, 2782–2789. [CrossRef] [PubMed]

27. Castro-Puyana, M.; Pérez-Sánchez, A.; Valdés, A.; Ibrahim, O.H.M.; Suarez-Álvarez, S.; Ferragut, J.A.; Micol, V.; Cifuentes, A.; Ibáñez, E.; García-Cañas, V. Pressurized liquid extraction of Neochloris oleoabundans for the recovery of bioactive carotenoids with anti-proliferative activity against human colon cancer cells. *Food Res. Int.* **2017**, *99*, 1048–1055. [CrossRef] [PubMed]

28. Sozgen Baskan, K.; Tutem, E.; Ozer, N.; Apak, R. Spectrophotometric and chromatographic assessment of contributions of carotenoids and chlorophylls to the total antioxidant capacities of plant foods. *J. Agric. Food Chem.* **2013**, *61*, 11371–11381. [CrossRef]

29. Gupta, P.; Sreelakshmi, Y.; Sharma, R. A rapid and sensitive method for determination of carotenoids in plant tissues by high performance liquid chromatography. *Plant Methods* **2015**, *11*, 5. [CrossRef]

30. Prior, R.L.; Wu, X.; Schaich, K. Standardized methods for the determination of antioxidant capacity and phenolics in foods and dietary supplements. *J. Agric. Food Chem.* **2005**, *53*, 4290–4302. [CrossRef]

31. Bender, C.; Graziano, S.; Zimmerman, B.F.; Weidlich, H.H. Antioxidant potential of aqueous plant extracts assessed by the cellular antioxidant activity assay. *Am. J. Biol. Life Sci.* **2014**, *2*, 72–79.
32. Gruz, J.; Novák, O.; Strnad, M. Rapid analysis of phenolic acids in beverages by UPLC-MS/MS. *Food Chem.* **2008**, *111*, 789–794. [CrossRef]
33. Parejo, I.; Jauregui, O.; Sánchez-Rabaneda, F.; Viladomat, F.; Bastida, J.; Codina, C. Separation and characterization of phenolic compounds in fennel (Foeniculum vulgare) using liquid chromatography-negative electrospray ionization tandem mass spectrometry. *J. Agric. Food Chem.* **2004**, *52*, 3679–3687. [CrossRef]
34. Francišković, M.; Gonzalez-Pérez, R.; Orčić, D.; de Medina, F.S.; Martínez-Augustin, O.; Svirčev, E.; Simin, N.; Mimica-Dukić, N. Chemical Composition and Immuno-Modulatory Effects of *Urtica dioica* L. (Stinging Nettle) Extracts. *Phyther. Res.* **2017**, *31*, 1183–1191. [CrossRef]
35. Zeković, Z.; Cvetanović, A.; Švarc-Gajić, J.; Gorjanović, S.; Sužnjević, D.; Mašković, P.; Savić, S.; Radojković, M.; Đurović, S. Chemical and biological screening of stinging nettle leaves extracts obtained by modern extraction techniques. *Ind. Crop. Prod.* **2017**, *108*, 423–430. [CrossRef]
36. Chen, Y.; Yu, H.; Wu, H.; Pan, Y.; Wang, K.; Jin, Y.; Zhang, C. Characterization and Quantification by LC-MS/MS of the Chemical Components of the Heating Products of the Flavonoids Extract in Pollen Typhae for Transformation Rule Exploration. *Molecules* **2015**, *20*, 18352–18366. [CrossRef] [PubMed]
37. Bucar, F.; Britzmann, B.; Streit, B.; Weigend, M. LC-PDA-MS-profiles of phenolic compounds in extracts of aerial parts of Urtica species. *Planta Med.* **2006**, *72*. [CrossRef]
38. Pinheiro, P.F.; Justino, G.C. Structural Analysis of Flavonoids and Related Compounds—A Review of Spectroscopic Applications. In *Phytochemicals—A Global Perspective of Their Role in Nutrition and Health*; Rao, V., Ed.; InTech: Rijeka, Croatia, 2012; pp. 33–56.
39. Farag, M.A.; Weigend, M.; Luebert, F.; Brokamp, G.; Wessjohann, L.A. Phytochemical, phylogenetic, and anti-inflammatory evaluation of 43 Urtica accessions (stinging nettle) based on UPLC–Q-TOF-MS metabolomic profiles. *Phytochemistry* **2013**, *96*, 170–183. [CrossRef] [PubMed]
40. Proestos, C.; Boziaris, I.S.; Nychas, G.-J.; Komaitis, M. Analysis of flavonoids and phenolic acids in Greek aromatic plants: Investigation of their antioxidant capacity and antimicrobial activity. *Food Chem.* **2006**, *95*, 664–671. [CrossRef]
41. Nencu, I.; Vlase, L.; Istudor, V.; Duțu, L.E.; Gird, C.E. Preliminary research regarding the therapeutic uses of *Urtica dioica* L. Note I. The polyphenols evaluation. *Farmacia* **2012**, *60*, 493–500.
42. Wang, F.; Jiang, K.; Li, Z. Purification and Identification of Genistein in Ginkgo biloba Leaf Extract. *Chin. J. Chromatogr.* **2007**, *25*, 509–513. [CrossRef]
43. Mazimba, O. Umbelliferone: Sources, chemistry and bioactivities review. *Bull. Fac. Pharm. Cairo Univ.* **2017**, *55*, 223–232. [CrossRef]
44. Ioana, N.; Viorica, I.; Diana-Carolina, I.; Valeria, R. Preliminary research regarding the therapeutic uses of *Urtica dioica* l note ii. The dynamics of accumulation of total phenolic compounds and ascorbic acid. *Farmacia* **2013**, *61*, 276–283.
45. Roslon, W.; Weglarz, Z. Polyphenolic acids of female and male forms of Urtica dioica. In Proceedings of the International Conference on Medicinal and Aromatic Plants (Part II), Budapest, Hungary, 8–11 July 2001; pp. 101–104.
46. Biesiada, A.; Kucharska, A.; Sokół-Łętowska, A.; Kuś, A. Effect of the Age of Plantation and Harvest Term on Chemical Composition and Antioxidant Avctivity of Stinging Nettle (*Urtica dioica* L.). *Ecol. Chem. Eng. A* **2010**, *17*, 1061–1068.
47. Biesiada, A.; Wołoszczak, E.; Sokół-Łętowska, A.; Kucharska, A.Z.; Nawirska-Olszańska, A. The effect of nitrogen form and dose on yield, chemical composition and antioxidant activity of stinging nettle (*Urtica dioica* L.). *Herba Pol.* **2009**, *55*, 84–93.
48. Naser Aldeen, M.G.; Mansoor, R.; AlJoubbeh, M. Fluctuations of phenols and flavonoids in infusion of lemon verbena (Lippia citriodora) dried leaves during growth stages. *Nutr. Food Sci.* **2015**, *45*, 766–773. [CrossRef]
49. Zeipiņa, S.; Alsiņa, I.; Lepse, L. Stinging nettle—The source of biologically active compounds as sustainable daily diet supplement. *Res. Rural Dev.* **2014**, *20*, 34–38.
50. Kong, D.X.; Li, Y.Q.; Wang, M.L.; Bai, M.; Zou, R.; Tang, H.; Wu, H. Effects of light intensity on leaf photosynthetic characteristics, chloroplast structure, and alkaloid content of Mahonia bodinieri (Gagnep.) Laferr. *Acta Physiol. Plant.* **2016**, *38*, 120. [CrossRef]
51. Pajevic, S.; Krstic, B.; Katic, S.; Nikolic, N.; Mihailovic, V. Some photosynthetic parameters of alfalfa (*Medicago sativa* L.) leaves at different phenological stages and in different cuttings. *Matica Srp. Proc. Nat. Sci.* **1999**, *35*. Available online: http://www.maticasrpska.org.rs/stariSajt/casopisi/prirodne_nauke_097.pdf#page=38 (accessed on 18 January 2021).
52. Cândido, T.L.N.; Silva, M.R.; Agostini-Costa, T.S. Bioactive compounds and antioxidant capacity of buriti (*Mauritia flexuosa* Lf) from the Cerrado and Amazon biomes. *Food Chem.* **2015**, *177*, 313–319. [CrossRef]
53. Česlová, L.; Šilarová, P.; Fischer, J. Effect of harvesting and processing of stinging nettle on the antioxidant capacity of its infusions. *Sci. Pap. Univ. Pardubice Ser. A Fac. Chem. Technol.* **2016**, *22*, 23–33.
54. Kırca, A.; Arslan, E. Antioxidant capacity and total phenolic content of selected plants from Turkey. *Int. J. Food Sci. Technol.* **2008**, *43*, 2038–2046. [CrossRef]

Article

Investigation of Phenolic Composition and Anticancer Properties of Ethanolic Extracts of Japanese Quince Leaves

Vaidotas Zvikas [1], Ieva Urbanaviciute [2], Rasa Bernotiene [3], Deimante Kulakauskiene [1], Urte Morkunaite [1], Zbigniev Balion [1,3], Daiva Majiene [3], Mindaugas Liaudanskas [1,4], Pranas Viskelis [1,2], Aiste Jekabsone [1,5] and Valdas Jakstas [1,4,*]

[1] Institute of Pharmaceutical Technologies, Lithuanian University of Health Sciences, Sukilėlių av. 13, LT-50162 Kaunas, Lithuania; vaidotas.zvikas@lsmuni.lt (V.Z.); deimante.kulakauskiene@lsmu.lt (D.K.); urte.morkunaite@stud.lsmu.lt (U.M.); zbigniev.balion@lsmuni.lt (Z.B.); mindaugas.liaudanskas@lsmuni.lt (M.L.); pranas.viskelis@lsmu.lt (P.V.); aiste.jekabsone@lsmuni.lt (A.J.)
[2] Laboratory of Biochemistry and Technology, Institute for Horticulture, Lithuanian Research Centre for Agriculture and Forestry, Kauno str. 30, LT-54333 Babtai, Lithuania; ievaurbanaviciute@yahoo.com
[3] Neuroscience Institute, Lithuanian University of Health Sciences, Eivenių str. 4, LT-50161 Kaunas, Lithuania; rasa.bernotiene@lsmuni.lt (R.B.); daiva.majiene@lsmuni.lt (D.M.)
[4] Department of Pharmacognosy, Faculty of Pharmacy, Lithuanian University of Health Sciences, Sukilėlių av. 13, LT-50162 Kaunas, Lithuania
[5] Institute of Cardiology, Lithuanian University of Health Sciences, Sukilėlių av. 17, LT-50009 Kaunas, Lithuania
* Correspondence: valdas.jakstas@lsmuni.lt; Tel.: +370-672-00844

Abstract: Glioblastoma multiforme is an aggressive and invasive disease with no efficient therapy available, and there is a great need for finding alternative treatment strategies. This study aimed to investigate anticancer activity of the extracts of the Japanese quince (JQ) cultivars 'Darius', 'Rondo', and 'Rasa' leaf extracts on glioblastoma C6 and HROG36 cells. As identified by ultra high performance liquid chromatography electrospray ionization tandem mass spectrometry, the extracts contained three prevailing groups of phenols: hydroxycinnamic acid derivatives; flavan-3-ols; and flavonols. Sixteen phenols were detected; the predominant compound was chlorogenic acid. The sum of detected phenols varied significantly between the cultivars ranging from 9322 µg/g ('Rondo') to 17,048 µg/g DW ('Darius'). Incubation with the extracts decreased the viability of glioblastoma HROG36 cells with an efficiency similar to temozolomide, a drug used for glioblastoma treatment. In the case of C6 glioblastoma cells, the extracts were even more efficient than temozolomide. Interestingly, primary cerebellar neuronal-glial cells were significantly less sensitive to the extracts compared to the cancer cell lines. The results showed that JQ leaf ethanol extracts are rich in phenolic compounds, can efficiently reduce glioblastoma cell viability while preserving non-cancerous cells, and are worth further investigations as potential anticancer drugs.

Keywords: *Chaenomeles japonica* leaves; phenolic compounds; glioblastoma; anticancer activity

Citation: Zvikas, V.; Urbanaviciute, I.; Bernotiene, R.; Kulakauskiene, D.; Morkunaite, U.; Balion, Z.; Majiene, D.; Liaudanskas, M.; Viskelis, P.; Jekabsone, A.; et al. Investigation of Phenolic Composition and Anticancer Properties of Ethanolic Extracts of Japanese Quince Leaves. *Foods* **2021**, *10*, 18. https://dx.doi.org/10.3390/foods10010018

Received: 30 November 2020
Accepted: 20 December 2020
Published: 23 December 2020

1. Introduction

Japanese quince (*Chaenomeles japonica* (Thunb.) Lindl. ex Spach), a representative of the *Rosaceae* Juss. family, has already been known in oriental folk medicine for about 3000 years [1]. This plant is a great source of secondary metabolites possessing various biological effects including anticancer activity [2,3]. For example, quince extracts have high amounts of triterpenes (such as ursolic and oleanolic acids) that are reported to decrease the viability of colon, breast, melanoma, lung, hepatic carcinoma, and other cancer cell types [4–7]. Furthermore, Japanese quince fruit extracts are rich in phenolic compounds, mostly flavonoids [8], that are also known for preventive and therapeutic anticancer potential [9]. Procyanidins and flavanols from Japanese quince fruits induce apoptosis and suppress invasiveness in human colon, prostate, and breast cancer cell cultures [10–12]. A recent study revealed that the extract of Japanese quince leaves reduces viability of colon cancer cells SW-480 and HT-29 to a greater extent compared to normal intestinal cells CCD-18 Co and CCD 841 CoN [13]. Such results encouraged us to investigate quince leaf extract efficiency on other cancer cell types.

Glioblastoma multiforme is one of the most aggressive and invasive cancerous diseases, and there are no efficient treatment options available [14]. The most common therapy is temozolomide, however the treatment is accompanied by severe side effects and the efficiency is poor [14]. Some promising results are achieved by applying plant-derived anticancer substances [15]. Therefore, it is important to continue investigating the new plant sources in order to find more efficient treatment or therapy complement for glioblastoma. Our previous research revealed that the leaves of three Japanese quince cultivars 'Darius', 'Rondo', and 'Rasa' are rich in phenols and triterpenes suggesting that the extracts might possess anticancer activity [16]. The current study aimed to perform a broader phenol analysis of the leaves of the same cultivars, and to investigate the effect of the extracts on the viability of glioblastoma HROG36 and C6 cells. In addition, to predict the potential level of cytotoxicity on healthy brain tissue, the effect of the extracts on viability of primary non-cancerous cultured brain cells was evaluated.

2. Materials and Methods

2.1. Chemicals

All the solvents, reagents, and standards used were of analytical grade. The following substances were used in the study: Ethanol 96% (*v*/*v*) (AB Strumbras, Kaunas, Lithuania), procyanidin C1, procyanidin B2, quercetin, hyperoside, avicularin, quercitrin, kaempherol 3-O-glucoside, luteolin 7-O-glucoside, phloridzin, formic acid, acetonitrile, (+)-catechin, (-)-epicatechin, rutin, isoquercitrin, chlorogenic acid, p-coumaric acid, caffeic acid, hydrochloric acid, Hoechst33342, propidium iodide, glucose, temozolomide, DMSO and KCl (Sigma-Aldrich, Steinheim, Germany), Dulbecco's Modified Eagle Medium (DMEM) with Glutamax, foetal bovine serum, penicillin-streptomycin, Versene solution, antibiotic-antimycotic solution (Anti-Anti) were of Gibco brand and purchased from Thermo Fisher Scientific, Waltham, MA, USA. During the study, we used purified de-ionized water prepared with the Milli–Q® (Millipore, Bedford, MA, USA) water purification system.

2.2. Plant Material and Extract Preparation

Japanese quince (*C. japonica*) leaves were collected in September 2018, after ripe fruits were harvested, from the garden of the Institute of Horticulture, Lithuanian Research Centre for Agriculture and Forestry, Babtai (55°60′ N, 23°48′ E). The leaves of each cultivar were collected from five shrubs, and frozen (at −40 °C) in a freezer with air circulation, and then lyophilized with a sublimator Zirbus 3 × 4 × 5 (ZIRBUS technology GmbH, Bad Grund, Germany), at a pressure of 0.01 mbar (temperature of condenser −85 °C) for 24 h. The lyophilized leaves were grounded to a fine powder with a knife mill GM (Retsch GmbH, Haan, Germany). Powdered leaf sample of each cultivar (2.5 g) was mixed with 50 mL of 40% ethanol, and extracted with ultrasonic bath Sonorex Digital 10 P (Bandelin Electronic GmbH & Co. KG, Berlin, Germany) for 40 min, at 60 °C, using 480 W ultrasonic power. The extracted samples were centrifuged and then filtered through filter paper (Watman no. 1). The Japanese quince ethanolic extracts (5 g/100 mL) were kept in a freezer (at −40 °C) in hermetically sealed containers for one week until further tests.

2.3. Evaluation of Phenolic Compound Composition (UPLC-ESI-MS/MS Conditions)

The variability in the qualitative and quantitative content of phenolic compounds in Japanese quince leaf samples was evaluated by applying validated UPLC-ESI-MS/MS method [17]. Samples were analyzed with Acquity H-class UPLC system (Waters Corporation, Milford, MA, USA) coupled with triple quadrupole tandem mass spectrometer (Xevo, Waters Corporation, Milford, MA, USA). To obtain MS/MS data an electrospray ionization source (ESI) was used. Compounds of interest were separated with YMC Triart C18 (100 × 2.0 mm; 1.9 μm) column (YMC Europe Gmbh, Dislanken, Germany). Constant temperature of 40 °C and flow rate of 0.5 mL·min^{-1} were maintained during analysis. Mobile phase consisted of 0.1% formic acid solution in water (solvent A) and acetonitrile (solvent B). Gradient profile was applied with following proportions of solvent A: Initially 95% for 1 min followed by linear increase to 70% over 4 min; 50% over next 3 min and to 95% over last 2 min. Analysis was performed in negative electrospray ionization mode. Capillary voltage was set to negative 2 kV. Temperature in ion source was maintained at 150 °C. Nitrogen gas temperature was set to 400 °C and flow rate to 700 L·h^{-1}. Cone gas flow rate was set to 20 L·h^{-1}. Each compound of interest had a specific collision energy and cone voltage selected. The selected mass spectrometry parameters for this method are presented in Table 1. The validation characteristics of the developed method are presented as supplementary Table S1.

Table 1. Mass spectrometry parameters for the analysis of phenolic compounds.

Compound	Retention Time, min	Molecular Ion (*m/z*)	Production (*m/z*)	Cone Voltage, V	Collision Chamber Energy, eV
(+)-Catechin	3.50	289	123	60	34
Chlorogenic acid	3.52	353	191	32	14
Caffeic acid	3.89	179	107	36	22
Procyanidin B2	4.02	577	407	50	20
(-)-Epicatechin	4.13	289	123	60	34
Procyanidin C1	4.38	865	125	56	60
p-Coumaric acid	4.73	163	93	28	22
Rutin	5.06	609	300	70	38
Hyperoside	5.20	463	300	50	26
Isoquercitrin	5.30	463	301	52	28
Luteolin 7-O-glucoside	5.31	447	285	66	26
Avicularin	5.59	433	301	50	20
Kaempferol 3-O-glucoside	5.64	447	284	54	28
Quercitrin	5.68	447	300	50	26
Phloridzin	5.88	435	273	42	14
Quercetin	6.86	301	151	48	20

2.4. C6 and HROG36 Cell Culture

The C6 and HROG36 cell lines were purchased from the Cell Lines Service GmbH (Germany). The cells suspended in DMEM with 10% of foetal bovine serum, 100 U/mL penicillin and streptomycin, seeded in 75 cm^2 flasks, and incubated at 37 °C, with 5% CO$_2$ and saturated humidity. The cells were reseeded to new flasks every 3 days. Twenty-four hours prior to the treatment with quince leaf extracts, the cells were transferred to 96 well plates (VWR) at density of 0.2 × 10^6 cells/cm^2.

2.5. Primary Neuronal-Glial Cell Culture

Primary rat cerebellar-glial cell culture was prepared as described previously [18]. Briefly, the cerebella were isolated, minced, and triturated in Versene solution (1:5000) to a single-cell suspension. The suspension was centrifuged at 270× *g* for 5 min and resuspended in DMEM with Glutamax supplemented with 5% horse serum, 5% foetal calf serum, 38 mM glucose, 25 mM KCl, and antibiotic-antimycotic solution. The cells were plated at a density of 0.25 × 10^6 cells/cm^2 in 96-well plates (VWR) coated with 0.0001% poly-L-lysine and kept in a humidified incubator containing 5% CO$_2$ at 37 °C. The cultures were subjected to treatment after 7 days in vitro.

2.6. Treatments of the Cells with Quince Leaf Extracts

HROG36, C6, and primary cerebellar neuronal glial cells were treated with 0.88, 1.25, 1.63, 2.00, 2.38, 2.75, 3.13 and 3.75 mg/mL ethanolic extracts made from leaves of Japanese quince cultivars 'Rasa', 'Darius' or 'Rondo' for 24 h. The controls with the same volume of the solvent (ethanol) were made in parallel. In addition, the C6 and HROG36 cells were treated with temozolomide concentration range from 0.02 to 4.85 mg/mL, chlorogenic acid (range 5–500 g/mL), epicatechin (5–300 g/mL), hyperoside (5–333 g/mL), and quercitrin (2–220 g/mL). After treatment, the cells were subjected to viability evaluation.

2.7. Evaluation of Cellular Viability

Viability of C6, HROG36, and primary cerebellar neuronal glial cells after treatments was evaluated according to metabolic activity by means of PrestoBlue™ Cell Viability Reagent (Thermo Fisher Scientific). The fluorescence of resorufin produced after PrestoBlue reagent cleavage was measured in a plate reader Infinite M Plex (Tecan Austria, Salzburg, Austria) at excitation and emission wavelengths of 560 and 590 nm, respectively. The results were expressed as percentage of the untreated control fluorescence level.

In addition, C6 and primary cerebellar neuronal-glial cells were evaluated for necrosis by double-staining with Hoechst 33,342 (15 µg/mL, Merck) and propidium iodide (PI; 5 µg/mL, Merck). After 15 min incubation with the dyes in dark at room temperature, the nuclear fluorescence was assessed under fluorescent microscope OLYMPUS IX71SIF-3 (Olympus Corporation, Tokyo, Japan). Hoechst33342-only-positive nuclei exhibiting blue fluorescence were considered viable, and Hoechst3334-plus-PI-positive nuclei stained magenta (because of blue and red signal overlay) were identified as necrotic.

2.8. Statistical Analysis

The phenolic compound content of each cultivar was expressed as means $\pm$ SD (standard deviation) of three replicates. The significant differences ($p \leq 0.05$) between means were evaluated using Tukey's HSD (Honest Significant Difference test). Cellular viability and metabolic activity results are presented as means $\pm$ standard deviation of 5 experimental repeats, each of 3 technical repeats. The data are expressed as percentage of the untreated control. Statistical analysis was performed by one-way analysis of variance (ANOVA) with the Dunnett's post-test by SigmaPlot 13.0 software (Systat Software Inc., Surrey, UK). A value of $p < 0.05$ was taken as the level of significance. EC$_{50}$ was calculated by SigmaPlot 13.0 (Systat Software Inc., Surrey, UK) software by means of four-parameter logistic function. Correlations were analyzed by Microsoft Office Excel 2010 (Microsoft, Redmond, WA, USA) software Correlation function.

3. Results

3.1. Phenolic Compound Composition of Japanese Quince Leaves

The sum of detected phenols varied significantly between cultivars, the highest amount was found in 'Darius', and the lowest in 'Rondo' leaves (Table 2). Sixteen phenolic compounds were identified in the leaves of 'Rondo', while 15 in 'Darius', and 14 in 'Rasa'. The majority of the identified phenols belong to three groups: Hydroxycinnamic acid derivatives, flavonols, and flavan-3-ols. There were also others, such as flavone luteolin 7-O-glucoside, and dihydrochalcone phloridzin. Total amount of hydroxycinnamic acid derivatives ranged from 5533 ('Darius') to 5839 ('Rasa') µg/g, and consisted of chlorogenic acid, p-coumaric acid, and caffeic acid. The flavan-3-ol group members were (-)-epicatechin, procyanidin B2, procyanidin C1, and (+)-catechin, and their total amount ranged from 700.4 ('Rondo') to 6426 ('Darius') µg/g. The flavonols found in the extracts were rutin, isoquercitrin, avicularin, kaempferol 3-O-glucoside, hyperoside, and quercetin. The total amount of the flavonols ranged from 2506 ('Rondo') to 4872 ('Darius') µg/g. The predominant group of phenols in 'Rondo' and 'Rasa' was the hydroxycinnamic acids, but in 'Darius' was flavan-3-ols. However, the total amount of hydroxycinnamic acids did not differ between the cultivars.

Table 2. Quantitative composition of phenols in Japanese quince leaves, µg/g DW.

Compound, µg/g DW	Japanese Quince Cultivars		
	'Darius'	'Rondo'	'Rasa'
Hydroxycinnamic Acids			
Chlorogenic acid	5373 ± 244 [a]	5773 ± 271 [a]	5737 ± 269 [a]
p-Coumaric acid	155.4 ± 10.2 [a]	58.9 ± 3.3 [c]	102.9 ± 7.5 [b]
Caffeic acid	4.8 ± 0.2 [a]	0.84 ± 0.03 [b]	ND
Total, µg·g^{-1}	5533 ± 202 [a]	5833 ± 307 [a]	5840 ± 188 [a]
Flavonols			
Isoquercitrin	2131.1 ± 54.4 [a]	418.6 ± 11.5 [c]	1700.7 ± 49.6 [b]
Hyperoside	907.7 ± 40.3 [b]	1124.5 ± 51.4 [a]	1107.8 ± 50.3 [a]
Quercitrin	1314.3 ± 39.2 [a]	248.5 ± 10.1 [c]	350.0 ± 20.2 [b]
Rutin	189.7 ± 9.0 [c]	289.1 ± 13.2 [b]	511.2 ± 20.0 [a]
Kaempherol 3-O-glucoside	318.0 ± 14.3 [b]	407.9 ± 18.3 [a]	226.9 ± 10.3 [c]
Avicularin	9.34 ± 0.32 [b]	14.97 ± 0.63 [a]	5.69 ± 0.20 [b]
Quercetin	2.14 ± 0.09 [a]	2.42 ± 0.10 [a]	1.23 ± 0.04 [b]
Total, µg·g^{-1}	4872 ± 140 [a]	2506 ± 56 [c]	3903 ± 14.3 [b]
Flavan-3-Ols			
(-)-Epicatechin	3963 ± 112 [a]	314.3 ± 9.4 [c]	2038 ± 88 [b]
Procyanidin C1	1333.8 ± 53.5 [a]	176.1 ± 8.4 [c]	792.4 ± 21.1 [b]
Procyanidin B2	1129.3 ± 50.3 [a]	209.4 ± 15.6 [c]	820.9 ± 33.6 [b]
(+)-Catechin	ND	0.6 ± 0.02	ND
Total, µg·g^{-1}	6426 ± 145 [a]	700.4 ± 10.7 [c]	3652 ± 73.6 [b]
Others Phenols			
Luteolin 7-O-glucoside	213.7 ± 8.6 [b]	277.9 ± 11.3 [a]	155.5 ± 6.9 [c]
Phloridzin	3.65 ± 0.16 [b]	5.32 ± 0.23 [a]	5.75 ± 0.25 [a]
Sum of all compounds, µg/g	17048 ± 461 [a]	9322 ± 287 [c]	13556 ± 375 [b]

Value is average ± SD (n = 3); Different letters in the same line indicate a statistically significant difference ($p \leq 0.05$); DW: dry weight; ND: not detected.

The total flavan-3-ol levels varied significantly between the cultivars. The highest amount was detected in 'Darius' leaves (6426 ± 145 µg/g), while 'Rasa' and 'Rondo' had around two and nine-fold less of the flavan-3-ols (3652 ± 73.6 µg/g and 700.4 ± 10.7 µg/g, respectively).

The total amount of flavonols was also significantly different between the cultivars. In addition, the distribution of individual compounds of this phenol group was different, too. The leaves of 'Darius' had significantly more isoquercitrin and quercitrin, 'Rondo' had more kaempferol 3-O-glucoside and luteolin 7-O-glucoside, and 'Rasa' had higher rutin amount. The main phenolic glycosides found in the extracts were isoquercitrin, hyperoside, and quercitrin.

3.2. The Effect of Quince Leaf Extracts on Viability of C6 and HROG36 Glioblastoma Cells

Next in the study, we evaluated the effect of the extracts from leaves of Japanese quince cultivars 'Darius', 'Rasa', and 'Rondo' on metabolic activity of rat C6 and human HROG36 glioblastoma cells (Figure 1).

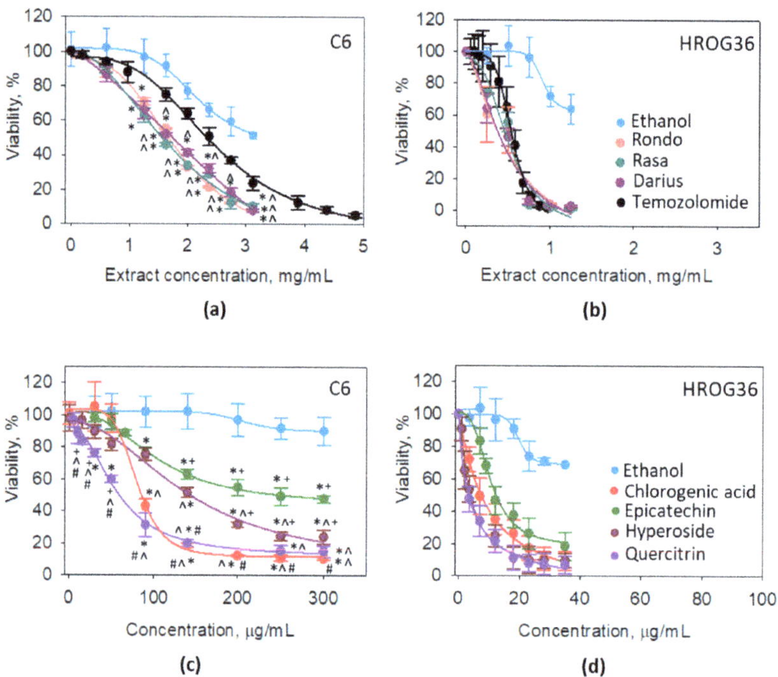

Figure 1. The effect of the extracts from leaves of different Japanese quince cultivars (**a**,**b**) and some phenolic compounds found in the extracts (**c**,**d**) on viability of glioblastoma C6 (**a**,**c**) and HROG36 (**b**,**d**) cells evaluated by metabolic activity assay with PrestoBlue reagent. For Ethanol, the concentrations were the same as in other samples at the indicated concentration point, i.e., 35, 50, 65, 80, 95, 110, 125, and 150 L/mL. In both (**a**,**c**), * indicates significant difference compared with ethanol-only treated samples; in (**a**), ˆ significant difference compared with temozolomide; in (**c**), ˆ significant difference compared with Epicatechin, + with Chlorogenic acid, # with Hyperoside. In (**b**), all extract-treated samples were statistically significantly different from Ethanol starting from concentration 0.25 mg/mL, and Temozolomide–starting from 0.5 mg/mL. in (**d**), all the samples were statistically significantly different from Ethanol starting from the concentration of 7 g/mL, and there was statistically significant difference between Epicatechin and other samples at 7 and 9 g/mL. The level of significance $p < 0.05$.

After 24-h application of 1.25 mg/mL quince leaf extracts, the viability of C6 cells (assessed as cellular metabolic activity by PrestoBlue assay) was significantly decreased compared to ethanol control. The viability was by 14%, 11%, and 13% lower after treatment with 'Rondo', 'Rasa', and 'Darius', respectively (Figure 1a). The metabolic activity of the cells continued to decrease with increase in the concentration of the extracts and reached 8% of control in 3.125 mg/mL 'Rondo'- and 'Darius'-treated samples, and 10% in 3.125 mg/mL 'Rasa"-treated samples. The difference between ethanol control and the 3.125 mg/mL extract-treated samples were 43%, 41%, and 43% for 'Rondo', 'Rasa', and 'Darius', respectively. The metabolic activity of C6 cells was more sensitive to the extracts compared to the effect of temozolomide, the drug used for glioblastoma treatment. The metabolic activity of C6 cells treated with extracts at 1.25 mg/mL and higher concentration was significantly lower than in the C6 samples treated by the same concentrations of temozolomide. This was also reflected in calculated EC_{50}; the values of 'Rondo', 'Rasa', and 'Darius' were 71%, 67%, and 74% lower compared to the EC_{50} value of temozolomide for C6 cells (Table 3).

Table 3. Calculated EC_{50} of extracts from leaves of quince cultivars 'Rondo', 'Rasa', and 'Darius' and of some phenolic compounds found in the extracts for C6 and HROG36 cells.

Substance, µg/mL	'Rondo'	'Rasa'	'Darius'	Chlorogenic Acid	Epicate Chin	Hypero Side	Querci Trin	Temozo Lomide
C6	1660.2	1560.0	1738.9	85.5	252.2	143.1	59.3	2341.5
HROG36	373.8	378.5	373.5	7.3	12.3	3.8	3.4	535.8

Human glioblastoma cells HROG36 were more sensitive to quince leaf extract treatment compared to C6 cells (Figure 1b). 'Rondo', 'Rasa', and 'Darius' applied at 0.25 mg/mL significantly decreased metabolic activity of HROG36 cells compared to ethanol control by 37%, 24%, and 34%, respectively. After treatment with 0.75 mg/mL extracts, metabolic activity of HROG36 cells was less than 5% and about 91% lower compared to the ethanol control. The effect of temozolomide on HROG36 cell metabolic activity was similar to that of the extracts, and there were no statistically significant differences detected. However, the EC_{50} value of temozolomide calculated from the average titration data was by 157.3–162.0 µg/mL higher compared to the EC_{50} of the extracts (Table 3).

For the next step in the study, we have investigated the toxicity of the phenolic compounds identified in the quince leaf extracts on C6 and HROG36 cells. Chlorogenic acid was selected as a representative of hydroxycinnamic acids, hyperoside and quercitrin from flavonols, and epicatechin from flavan-3-ols. C6 cells were most sensitive to quercitrin and chlorogenic acid, as presented in the titration curves in Figure 1c and EC_{50} values in Table 3. The next least toxic compound was hyperoside, and the least toxic was epicatechin. Similarly to the case of extract treatment, HROG36 cells were more sensitive to the phenolic compounds compared to the C6 cells (Figure 1c,d). After treatment with 23 µg/mL of each compound, the metabolic activity of HROG36 cells was 8% (for quercitrin)–26% (for epicatechin) of the untreated control. For comparison, after similar 30 µg/mL treatment in C6 cell samples, the metabolic activity was either unchanged (chlorogenic acid, epicatechin), or decreased only to 90% (hyperoside) and 75% (quercitrin) of untreated control. The most toxic for HROG36 cells were quercitrin and hyperoside, although chlorogenic acid was also very close to that level. A slightly lower toxicity was caused by epicatechin; there was statistically significant difference between epicatechin and other investigated phenolic compounds at 7 and 9 g/mL.

The potential input of each group of phenolic compounds and some individual phenols in the extracts was estimated by correlation analysis (Table 4).

Table 4. The values of coefficient for correlation between viability (metabolic activity) of C6 or HROG36 cells and amount of phenolic compounds in the extracts from leaves of quince cultivars 'Rondo', 'Rasa', and 'Darius'.

Compound	Hydroxycinnamic Acids	Flavonols	Flavan-3-Ols	Total Phenols	Chlorogenic Acid	Epicate Chin	Hypero Side	Querci Trin
C6	−0.99	0.46	0.21	0.39	−0.99	0.33	−0.75	−0.56
HROG36	−0.86	−0.38	0.28	−0.42	−0.81	−0.21	−0.31	−0.81

Strong negative correlation with r value close to −1 was found between viability level of both glioblastoma cell types and amounts of hydroxycinnamic acids and chlorogenic acid. In addition, strong negative correlation was between metabolic activity of C6 and hyperoside, and between metabolic activity of HROG36 and quercitrin. Moderate negative correlation was between C6 viability and the amount of quercitrin, and low negative— between HROG36 viability and total contents of flavonols and phenols. The analysis allows to suggest that the toxicity of the extracts was most likely mediated by chlorogenic acid.

Metabolic activity assays such as PrestoBlue reflects not only changes in cell viability, but also differences in proliferation rate and metabolic disturbances. Therefore, we have additionally investigated viability of C6 cells after quince leaf extract treatments by double nuclear fluorescence staining that allows to detect necrotic cells with lost membrane integrity (Figure 2).

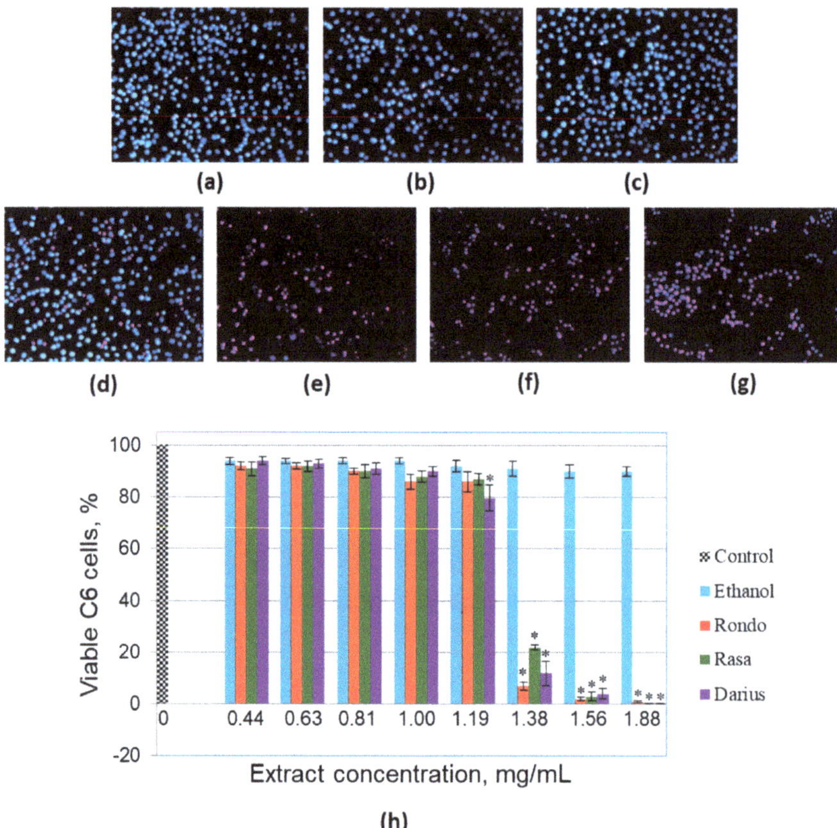

Figure 2. The effect of the extracts from leaves of different Japanese quince cultivars on viability of glioblastoma C6 cells. (a–g) Characteristic images of Hoechst/propidium iodide staining of the C6 cell nuclei after extract treatments; (a) samples treated with 125 µL/mL of the solvent ethanol (amount corresponds to 1.88 mg/mL extract treatment); (b–d) treated with 1.19 mg/mL extracts from 'Rondo', 'Rasa', and 'Darius' cultivars, respectively; (e–g) samples, after treatment with 1.56 mg/mL extracts from 'Rondo', 'Rasa', and 'Darius' cultivars, respectively; (h) quantitative summary of viability data presented as averages with standard deviation, * indicates significant difference compared with ethanol-only treated samples, $p < 0.05$.

The extracts did not significantly affect C6 cell viability up to the concentration of 1.19 mg/mL (Figure 2b–d,h). Treatment with 1.19 mg/mL extract from leaves of 'Darius' induced small yet significant decrease in percentage of viable C6 cells (Figure 2d,h). The average level of viable cells in 'Darius' extract-treated samples was by 14% lower compared with samples treated with the same amount of ethanol. Further increase in extract concentration to 1.38 mg/mL caused a remarkable drop in C6 viability in all three cultivar groups (Figure 2h). The average level of viable cells decreased by 93%, 78%, and 88% after treatment with extracts from 'Rondo', 'Rasa', and 'Darius', respectively. After treatment with 1.88 mg/mL extracts, the number of viable cells in C6 samples was close to 'zero' in all three cultivar groups. There was no significant decrease in C6 cell viability, observed after treatment with ethanol up to 1.88 mg/mL. Calculated levels of EC_{50} from double nuclear staining experiments were 1.26 mg/mL, 13.0 mg/mL, and 1.26 mg/mL for 'Rondo', 'Rasa', and 'Darius', respectively; the values slightly lower yet similar to those revealed by metabolic activity assay.

3.3. The Effect of Quince Leaf Extracts on Viability of Primary Non-Cancerous Brain Cells

All the Japanese quince leaf extracts investigated in the study were toxic to human rat glioblastoma C6 cells at concentrations equal to or higher than 1.38 mg/mL. Human glioblastoma HROG36 cells were even more sensitive to the treatments. The next step in this study was to investigate whether primary non-cancerous brain cells have similar susceptibility to the same extract treatment (Figure 3).

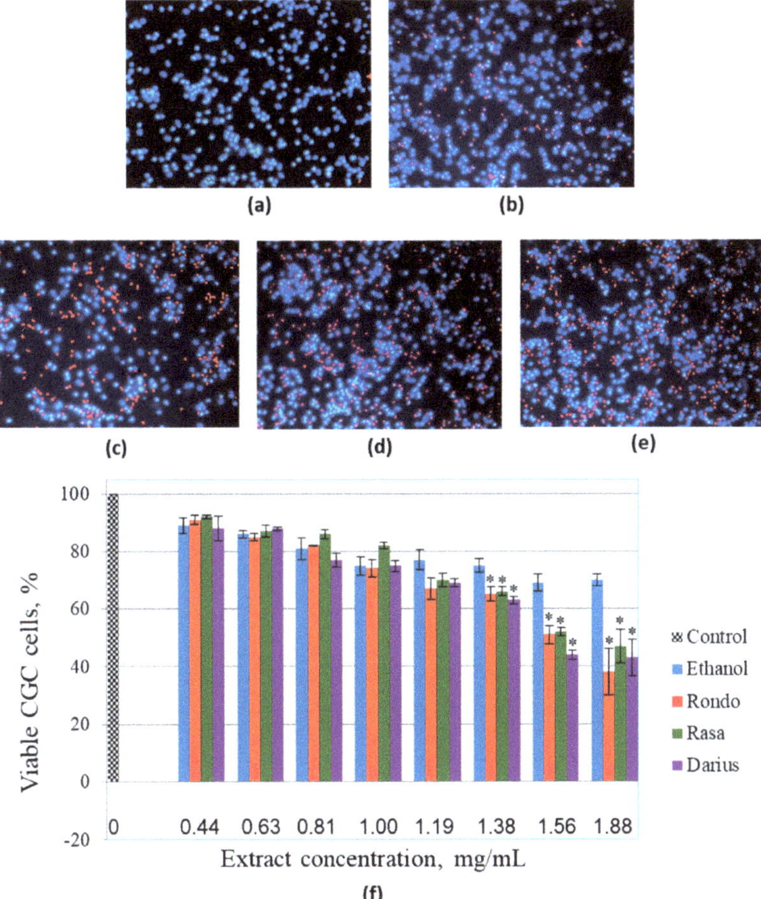

Figure 3. The effect of the extracts from leaves of different Japanese quince cultivars on viability of cultivated primary rat cerebellar neuronal-glial cells. (**a**–**e**) Representative images of nuclei stained with Hoechst33342 and propidium iodide after extract treatments: (**a**) Untreated control, (**b**) sample treated with 125 µL/mL of the solvent ethanol; (**c**–**e**) samples treated with 1.19 mg/mL extracts from 'Rondo', 'Rasa', and 'Darius' cultivars, respectively; (**f**) quantitative summary of viability data presented as averages with standard deviation, * indicates significant difference compared with the samples treated with the same amount of ethanol.

For these experiments, rat cerebellar neuronal-glial cell cultures consisting of approximately of 81 ± 4% granule neurons, 14 ± 3% astrocytes, and 6 ± 2% microglial cells [18] were used. The extracts up to the concentration of 1.0 mg/mL did not cause significant decrease in viability of primary cerebellar cells compared to the respective ethanol control (Figure 3f). However, increase in concentration up to 1.19 mg/mL caused viability drop by 13% ('Rondo'), 9% ('Rasa'), and 10% ('Darius') compared with the respective ethanol treatment. Further increase in extract concentration continued to lower the number of viable cells, and after treatment with 1.88 mg/mL extracts, the average numbers such cells in primary neuronal-glial cultures were 38%, 47%, and 43% for 'Rondo', 'Rasa', and 'Darius', respectively. In the case of C6 glioblastoma cells, the extracts applied at concentration 1.56 mg/mL induced 80% and higher loss of viable cells, and there were almost no viable cells left after treatment with 1.88 mg/mL extracts (Figure 2). In primary brain cell cultures, the level of viable cells remained higher than 50% after treatment with 1.56 mg/mL extracts and close to 50% after treatment with 1.88 mg/mL extracts. Thus, primary neuronal-glial cells from rat cerebella were less sensitive to the toxic effect of the 1.56–1.88 mg/mL extracts from quince leaves compared to the glioblastoma C6 cells. This was also confirmed by the calculated EC_{50} of the extracts for primary cerebellar cells; the values were 1.58 mg/mL, 1.72 mg/mL, and 1.64 mg/mL for 'Rondo', 'Rasa', and 'Darius', respectively, and they were higher than the values for C6 cells calculated from the double nuclear staining assay data. However, primary neuronal-glial cells were more sensitive to the solvent ethanol compared to C6 cell line. There was significant decrease in primary cerebellar cell viability observed after treatment with 80 μL/mL ethanol; the average level of viable cells after this treatment was by 25% lower compared to the untreated control. After treatment with 150 μL/mL ethanol, the percentage of viable cells were by 30% lower than in control samples.

4. Discussion

Japanese quince leaf extract consisted of three major phenol groups: Hydroxycinnamic acids, flavan-3-ols, and flavonols. However, distribution of these groups between cultivars was significantly different. For example, in 'Rondo', more than half of total phenols consisted of hydroxycinnamic acid derivatives (62.6%), while 'Darius' and 'Rasa' contained only 32.5%, and 43% of these compounds, respectively. Similar amounts (from 42.90% to 50.90%) of hydroxycinnamic acids in quince leaves were found in a recent study of Chojnacka and co-authors [13]. The predominant compound of this group in all cultivars was chlorogenic acid, and this is in agreement with other studies [13,19]. The amount of flavonols in all cultivars was around 30% of total phenols. The majority of the flavonols consisted of isoquercitrin, hyperoside, and quercitrin, all the three are known for anti-cancer properties [20–22]. Saccharide moiety of the compounds mediates the toxicity interacting with membranes of cancer cells and promoting active uptake of the compounds [23,24]. The distribution of flavan-3-ols between cultivars was most diverse and varied from 7.5% ('Rondo') to 37.7% ('Darius').

In the present study, we have found that ethanolic quince leave extracts were cytotoxic similarly as (in the case of human glioblastoma HROG36 cells) or even more than (in rat C6 cell case) temozolomide, the drug used for glioblastoma treatment in the clinical practice. The viability of both investigated cell types had strong negative correlation with the amount of hydroxycinnamic acid derivatives and chlorogenic acid, and the cytotoxicity of the chlorogenic acid was also demonstrated in both C6 and HROG36 cell cultures. Although other investigated compounds were also toxic to the cells, especially flavonols quercitrin and hyperoside, the amounts of the compounds present in the extracts were far too small to mediate the toxicity for C6 cells. However, for HROG36 cells, the flavonols could contribute to the toxicity of chlorogenic acid in the extracts because the EC_{50} values of the compounds for the cells were close to the levels found in the extracts. The anticancer activity of hydroxycinnamic acids is also reported by others. These compounds promote apoptosis, arrest cell cycle, and prevent metastasis of different types of breast, lung, colon, gastric, liver, pancreatic, and prostate cancer cells [25–27]. Ekbatan and colleagues have shown that chlorogenic acid and its metabolites caffeic acid, 3-phenyl propionic acid, and benzoic acid cause cell cycle arrest and apoptosis of colon cancer cells Caco 2 [28].

Another group of scientists has demonstrated that chlorogenic acid disrupts cytoskeleton organization and mTORC2 signalling of both adenocarcinomic human alveolar basal epithelial cells A549 and human hepatocyte carcinoma HepG2 [29]. The above-mentioned findings are in line with another extensive study performed by Huang and co-workers, who investigated anticancer activity of chlorogenic acid on human cancerous lung, liver, kidney, colon, and brain cells including human glioblastoma lines U87MG and M059J, rat C6, and mouse G422 [30]. The study revealed that chlorogenic acid promotes all cancer cell (including glioblastoma) cycle arrest and differentiation to maturation phenotype via miR-17 family downregulation, p21 upregulation and mitochondrial suppression. The efficiency of chlorogenic acid was comparable to that of temozolomide.

Comparison of viability of primary rat non-cancerous brain cells and rat glioblastoma C6 cells evaluated by double nuclear staining revealed the significantly higher sensitivity of glioblastoma for the quince leave extracts. Although the EC_{50} values of the extracts for the non-cancerous cells were similar to those calculated for C6 cells from the metabolic PrestoBlue evaluation data, to our opinion, it would not be relevant to compare the data obtained from different viability assessment assays. PrestoBlue assay monitors the rate of cellular metabolic activity, which is proportional to the number of viable cells. However, the metabolic activity might be directly influenced by the investigated compound without causing cell death, e.g., chlorogenic acid, the main component of the quince leaf extract, is reported to decrease mitochondrial activity of glioblastoma [30]. The number of viable cells might be lower not, or not only, due to the increase in cell death, but also due to the suppression of proliferation, because hydroxycinnamic acids present in the extracts might induce cell cycle arrest [26]. Thus, the metabolic assay is different from the evaluation of viability by double nuclear staining which gives information about percentage of necrotic and viable cells without sensing the metabolic activity or proliferation rate of them.

Similar finding about lower susceptibility to quince leaf extract of non-cancerous cells compared to cancer cells was recently described by Chojnacka and co-authors [13]. Such higher sensitivity of cancer cells to bioactive compounds of quince leaf extracts could be related to specific biology of these cells. Usually, cancer cells have higher proliferation rate, migration capacity and altered energy metabolism compared to the surrounding non-cancerous cells [31,32]. Such a difference opens a niche for selective targeting of cancer cells with lower risk to destroy healthy non-cancerous cells. Ethanolic extracts from Japanese quince leaves have several compounds that interfere with cancer cell-specific pathways related to proliferation, migration and energy metabolism [4,29,30]. This might at least partially explain cancer cell-selective cytotoxicity of quince leaf extracts.

Thinking about quince leaf extracts as complementary therapy for glioblastoma, it is important to evaluate the ability of the extract compounds to cross the blood-brain barrier (BBB). One of the best studied pathways for plant phenolic metabolites to cross BBB is passive permeation, however active transport might be also possible. Lee and co-authors have found that intraperitoneally administered chlorogenic acid ameliorates brain damage and oedema after cerebral ischaemia in rats [33]. In another study, chlorogenic acid was found both in the blood and brain after intraperitoneal administration in mice, and was safe even at very high doses (1000 mg/kg) [30]. A pharmacokinetics and brain penetration study shows that chlorogenic acid is rapidly absorbed in plasma after both intranasal and intravenous administration of 10 mg/kg and reaches brain and cerebrospinal fluid [34]. The concentration of chlorogenic acid in the brain after 30 min of intranasal administration was about 250 g/mL and remained about 25 g/mL after 6 h. Such data allow to suggest that chlorogenic acid and other compounds from quince leave extracts might be applied as complementary therapy for glioblastoma. However, future research should focus on an effective and safe dose, biologically active compound absorption, distribution, and excretion.

5. Conclusions

The main compound in ethanolic extracts from Japanese quince cultivars 'Rondo', 'Rasa', and 'Darius' is chlorogenic acid, and the amount of this compound is similar in all three cultivars. The amount of other phenolic compounds is more variable between the cultivars. The extracts of the leaves of all three cultivars significantly decrease viability of C6 and HROG36 glioblastoma cells; in the case of the HROG36 cells, the extracts are equally toxic, but in the C6 cells, the extracts are more toxic than glioblastoma drug temozolomide. The effect on viability has strong correlation with the level of chlorogenic acid for both cell types. In addition, quince leaf extracts exert significantly higher toxicity on rat C6 glioblastoma cells compared to primary rat neuronal-glial cerebellar cells. This finding suggests that Japanese quince leaves could be further investigated as anticancer drugs for glioblastoma treatment.

Supplementary Materials: The following are available online at https://www.mdpi.com/2304-815 8/10/1/18/s1, Table S1: Validation characteristics of developed UPLC-ESI-MS/MS method.

Author Contributions: Conceptualization, P.V., V.J., and A.J.; methodology, V.Z., I.U., R.B., D.M., and Z.B.; formal analysis, I.U., R.B., D.K., U.M., D.M., and M.L.; investigation, V.Z., I.U., R.B., U.M., and Z.B.; resources, P.V., A.J., and V.J.; writing—original draft preparation, I.U. and A.J.; writing—review and editing, I.U., M.L., P.V., A.J., and V.J.; visualization, I.U., R.B., M.L., and A.J.; supervision, P.V., A.J., and V.J.; All authors have read and agreed to the published version of the manuscript.

Funding: This study was financed by the Institute of Horticulture, Lithuanian Research Centre for Agriculture and Forestry and the EUREKA Network Project E! 13496 "OHMDRINKS" (No. 01.2.2-MITA-K-702-08-003).

Data Availability Statement: The data presented in this study are available on request from the corresponding author. The data are not publicly available due to the institutional data policy.

Conflicts of Interest: The authors declare no conflict of interest.

References

1. Bieniasz, M.; Dziedzic, E.; Kaczmarczyk, E. The effect of storage and processing on Vitamin C content in Japanese quince fruit. *Folia Hortic.* **2017**, *29*, 83–93. [CrossRef]
2. Du, H.; Wu, J.; Li, H.; Zhong, P.-X.; Xu, Y.-J.; Li, C.-H.; Ji, K.-X.; Wang, L.-S. Polyphenols and triterpenes from Chaenomeles fruits: Chemical analysis and antioxidant activities assessment. *Food Chem.* **2013**, *141*, 4260–4268. [CrossRef] [PubMed]
3. Urbanavičiūtė, I.; Liaudanskas, M.; Bobinas, Č.; Šarkinas, A.; Rezgienė, A.; Viskelis, P. Japanese Quince (*Chaenomeles japonica*) as a Potential Source of Phenols: Optimization of the Extraction Parameters and Assessment of Antiradical and Antimicrobial Activities. *Foods* **2020**, *9*, 1132. [CrossRef] [PubMed]
4. Feng, X.M.; Su, X.L. Anticancer effect of ursolic acid via mitochondria-dependent pathways (Review). *Oncol. Lett.* **2019**, *17*, 4761–4767. [CrossRef] [PubMed]
5. Liu, P.; Du, R.; Yu, X. Ursolic acid exhibits potent anticancer effects in human metastatic melanoma cancer cells (SK-MEL-24) via apoptosis induction, inhibition of cell migration and invasion, cell cycle arrest, and inhibition of mitogen-activated protein kinase (MAPK)/ERK signaling pathway. *Med. Sci. Monit.* **2019**, *25*, 1283–1290. [CrossRef] [PubMed]
6. Shyu, M.H.; Kao, T.C.; Yen, G.C. Oleanolic acid and ursolic acid induce apoptosis in HuH7 human hepatocellular carcinoma cells through a mitochondrial-dependent pathway and downregulation of XIAP. *J. Agric. Food Chem.* **2010**, *48*, 3396–3402. [CrossRef] [PubMed]
7. Zhu, Y.Y.; Huang, H.Y.; Wu, Y.L. Anticancer and apoptotic activities of oleanolic acid are mediated through cell cycle arrest and disruption of mitochondrial membrane potential in HepG2 human hepatocellular carcinoma cells. *Mol. Med. Rep.* **2015**, *12*, 5012–5018. [CrossRef]
8. Silva, B.M.; Andrade, P.B.; Ferreres, F.; Domingues, A.L.; Seabra, R.M.; Ferreira, M.A. Phenolic profile of quince fruit (Cydonia oblonga Miller) (pulp and peel). *J. Agric. Food Chem.* **2002**, *50*, 4615–4618. [CrossRef]
9. Kopustinskiene, D.M.; Jakstas, V.; Savickas, A.; Bernatoniene, J. Flavonoids as anticancer agents. *Nutrients* **2020**, *12*, 457. [CrossRef]
10. Gorlach, S.; Wagner, W.; Podsedek, A.; Szewczyk, K.; Koziokiewicz, M.; Dastych, J. Procyanidins from Japanese quince (*Chaenomeles japonica*) fruit induce apoptosis in human colon cancer caco-2 cells in a degree of polymerization—Dependent manner. *Nutr. Cancer* **2011**, *63*, 1348–1360. [CrossRef]
11. Lewandowska, U.; Szewczyk, K.; Owczarek, K.; Hrabec, Z.; Podsędek, A.; Koziołkiewicz, M.; Hrabec, E. Flavanols from Japanese quince (*Chaenomeles Japonica*) fruit inhibit human prostate and breast cancer cell line invasiveness and cause favorable changes in Bax/Bcl-2 mRNA Ratio. *Nutr. Cancer* **2013**, *65*, 273–285. [CrossRef] [PubMed]

12. Owczarek, K.; Hrabec, E.; Fichna, J.; Sosnowska, D.; Koziołkiewicz, M.; Szymański, J.; Lewandowska, U. Flavanols from Japanese quince (*Chaenomeles japonica*) fruit suppress expression of cyclooxygenase-2, metalloproteinase-9, and nuclear factor-kappaB in human colon cancer cells. *Acta Biochim. Pol.* **2017**, *64*, 567–576. [CrossRef] [PubMed]

13. Chojnacka, K.; Sosnowska, D.; Polka, D.; Owczarek, K.; Gorlach-Lira, K.; Oliveira De Verasa, B.; Lewandowska, U. Comparison of phenolic compounds, antioxidant and cytotoxic activity of extracts prepared from Japanese quince (*Chaenomeles japonica* L.) leaves. *J. Physiol. Pharmacol.* **2020**, *71*, 1–10. [CrossRef]

14. Shergalis, A.; Bankhead, A.; Luesakul, U.; Muangsin, N.; Neamati, N. Current challenges and opportunities in treating glioblastomas. *Pharmacol. Rev.* **2018**, *70*, 412–445. [CrossRef] [PubMed]

15. Vengoji, R.; Macha, M.A.; Batra, S.K.; Shonka, N.A. Natural products: A hope for glioblastoma patients. *Oncotarget* **2018**, *9*, 22199–22224. [CrossRef] [PubMed]

16. Urbanaviciute, I.; Liaudanskas, M.; Seglina, D.; Viskelis, P. Japanese Quince *Chaenomeles Japonica* (Thunb.) Lindl. ex Spach Leaves a New Source of Antioxidants for Food. *Int. J. Food Prop.* **2019**, *22*, 795–803. [CrossRef]

17. González-Burgos, E.; Liaudanskas, M.; Viškelis, J.; Žvikas, V.; Janulis, V.; Gómez-Serranillos, M.P. Antioxidant activity, neuroprotective properties and bioactive constituents analysis of varying polarity extracts from Eucalyptus globulus leaves. *J. Food Drug Anal.* **2018**, *26*, 1293–1302. [CrossRef]

18. Balion, Z.; Cėpla, V.; Svirskiene, N.; Svirskis, G.; Druceikaitė, K.; Inokaitis, H.; Rusteikaitė, J.; Masilionis, I.; Stankevičienė, G.; Jelinskas, T.; et al. Cerebellar cells self-assemble into functional organoids on synthetic, chemically crosslinked ECM-mimicking peptide hydrogels. *Biomolecules* **2020**, *10*, 754. [CrossRef]

19. Kikowska, M.; Włodarczyk, A.; Rewers, M.; Sliwinska, E.; Studzińska-Sroka, E.; Witkowska-Banaszczak, E.; Stochmal, A.; Żuchowski, J.; Dlugaszewska, J.; Thiem, B. Micropropagation of *Chaenomeles japonica*: A step towards production of polyphenol-rich extracts showing antioxidant and antimicrobial activities. *Molecules* **2019**, *24*, 1314. [CrossRef]

20. Di Camillo Orfali, G.; Duarte, A.C.; Bonadio, V.; Martinez, N.P.; De Araújo, M.E.M.B.; Priviero, F.B.M.; Carvalho, P.O.; Priolli, D.G. Review of anticancer mechanisms of isoquercitin. *World J. Clin. Oncol.* **2016**, *7*, 189–199. [CrossRef]

21. Cincin, Z.B.; Unlu, M.; Kiran, B.; Bireller, E.S.; Baran, Y.; Cakmakoglu, B. Apoptotic Effects of Quercitrin on DLD-1 Colon Cancer Cell Line. *Pathol. Oncol. Res.* **2015**, *21*, 333–338. [CrossRef] [PubMed]

22. Raza, A.; Xu, X.; Sun, H.; Tang, J.; Ouyang, Z. Pharmacological activities and pharmacokinetic study of hyperoside: A short review. *Trop. J. Pharm. Res.* **2017**, *16*, 483–489. [CrossRef]

23. Chen, X.; Hui, L.; Foster, D.A.; Drain, C.M. Efficient synthesis and photodynamic activity of porphyrin-saccharide conjugates: Targeting and incapacitating cancer cells. *Biochemistry* **2004**, *43*, 10918–10929. [CrossRef] [PubMed]

24. Salucci, M.; Bugianesi, R.; Maiani, G.; Stivala, L.A.; Vannini, V. Flavonoids uptake and their effect on cell cycle of human colon adenocarcinoma cells (Caco2). *Br. J. Cancer* **2002**, *86*, 1645–1651. [CrossRef] [PubMed]

25. Deka, S.; Gorai, S.; Manna, D.; Trivedi, V. Evidence of PKC Binding and Translocation to Explain the Anticancer Mechanism of Chlorogenic Acid in Breast Cancer Cells. *Curr. Mol. Med.* **2017**, *17*, 79–89. [CrossRef] [PubMed]

26. Rocha, L.D.; Monteiro, M.C.; Teodoro, A.J. Anticancer Properties of Hydroxycinnamic Acids—A Review. *Cancer Clin. Oncol.* **2012**, *1*, 109. [CrossRef]

27. Yamaguchi, M.; Murata, T.; El-Rayes, B.F.; Shoji, M. The flavonoid p-hydroxycinnamic acid exhibits anticancer effects in human pancreatic cancer MIA PaCa-2 cells in vitro: Comparison with gemcitabine. *Oncol. Rep.* **2015**, *34*, 3304–3310. [CrossRef]

28. Ekbatan, S.S.; Li, X.Q.; Ghorbani, M.; Azadi, B.; Kubow, S. Chlorogenic acid and its microbial metabolites exert anti-proliferative effects, S-phase cell-cycle arrest and apoptosis in human colon cancer caco-2 cells. *Int. J. Mol. Sci.* **2018**, *19*, 723. [CrossRef]

29. Tan, S.; Dong, X.; Liu, D.; Hao, S.; He, F. Anti-tumor activity of chlorogenic acid by regulating the mTORC2 signaling pathway and disrupting F-actin organization. *Int. J. Clin. Exp. Med.* **2019**, *12*, 4818–4828.

30. Huang, S.; Wang, L.L.; Xue, N.N.; Li, C.; Guo, H.H.; Ren, T.K.; Zhan, Y.; Li, W.B.; Zhang, J.; Chen, X.G.; et al. Chlorogenic acid effectively treats cancers through induction of cancer cell differentiation. *Theranostics* **2019**, *9*, 6745–6763. [CrossRef]

31. Danhier, P.; Bański, P.; Payen, V.L.; Grasso, D.; Ippolito, L.; Sonveaux, P.; Porporato, P.E. Cancer metabolism in space and time: Beyond the Warburg effect. *Biochim. Biophys. Acta Bioenerg.* **2017**, *1858*, 556–572. [CrossRef] [PubMed]

32. Gallaher, J.A.; Brown, J.S.; Anderson, A.R.A. The impact of proliferation-migration tradeoffs on phenotypic evolution in cancer. *Sci. Rep.* **2019**, *9*, 1–10. [CrossRef] [PubMed]

33. Lee, K.; Lee, J.S.; Jang, H.J.; Kim, S.M.; Chang, M.S.; Park, S.H.; Kim, K.S.; Bae, J.; Park, J.W.; Lee, B.; et al. Chlorogenic acid ameliorates brain damage and edema by inhibiting matrix metalloproteinase-2 and 9 in a rat model of focal cerebral ischemia. *Eur. J. Pharmacol.* **2012**, *689*, 89–95. [CrossRef] [PubMed]

34. Kumar, G.; Paliwal, P.; Mukherjee, S.; Patnaik, N.; Krishnamurthy, S.; Patnaik, R. Pharmacokinetics and brain penetration study of chlorogenic acid in rats. *Xenobiotica* **2019**, *49*, 339–345. [CrossRef] [PubMed]

Article

Biomanufacturing of Tomato-Derived Nanovesicles

Ramesh Bokka [1], Anna Paulina Ramos [1], Immacolata Fiume [1], Mauro Manno [2], Samuele Raccosta [2], Lilla Turiák [3], Simon Sugár [3], Giorgia Adamo [4], Tamás Csizmadia [5] and Gabriella Pocsfalvi [1,*]

[1] Extracellular Vesicles and Mass Spectrometry Group, Institute of Biosciences and BioResources, National Research Council of Italy, 80131 Naples, Italy; ramesh.chem2008@gmail.com (R.B.); a.paulina.ramos@gmail.com (A.P.R.); immacolata.fiume@ibbr.cnr.it (I.F.)
[2] Institute of Biophysics, National Research Council of Italy, 90146 Palermo, Italy; mauro.manno@cnr.it (M.M.); samuele.raccosta@ibf.cnr.it (S.R.)
[3] MS Proteomics Research Group, Hungarian Academy of Sciences, Research Centre for Natural Sciences, 1117 Budapest, Hungary; liliat7@gmail.com (L.T.); sugarsimi@gmail.com (S.S.)
[4] Institute for Biomedical Research and Innovation, National Research Council of Italy, 90146 Palermo, Italy; giorgia.adamo@gmail.com
[5] Department of Anatomy, Cell and Developmental Biology, Eötvös Loránd University, 1117 Budapest, Hungary; aldhissla1987a@gmail.com
* Correspondence: gabriella.pocsfalvi@ibbr.cnr.it

Received: 1 November 2020; Accepted: 7 December 2020; Published: 11 December 2020

Abstract: Micro- and nano-sized vesicles (MVs and NVs, respectively) from edible plant resources are gaining increasing interest as green, sustainable, and biocompatible materials for the development of next-generation delivery vectors. The isolation of vesicles from complex plant matrix is a significant challenge considering the trade-off between yield and purity. Here, we used differential ultracentrifugation (dUC) for the bulk production of MVs and NVs from tomato (*Solanum lycopersicum* L.) fruit and analyzed their physical and morphological characteristics and biocargo profiles. The protein and phospholipid cargo shared considerable similarities between MVs and NVs. Phosphatidic acid was the most abundant phospholipid identified in NVs and MVs. The bulk vesicle isolates were further purified using sucrose density gradient ultracentrifugation (gUC) or size-exclusion chromatography (SEC). We showed that SEC using gravity column efficiently removed co-purifying matrix components including proteins and small molecular species. dUC/SEC yielded a high yield of purified vesicles in terms of number of particles (2.6×10^{15} particles) and protein quantities (6.9 ± 1.5 mg) per kilogram of tomato. dUC/gUC method separated two vesicle populations on the basis of buoyant density. Proteomics and in silico studies of the SEC-purified MVs and NVs support the presence of different intra- and extracellular vesicles with highly abundant lipoxygenase (LOX), ATPases, and heat shock proteins (HSPs), as well as a set of proteins that overlaps with that previously reported in tomato chromoplast.

Keywords: nanovesicles; tomato; size-exclusion chromatography; biomanufacturing; proteomics; phospholipids; *Solanum lycopersicum* L.; gradient ultracentrifugation; differential ultracentrifugation

---

1. Introduction

The study of nanometer-sized vesicles (NVs) isolated from whole plants or plant organs has opened a new branch of research in the field of extracellular vesicles (EVs). [1–3] Several recent reports and review articles describe the successful isolation of NVs from a great variety of edible fruits and vegetables (Table 1) [1–15]. Plant-derived exosome-like NVs are morphologically similar to the small EVs (sEVs) or exosomes isolated from mammalian cell cultures and biofluids [16], and the methods used for their isolation and characterization are also the same or similar. Typically, homogenized plant is the starting material for the isolation of NVs, which is a very complex matrix that makes the isolation process very challenging. Plant-derived vesicle isolates are more complex than mammalian cell-derived EVs and contain both intra- and extracellular vesicles [7]. There is a growing interest in advanced strategies for the production of NVs from plant resources due to their numerous promising applications, especially in the nutraceutical [6], cosmeceutical [8], and therapeutic fields [9]. Due to their inherent role in intracellular trafficking, native NVs are efficiently taken up by recipient cells to which they transfer their lipids, mRNAs, microRNAs, and protein biocargo [10,11]. Interestingly, some native NVs have been shown to possess antitumor [12], anti-inflammatory [13–16], anti-aging [8], and anti-Alzheimer [17] properties. For example, Zhang et. al. reported that ginger-derived NVs reduce inflammation in inflammatory bowel disease and colitis-associated cancer in mice [13]. In another work, ginger-derived NVs were proven to be efficient against Alzheimer's disease in a rat model [17]. Both grape- and broccoli-derived NVs were shown to inhibit colitis [14] and to protect from dextran sulfate sodium-induced colitis (DSS-induced colitis) [16] using mouse models. Moreover, turmeric (*Curcuma longa* L.)-derived NVs were shown to reduce colitis and promote intestinal wound repair [15]. Interestingly, NVs from the fruit juice of *Citrus limon* L. strongly suppressed tumor growth in rats [12]. NVs extracted from *Dendropanax morbifera* were shown to have strong inhibitory effect on melanin production in a human epidermis model, which promotes their future cosmeceutical applications [18]. The use of plant-derived NVs as novel drug delivery systems is boosted by their intrinsic resistance to the acidic gastric environment of the stomach [19], efficient uptake at target site, and low cost and sustainable production [18]. Moreover, NVs can be loaded with exogenous molecules such as drugs or other health-promoting substances or modified for engineered targeting. For example, Wang et al. loaded grapefruit-derived NVs with curcumin, folic acid, and zymosan A [19], and in another work ginger-derived NVs were loaded with doxorubicin anti-cancer drug [20].

Table 1. Methods used for the isolation of plant-derived microvesicles (MVs), nanovesicles (NVs), and apoplastic vesicles (AVs) from different organs, such as fruit, flower, seed, rhizome, and leaf, and the yields obtained.

Resource	Organ	Isolation Method	Vesicle Type(s) Isolated	Yield (g/L or g/kg of Starting Plant Material)	Particle Number (Particles/kg or Particles/L of Starting Plant Material)	Ref.
Ginger	rhizome	dUC/gUC	NVs	0.25–1.25 g/L	n.r.	[21]
Ginger	rhizome	dUC/precipitation	NVs	2–3.8	n.r.	[22]
Ginger	rhizome	dUC	NVs	4	n.r.	[22]
Ginger	rhizome	dUC	NVs	$48.5 \pm 4.8 \times 10^{-3}$	n.r.	[20]
Ginger	rhizome	dUC/gUC	NVs	0.890	4.2×10^{12}	[23]
Ginger	rhizome	dUC	NVs	n.r.	$0.5 - 2 \times 10^{14}$	[17]
Ginger	rhizome	dUC/gUC	NVs	Three bands each containing ≅0.05	n.r.	[13]
Grape	fruit	dUC/gUC	NVs	1.76 ± 0.15	n.r.	[19]
Grapefruit	fruit	dUC/gUC	NVs	2.21 ± 0.044	n.r.	[19]
Tomatoes	fruit	dUC/gUC	NVs	0.44 ± 0.02	n.r.	[19]
Grape	fruit	dUC/gUC	NVs	n.r.	n.r.	[16]
Broccoli	flower	dUC/gUC	NVs, MVs	n.r.	n.r.	[14]
Apple	fruit	dUC	NVs	n.r.	1.6×10^{13} particles/L	[24]
Coconut	fruit	dUC/MF	NVs	n.r.	n.r.	[25]
Citrus clementina	fruit	dUC/gUC	NVs	1.67×10^{-3} g/L (protein)	1.16×10^{12} particles/L	[26]
Citrus sinensis	fruit	dUC	NVs	0.178 g/L (protein)	n.r.	[27]
Citrus paradisi	fruit	dUC	NVs	0.134 g/L (protein)	n.r.	[27]

Table 1. *Cont.*

Resource	Organ	Isolation Method	Vesicle Type(s) Isolated	Yield (g/L or g/kg of Starting Plant Material)	Particle Number (Particles/kg or Particles/L of Starting Plant Material)	Ref.
Citrus aurantium	fruit	dUC	NVs	0.161 g/L (protein)	n.r.	[27]
Citrus limon	fruit	dUC	NVs	0.409 g/L (protein)	n.r.	[27]
Citrus limon	fruit	dUC/MF/gUC	NVs	2.5×10^{-3} g/L	n.r.	[12]
Carrot	root	dUC/gUC	NVs	0.298	n.r.	[28]
Blueberry	fruit	dUC/MF	NVs	n.r.	n.r.	[29]
Hami melon	fruit	dUC/MF	NVs	n.r.	n.r.	[29]
Pea	seed	dUC/MF	NVs	n.r.	n.r.	[29]
Pear	fruit	dUC/MF	NVs	n.r.	n.r.	[29]
Soybean	seed	dUC/MF	NVs	n.r.	n.r.	[29]
Orange	fruit	dUC/MF	NVs	n.r.	n.r.	[29]
Kiwifruit	fruit	dUC/MF	NVs	n.r.	n.r.	[29]
Arabidopsis thaliana L.	leaf	dUC/gUC	EVs	n.r.	n.r.	[30]
Sunflower	seed	MF/dUC	AVs	n.r.	n.r.	[31]
Nicotiana tabacum L.	leaf	dUC	AVs	n.r.	n.r.	[32]
Vinca minor L.	leaf	dUC	AVs	n.r.	n.r.	[32]
Viscum album L.	leaf	dUC	AVs	n.r.	n.r.	[32]
Phaseolus vulgaris L.	leaf	dUC	EVs	0.081 ± 0.03	n.r.	[33]
Oryza sativa L. (Rice)	leaf	dUC	AVs	n.r.	n.r.	[34]

Abbreviations are as follows: dUC: differential ultracentrifugation, gUC: gradient ultracentrifugation, MF: microfiltration, n.r.: not reported data, and Ref.: reference. Yield refers to the weight of the vesicle-containing pellet if not stated otherwise.

EVs can be isolated in several ways. The applied methods rely on centrifugation, filtration, precipitation, and chromatography-based separations. Intense research is undertaken to make improvements in this field by optimizing traditionally used methods or by finding new ways. The ISEV community for the efficient separation of EVs suggests the gradient ultracentrifugation (gUC) method. gUC employs an inert gradient medium in which the EV-containing sample is centrifuged at high centrifugal force to reach the equilibrium isodensity zone. The method generally achieves good separation of particles of different densities and is able to separate EVs from the soluble smaller components and to resolve EV subpopulations differing in buoyant densities. There are a number of different gradient media for the separation of EVs—sucrose and iodixanol being the most popular ones. Recently, other methods, such as tangential flow filtration (TFF), [35] polymer-based precipitation, and size-exclusion chromatography (SEC) are gaining field for the isolation or second step purification of vesicles [36]. Amongst these, SEC is one of the best performing methods for EV separation/purification, especially from biological fluids [37] and in combination with ultrafiltration or ultracentrifugation. SEC uses porous beads to separate EVs from other biopolymers (proteins, polysaccharides, proteoglycans, etc.) and small molecules on the basis of their hydrodynamic volume [38]. The separation takes place during the filtration of a sample solution through a gravity or HPLC column containing the porous beads with radii smaller than the EVs [37]. SEC columns can be packed with different stationary phases including Sepharose 2B, Sepharose CL-4B, Sepharose CL-2B, and Sephacryl S-400 for the gravity-driven separation of exosome-like vesicles; however, commercially available, ready-made columns (IZON qEV and ExoSpin) with proprietary resin bed are also available and they respond well to the inter-laboratory reproducibility challenges. The separation efficiency of the column depends on the chemical composition and structure of the stationary phase [38]. SEC is relatively fast and reproducible, providing relatively high yields [37]. Moreover, SEC was shown to separate EVs from soluble smaller molecules without effecting the integrity and biological activity of the vesicles [39].

Isolation of exosome-like vesicles from complex plant matrices is very challenging. The different organs such as fruit, leaf, seed, and root have different physical structures and tissue types. As is shown in (Table 1), in the plant field, differential ultracentrifugation (dUC) is the most frequently used method today for the purification of EV-like vesicles. The main drawback of the dUC method in the isolation of NVs from the highly complex matrix is the low efficiency to separate the vesicles from the co-sedimenting broken cells, insoluble polymers from the extracellular matrix, cell wall, etc. This usually negatively influences not only the reproducibility but also the downstream analysis and their applications in biotechnology. The combination of dUC/gUC generally solves this problem and results in purer fraction than dUC alone. Thus far, dUC/gUC has only been limitedly applied in the plant field (Table 1) [40] because it is time-consuming and includes multiple washing and pelleting steps that can negatively affect the final vesicle yields. Other methods, such as polyethylene glycol precipitation, have also been employed for the purification of ginger rhizome-derived vesicles [22]. Even though the precipitation method is easy and does not require specialized equipment, it has some drawbacks such as co-purification of non-vesicular proteins and requirement of pre- and post-clean-up steps [22]. Ultrafiltration using membrane filters with defined molecular weight or size limits [25] has also been used in combination with dUC. Ultrafiltration is quick and easy but the applied force during filtration may lead to deformation and breaking of large vesicles [40].

Here, micro- (MVs) and nanovesicles (NVs) were isolated by dUC from tomato (*Solanum lycopersicum* L.) fruit. Tomato is the second most important vegetable crop after potato [41]. Tomato fruit has been extensively studied for its health benefits such as antioxidant activity associated with lycopene. Tomato fruit is a potentially high-value resource of vesicles to be applied in future functional foods. Vesicles isolated by dUC were characterized by dynamic light scattering (DLS), nanoparticle tracking analysis (NTA), transmission electron microscopy (TEM), sodium dodecyl sulphate polyacrylamide gel electrophoresis (SDS-PAGE), and thin-layer chromatography (TLC) to prove the vesicle character and the biocargo complexity of the isolates. Vesicles were further purified using two different methods: gUC, which separates the components on the basis of their buoyant density, and SEC, which separates according to their size. dUC/SEC and dUC/gUC methods were compared in their ability to purify tomato-derived vesicles in terms of yield, purity, and number of vesicles. Finally, nanoHPLC–MS/MS-based shotgun proteomics was performed on the SEC-purified vesicles to obtain information about the complexity of protein biocargo they carry.

2. Materials and Methods

2.1. Plant Material and Isolation of Vesicles by Differential Ultracentrifugation

Tomato fruits (Piccadilly) were purchased from the local market (G.M Fruit, Sicily, Italy). Tomatoes (500 g) were washed with Milli-Q water and put into boiling water for a few seconds to remove exocarp. Fruits were transferred to mixture grinder containing a 1:1 weight to volume ratio extraction buffer (pH 8) composed of 100 mM phosphate, 10 mM ethylenediamine tetraacetic acid (EDTA), and protease inhibitor cocktail (0.25 mL leupeptine (1 mg/mL), 1.25 mL 100mM phenylmethylsulfonyl fluoride (PMSF), and 0.8 mL 1M sodium azide). The sample was homogenized 3 times at maximum velocity for 10 s. Homogenized sample was subjected to the dUC protocol. Briefly, sequential low velocity centrifugations were performed at $400\times g$, $800\times g$, and $2000\times g$ using a swinging-bucket rotor for 30 minutes for each step at 22 °C. Supernatant was centrifuged at $15,000\times g$ in a fixed-angle rotor for 30 minutes at 22 °C to collect the pellet containing the microvesicle (MV) fraction. The supernatant was ultracentrifuged at $100,000\times g$ for 120 minutes at 4 °C using a SW28 Beckman rotor in a Beckman

Coulter Optima L-90K ultracentrifuge. Pellet was solubilized in a small amount of the extraction buffer and protein quantity was measured using the Qubit Protein Assay Kit (Thermo Fisher Scientific, Rockford, IL USA).

2.2. Gradient Ultracentrifugation

Gradient ultracentrifugation (gUC) was performed on NV-enriched samples isolated by dUC using a continuous sucrose gradient 8–45% (*w/v*) in a polypropylene ultracentrifugation tube (Beckman Coulter, Brea, CA, USA). Sample (5 mg measured as protein content) was dispersed in 500 μL of extraction buffer by agitation and pipetting, then centrifuged at 100,000× g for 2 h at 4 °C using an SW28Ti rotor (Beckman Coulter, Brea, CA, USA). Six fractions were collected from top to bottom as follows: Fr1, 5 mL; Fr2, 7 mL; Fr3 (Band 1) 7.5 mL; Fr4, 4.5 mL; Fr 5 (band2), 3 mL; and Fr6, 11 mL. Each fraction was washed to remove sucrose by using centrifugation at 100,000× g for 1 h at 4 °C in extraction buffer. The resulting pellets were suspended in small volumes of extraction buffer and protein quantity was measured. The experiment was performed three times using NVs from 3 different dUC extractions.

2.3. Size-Exclusion Chromatography

2.3.1. Column Packing

The gravity size-exclusion chromatography column (SEC) was prepared using 15 mL of sepharose CL-2B (GE Healthcare, Cytiva, Uppsala, Sweden) particle size 60–200 μm, pore size 100,000–20,000,000 Da, and 2% cross-linked agarose gel filtration matrix. A total of 15 mL of the sepharose CL-2B ethanol suspension were equilibrated with phosphate-buffered saline (PBS; pH 7.4), which was previously degassed and filtered with a 0.22 μm filter (Millex-GP filter, Millipore, Burlington, MA, USA). Then, 10 mL of the sepharose CL-2B was packed in a 15 mL Chromabond column with a polyethylene (PE) frit integrated at the bottom (Chromabond, Macherey-Nagel, Düren, Germany) and another PE frit at the top to allow the loading of the sample uniformly. The dimensions of the column used in this work were 1.5 cm diameter and 5.6 cm height. The void volume was 2.5 mL determined using dextran blue. Bovine serum albumin (BSA, Sigma, Burbank, CA, USA) was used to check the elution profile of the SEC column. Another column of 5 mL volume was prepared similarly. After chromatography, the column was cleaned by 10 volumes of elution buffer followed by 1 volume 1% (*v/v*) Triton, 1 volume of 0.5 M NaOH, and 10 volumes of elution buffer before reuse.

2.3.2. Size-Exclusion Chromatography of Tomato Fruit-Derived Vesicles

NVs or MVs isolated using dUC and suspended in appropriate volume of extraction buffer (i.e., 500 μL for the 10 mL column and in 250 μL for the 5 mL volume column) were loaded on the SEC column. A total of 30 fractions (500 μL volume each in the case of 10 mL column and 250 μL volume each in case of the 5 mL volume column) were collected by using extraction buffer for elution. Protein quantity in each fraction was measured using Qubit protein assay (Invitrogen, Life Technology Corporation, Eugene, OR, USA). NVs isolated by dUC were mixed in a 1:1 ratio with BSA to determine the efficiency of the separation of vesicles from medium molecular mass soluble proteins.

2.4. Characterization of MVs and NVs

For the physicochemical and morphological characterization of NVs and MVs, we applied DLS, NTA, and TEM techniques.

2.4.1. Dynamic Light Scattering

Samples (0.33 mg/mL protein concentration) were centrifuged at 1000× g for 10 minutes, and were transferred by using clean pipettes into a quartz cell for DLS measurements. The intensity autocorrelation function g2(t) was measured at 20 °C by using a Brookhaven instrument BI-9000 correlator and a solid-state laser tuned at 532 nm. The autocorrelation function was fit to obtain the distribution of the diffusion coefficients by using the expression $g_2(t) - 1 = \beta \left[\int P(D) \exp(-Dq^2t) \, dD \right]^2$ and assuming a multi-peak Schultz distribution [33]. Therefore, the size distribution function, namely, the hydrodynamic diameter distribution function $P(D_h)$, was derived from the distribution of diffusion coefficients $P(D)$ by using the Stokes–Einstein relation $D_h = k_B T/(3\pi\eta D)$.

2.4.2. Nanoparticle Tracking Analysis

Particle number concentration was measured by NTA using a Nanosight NS300 (Malvern Panalytical, UK). The samples were diluted to obtain less than 100 particles per frame; then, 5×60 second measurements were performed with a moderate flow and analyzed by the built in software of the instrument (NanoSight NTA software 3.4 version 003).

2.4.3. Transmission Electron Microscope (TEM)

TEM analyses were performed on NVs and MVs isolated by dUC. Briefly, 5 µL samples at 1 µg/µL protein concentration in 0.1 M PBS (pH 7.6) were deposited onto the formvar and carbon-coated 300 mesh copper grids. After 1 minute, the droplets were removed and the grids were dried. Samples were negatively stained with 2% (*w/v*) aqueous uranyl acetate. TEM images were acquired using a Jeol JEM 1011 electron microscope operating at 60 kV and mounted with a Morada CCD camera (Olympus Soft Imaging Solutions, Münster, Germany).

2.4.4. Protein Profiling by SDS-PAGE

Protein profiles of the different vesicle isolates were obtained by SDS-PAGE. Samples (10 µg of protein based on Qubit protein assay) were electrophoretically separated under reducing conditions on a precast Novex Bolt 4–12% Bis-Tris Plus gel (Invitrogen, Carlsbad, CA, USA) using the Bolt MOPS SDS running buffer (Invitrogen, Carlsbad, CA, USA) according to the manufacturer's instructions. Gels were stained with colloidal Coomassie blue (Applichem GmbH, Darmstadt, Germany).

2.4.5. Lipid Profiling by Thin-Layer Chromatography

Total lipid extraction and TLC analysis were performed on dUC-isolated NV and MV samples. A total of 200 µg of sample (expressed in protein amount measured by the Qubit protein assay) (Invitrogen, Life Technology Corporation, Eugene, OR, USA) was mixed with 1 mL of methanol/water/chloroform (2.5:1:1) solution at −20 °C. After rigorous mixing for 1 minute, the sample was centrifuged at 15,000× *g* for 5 minutes at 4 °C. Supernatant was collected and pellet was dissolved in 0.5 mL methanol/chloroform (1:1), kept at −20 °C, and centrifuged again at 15,000× *g* for 5 minutes at 4 °C. The resulting 2 supernatants containing the lipid extracts were combined and 300 µL of water was added; then, the sample was centrifuged at 15,000× *g* for 5 minutes at 4 °C. The bottom layer was collected and dried under the stream of nitrogen. The dried sample was suspended in 50 µL of chloroform, and 10 µL was applied on the TLC plate. Lipids were separated on silica gel 60 F254 TLC plates (Merck KGaA, Darmstadt, Germany) by using chloroform/methanol/water (5/1.5/0.5, *v/v/v*) as mobile phase. After development, plate was dried at room temperature. For visualization, the plate was placed in a solution containing 10% copper sulfate (Carlo Erba, Milano, Italy), 8% phosphoric acid (Deltek, Pozzuoli, Italy), and 5% methanol (Romil, Deltek, Pozzuoli, Italy) for 10 seconds, and then the plate was placed in an oven at 150 °C for 10 minutes. Phospholipid (PL) standards, phosphatidylserine (PS), phosphatidic acid (PA), phosphatidylglycerol (PG), phosphatidylcholine (PC), and phosphatidylethanolamine (PE) from Larodan AB (Solna, Sweden) were used for the identification of the lipids. Retardation factors (RFs) of the PL standards were measured in 3 experiments and they were as follows: Rf(PS): 0.21 ± 0.01, Rf(PA): 0.30 ± 0.01, Rf(PG): 0.33 ± 0.01, Rf(PC): 0.40 ± 0.01, and Rf(PE): 0.5 ± 0.01. PLs in the NV and MV samples were tentatively identified on the basis of their RF values. Quantification of the major PLs in NVs and MVs was performed by image analysis scanning densitometry using a VersaDoc (Bio-Rad

Laboratories Inc., Munchen, Germany) imaging system in densitometry mode. Experiments were performed 5 times on each sample; mean values and standard deviations were calculated and reported.

2.4.6. Proteomic and Bioinformatics Analysis

Two biological replicates of both the MV and NV fractions isolated and purified by dUC/SEC were subjected to in-solution digestion and nanoLC–MS/MS analysis. Vesicles were lysed using 5 freeze–thaw cycles in the presence of Rapigest (Waters, Milford, MA, USA) detergent according to the manufacturer recommendations. Lysed vesicles were in-solution digested using Trypsin (Mass Spec grade, Promega Corporation, Madison, WI, USA) at a 1:100 ratio. A total of 1 μg tryptic digest was analyzed using a Dionex Ultimate 3000 nanoRSLC (Dionex, Sunnyvale, CA, USA) coupled to a Bruker Maxis II mass spectrometer (Bruker Daltonics GmbH, Bremen, Germany) via CaptiveSpray nanobooster ion source. Samples were desalted by 0.1% trifluoroacetic acid at a flow rate of 5 μL/min for 8 minutes using an Acclaim PepMap100 C-18 trap column (100 μm × 20 mm, Thermo Scientific, Sunnyvale, CA, USA). Peptides eluting from the precolumn were separated on the ACQUITY UPLC M-Class Peptide BEH C18 column (130 Å, 1.7 μm, 75 μm × 250 mm, Waters, Milford, MA, USA) at 300 nL/min flow rate at 48 °C column temperature using a linear gradient from 4% B to 50% B in 120 minutes. Solvent A was 0.1% formic acid, solvent B was acetonitrile with 0.1% formic acid. The cycle time for data-dependent acquisition, was 2.5 s. MS spectra were acquired at 3 Hz, while MS/MS spectra were acquired at 4 or 16 Hz, depending on the intensity of the precursor ion. Singly charged ions were excluded from the analysis. The default peak-picking settings were used to process the raw MS files in MaxQuant version 1.5.3.30 for label-free quantitation. Peptide identifications were performed within MaxQuant using its built-in Andromeda search engine. During the Andromeda search, proteins were examined against a focused database. The focused database was created following a Byonic (v3.6.0, Protein Metrics Inc, Cupertino, CA, USA) search on the merged mgf files against the *Solanum lycopersicum* L. taxonomy (Tax ID: 20161018) using loose criteria (20 ppm precursor and fragment ion mass tolerance, 2% false discovery rate (FDR), two missed cleavages, cysteine as fixed modification, with methionine oxidation and asparagine and glutamine deamidation as variable modifications). In terms of creating the fasta file for the Andromeda search, only active proteins remained, decreasing the number of identified proteins from around 1000 to 200. Only proteins with at least 2 identified peptides were accepted. For the list of active proteins identified and MaxQuant LFQ values (average of the 2 biological replicates), see Table S1. The Gene Ontology (GO) annotation of the top 20 proteins was performed with Perseus v1.6 using the publicly available *Solanum lycopersycum* L. database. The comparison of the protein content of the different types of vesicles was performed and visualized by the Venny 2.1.0 web app (https://bioinfogp.cnb.csic.es/tools/venny/).

3. Results and Discussion

MVs and NVs were isolated from tomato fruit (Piccadilly variety) as red color pellets obtained by dUC after the 15,000× g and 100,000× g centrifugation steps, respectively (Figure 1). The isolates (Figure 2A) were suspended in the extraction buffer and further purified by (i) gUC using sucrose gradient or (ii) SEC using Sepharose CL-2B-packed SEC columns (Figure 1). Since SEC has not yet been applied to the purification of plant-derived vesicles, we herein aimed to compare the performance of SEC to the more frequently used gUC method.

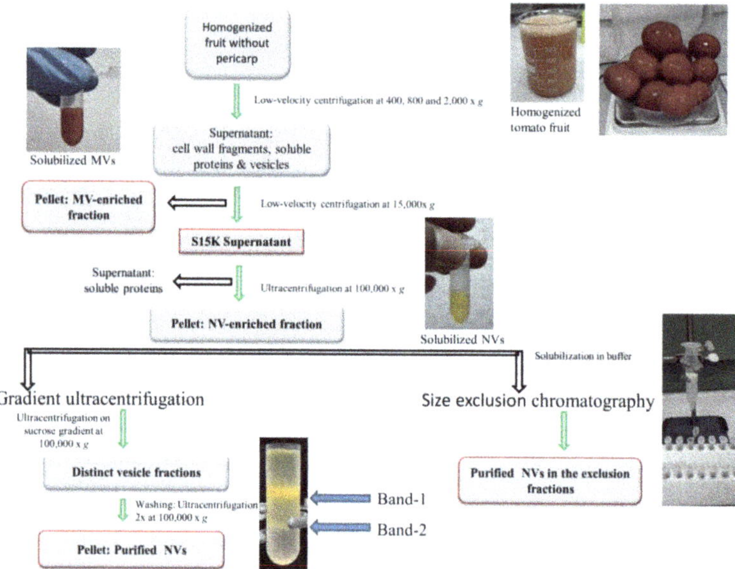

Figure 1. Schematic of the workflow used for the isolation and purification of microvesicles (MVs) and nanovesicles (NVs) from tomato fruit. Differential ultracentrifugation (dUC) was used to prepare samples enriched in MVs and NVs. Further separation and purification of dUC-isolated samples were performed by gradient ultracentrifugation (gUC) or size-exclusion chromatography (SEC). gUC resulted in two visible bands indicated as Band-1 and Band-2 in the insert. A total of 30 SEC fractions were collected in each sample.

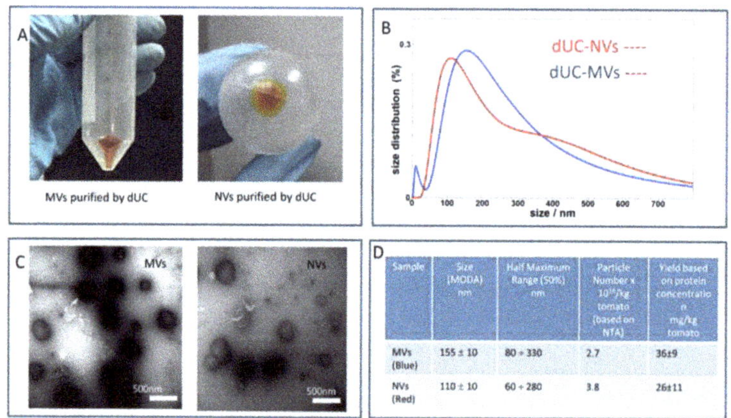

Sample	Size (MODA) nm	Half Maximum Range (50%) nm	Particle Number × 10¹⁶/kg tomato (based on NTA)	Yield based on protein concentration n mg/kg tomato
MVs (Blue)	155 ± 10	80 ÷ 330	2.7	36±9
NVs (Red)	110 ± 10	60 ÷ 280	3.8	26±11

Figure 2. Physical and morphological characteristics and yields of isolated microvesicles (MVs) and nanovesicles (NVs). (**A**) The 15,000× *g* (left image, MVs) and 100,000× *g* centrifugation steps (right image, NVs); (**B**) size distribution of crude MVs (blue line) and NVs (red line) measured in dynamic light scattering (DLS) experiments; (**C**) transmission electron microscopy (TEM) images of MVs and NVs; and (**D**) summary of the results on yield expressed both in protein amounts measured by the Qubit assay and particle numbers measured by nanoparticle tracking analysis (NTA) for kilogram of tomato fruit, and the size (MODA) measured by DLS.

3.1. dUC Isolation and Characterization of Tomato-Derived MVs and NVs

Figure 2 shows the physical characteristics of the crude vesicles obtained by dUC such as size distribution determined by DLS (Figure 2B); vesicle-like morphology acquired by TEM (Figure 2C); and particle and protein concentrations measured by NTA and the Qubit assay, respectively (Figure 2D). In the region below 200 nm, DLS showed size distributions peaked at 110 ± 10 nm and 155 ± 10 nm for NVs and MVs, respectively. In general, the size distributions of MVs and NVs were found to be quite similar. While NVs are on average smaller than MVs, one can notice the presence of a larger size vesicle population in NV samples as well as a population of smaller vesicles in the MV samples. In the case of MVs, we were able to clearly observe a population of very small sized objects, likely freely diffusible proteins (Figure 2B). TEM provided images of objects of vesicle appearance in the size range of 50–500 nm in both samples (Figure 2C). DLS, NTA, and TEM analyses confirmed the presence of heterogeneous vesicle-like objects with different shapes and broad size distribution that are typical characteristics of this kind of complex sample. The presence of small-sized vesicles in the NVs but not in the MVs was expected. In the MVs samples, this could have been due to their co-sedimentation during the $15,000 \times g$ centrifugation step or to the lysis of the larger vesicle containing organelles such as the chromoplast that are in high quantity in tomato fruit preparations. To determine protein concentration, we used two different assays, the micro bicinchoninic acid assay (BCA assay), which is widely used in EV research, and the Qubit assay. The latter utilizes target-selective dyes that emit fluorescence when bound to proteins. We found that the BCA assay, which is widely used in the determination of EV and NV protein concentration, overestimated the protein concentration in tomato-derived MVs and NVs samples by roughly one order of magnitude in comparison with the Qubit assay and on the basis of the protein profiles obtained by SDS-PAGE using Coomassie blue staining (not shown). Therefore, we show the results of Qubit assay throughout this work. The quantity of vesicles (based on protein concentration) isolated from one kilogram of tomato fruit and measured in three independent extractions (Table in Figure 2D) were 35.6 ± 8.6 mg proteins for MVs and 25.8 ± 11.4 mg proteins for NVs. The particle concentrations measured by NTA were 2.7×10^{16} particles for MVs and 3.8×10^{16} particles for NVs per 1 kg of tomato fruit. Ratio of particle to protein can be used as a mean of comparing sample purity [34], and these were 8.5×10^{11} vesicles/µg of protein and 2×10^{12} vesicles/µg of protein for MVs and NVs, respectively. This value for mammalian cell-derived EVs is usually in the range of 10^9. The higher particle-to-protein ratios for MVs and NVs may indicate that tomato vesicles contain less protein per vesicles than mammalian-cell derived EVs.

SDS-PAGE (Figure 3A) and TLC analyses (Figure 3B) confirmed the complex protein and lipid contents, respectively, a typical characteristic of phospholipid-enclosed vesicles. The SDS-PAGE protein profiles (Figure 3A) of MVs and NVs were similar but not identical. In both samples, we could observe numerous bands distributed evenly throughout the gel. PLs are the major lipids in plant-derived vesicles. Here, we extracted lipids from 200 µg of vesicles (expressed in protein content) from MVs and NVs and analyzed their PS, PA, PG, PC, and PE contents on the basis of their Rf and intensities values measured in the TLC densitometry analyses in five replicates (Figure 3B). The PL profiles of tomato-derived MVs and NVs were similar. Relative and absolute amount of PLs are reported in Figure 3B. NVs contained more PLs (30 µg) than MVs (20 µg). Similarly to ginger- [20] and grape-derived vesicles [16], the most abundant PL component was PA in both MVs (7.4 ± 0.5 µg) and NVs (10.3 ± 0.4 µg). A more than four times higher relative amount of PS was measured in NVs (4.8 ± 0.4 µg) in comparison with MVs (1 ± 0.7 µg). At high Rf, values other than phospholipids were also observed (Figure 3B left side image) on the TLC plate.

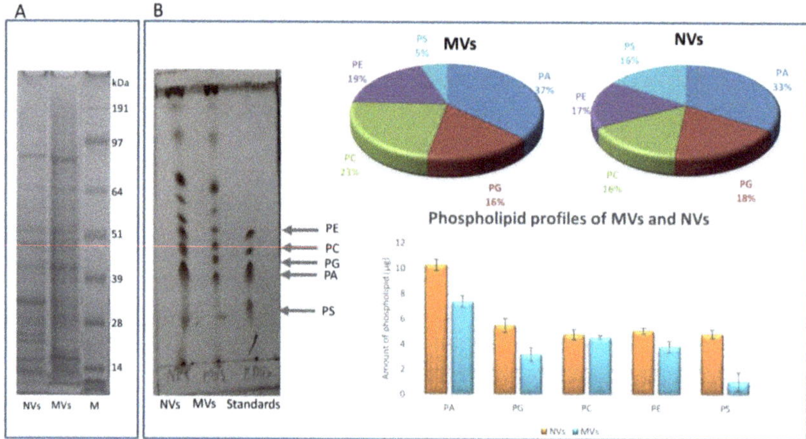

Figure 3. Tomato fruit-derived microvesicles (MVs) and nanovesicles (NVs) showed complex protein and phospholipid profiles. Proteins and lipids were extracted from MV and NV samples isolated by the differential centrifugation (dUC) method. Representative images of (**A**) protein and (**B**) phospholipid (PL) profiles. Phosphatidylserine (PS), phosphatidic acid (PA), phosphatidylglycerol (PG), phosphatidylcholine (PC), and phosphatidylethanolamine (PE) were used as standards for the tentative identification of PLs in 200 μg (expressed as protein quantity measured in the Qubit assay) of NVs and MVs (panel (**B**) left image). Relative (upper right image in panel (**B**)) and absolute PL quantitates (lower right image in panel (**B**)) were determined by densitometric image analysis scanning.

3.2. Gradient Ultracentrifugation of NVs Isolated by dUC

The NV sample was subjected to gUC using a stepwise 8/30/45% (*w/v*) sucrose gradient to remove co-purifying impurities associated with the vesicles and to separate the crude NVs into discrete EV-like fractions on the basis of density in three extraction experiments using different starting NV samples. Densities of mammalian EVs in sucrose were in the range of 1.13–1.19 g/mL that is obtainable at the 30–40% (*w/v*) sucrose concentration range. The protocol applied is similar to that previously used for ginseng vesicles [21]. Six fractions were collected from top to bottom. Two of the fractions contained visible bands: B1 (corresponds to Fraction 3) at low-density and B2 (corresponds to Fraction 5 in Figures 1 and 4A) at high density. SDS-PAGE protein profiles of the crude NV isolated by dUC and the density-separated six fractions are shown in Figure 4B.

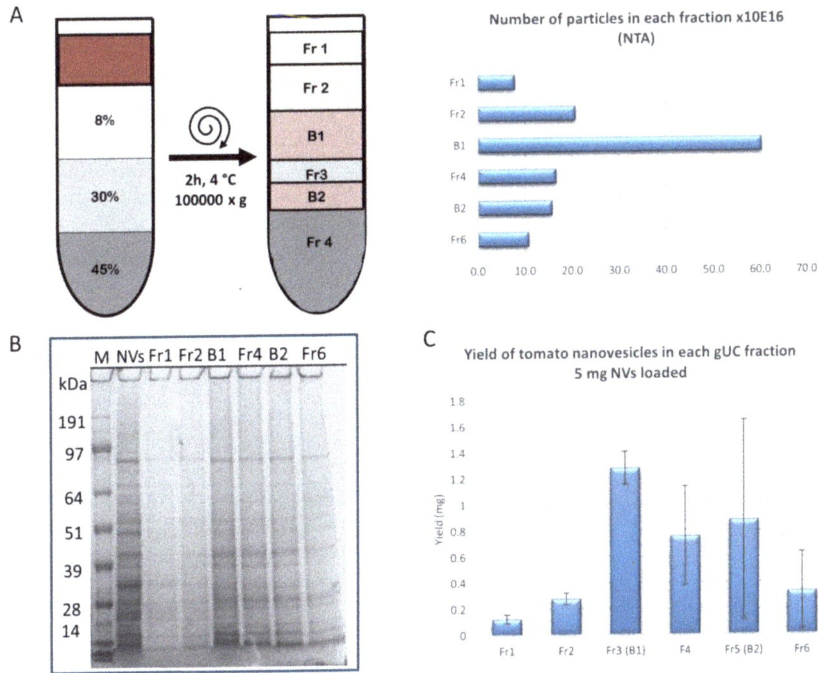

Figure 4. Tomato fruit-derived nanovesicles (NVs) isolated and separated by the differential ultracentrifugation (dUC) method followed by gradient ultracentrifugation (gUC) using 8%/30%/45% (w/v) stepwise sucrose gradient. dUC/gUC after 2 hours of centrifugation at 100,000× *g* yielded two visible bands, as shown in Figure 1: band 1 (B1) and band 2 (B2). Six fractions (F1, F2, B1, F3, B2, and F6) were collected and analyzed. (**A**) Results of nanoparticle tracking analysis (NTA) showing the number of particles in each fraction, (**B**) proteins of each fraction were separated by sodium dodecyl sulphate polyacrylamide gel electrophoresis (SDS-PAGE), and (**C**) yield expressed in protein amount calculated on the basis of Qubit assay in each fraction.

SDS-PAGE protein profiles of the gUC-separated NV fractions showed less background level and more resolved protein bands when compared to the dUC NVs sample. NTA confirmed the presence of particles in each gUC fraction. Distribution of particle numbers in the six fractions is shown in (Figure 4A). The highest number of particles were concentrated in B1. The separation of dUC-isolated NV sample into discrete vesicle populations indicates that this isolate contained vesicles not only of different shapes (as confirmed by TEM) and sizes (as confirmed by DLS) but also of different densities. Protein quantity obtained in each fraction is shown in Figure 4C. This dataset is related to the same experiment when NTA (Figure 4A) and SDS-PAGE (Figure 4B) were performed. The mean (*n* = 3) yields for the two visible bands were as follows: B1: 1.28 ± 0.13 mg, and B2: 0.88 ± 0.77 mg. The sum of these two bands corresponded to 43.2% of the total protein amount loaded (5 mg). We found that the relative protein amounts in these two bands were highly variable in the different experiments, and probably depended on the quality of the fruit used for the extraction.

3.3. SEC Purification of MVs and NVs Isolated by dUC

SEC was performed using a gravity column packed with Sepharose Cl-2B (Figure 1) and equilibrated with the extraction buffer to purify the crude NVs and MVs isolated by dUC. Two columns with bed volumes of 10 and 5 mL were packed to evaluate the influence of column bed volume on the separation and loading capacity. The effect of exogenously added soluble proteins on separation efficiency and the presence of vesicles in the supernatant obtained in the 15,000× g centrifugation step (Figure 1) were also studied.

MVs or NVs isolated by dUC were loaded on the SEC column and 30 volume fractions (0.5 mL) were collected. Vesicles and large molecules cannot enter the pores and are eluted in the column's void volume. Protein concentrations and particle numbers were measured and evaluated in each fraction. We obtained chromatograms with two broad peaks in both MVs and NVs (Figure 5A,B)—the first most abundant peak was eluted at fractions 4–6, and a second smaller and broader peak at fractions 15–17 (Figure 5C). The highest protein concentration was measured at fraction 5. These elution profiles indicate that most of the proteins were associated with the vesicles, and only a smaller portion of the signal could be attributed to the co-purifying proteins in the sample.

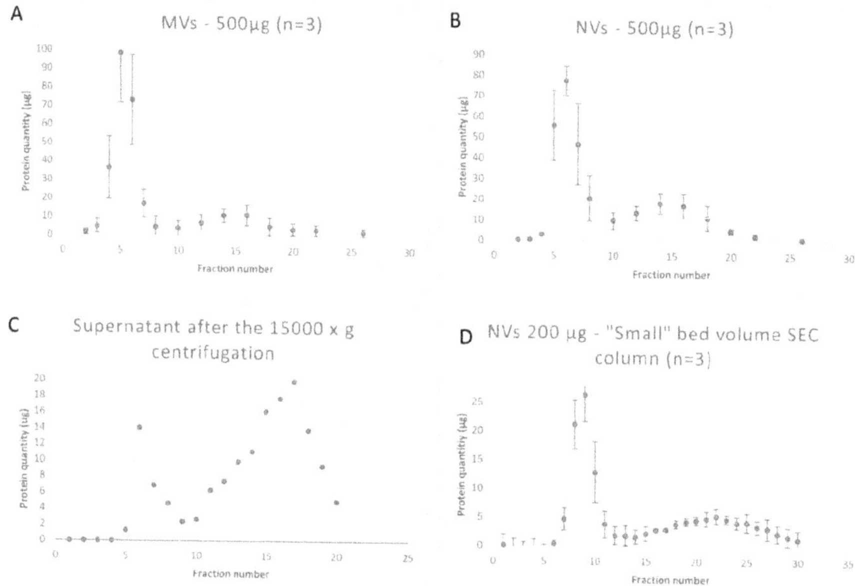

Figure 5. Tomato fruit-derived microvesicles (MVs) and nanovesicles (NVs) purified by size-exclusion chromatography (SEC); 30 fractions were collected in each sample. SEC chromatograms show the protein quantities determined in each fraction by the Qubit assay. (**A**) MVs pelleted after the 15,000× g centrifugation step; (**B**) NVs obtained as pellet after the 100,000× g centrifugation step; (**C**) supernatant after 15,000× g centrifugation step using a 10 mL bed volume SEC gravity column; (**D**) 5 mL bed volume columns were used to separate 200 µg of NVs.

The performance of SEC column was tested at varying loading quantities in the range of 50–500 µg. (Figure 6A) shows a linearly increasing yield with the increasing loading amounts in the studied range. Further, to prove the utility of SEC in removing exogenous soluble proteins from the vesicle samples, we performed experiments in which 200 µg bovine serum albumin (BSA) alone or mixed 1:1 with NVs were loaded. BSA in these experiments represented the soluble protein exogenously added to the vesicles. Figure 7B demonstrates that SEC efficiently separated the vesicles (fractions 4–6) from BSA, which eluted in later fractions (fractions 10–20), confirming a good efficiency in separation. SEC-based purification of NVs was performed using a smaller (5 mL) bed volume column with different quantities. One example is shown in Figure 5D when 200 µg of crude NVs were purified in three consecutive runs. Elution profiles of NVs obtained using the two columns with different bed volumes (10 mL and 5) were very similar. In one experiment, we loaded the filtered (0.45 µm followed by a 0.22

μm pore size filters) supernatant after the 15,000× *g* centrifugation step (S15k); thus, instead of the ultracentrifuge, SEC was used to isolate NVs directly. The chromatogram (Figure 5C) showed a peak (with a maximum at fraction 6) in the exclusion volume, which is typical of NVs. The calculated total amount of NVs purified by SEC from the S15K fraction was slightly lower than that obtained by the dUC/SEC purification.

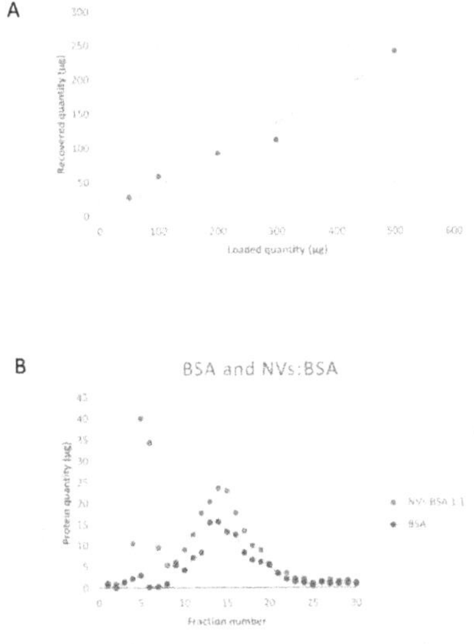

Figure 6. Performance of size-exclusion chromatography (SEC) in the purification of tomato-derived nanovesicles (NVs) isolated by differential ultracentrifugation. (**A**) Graph shows the quantities of SEC-purified NVs obtained at increasing loading quantities (50, 100, 200, 300, and 500 μg expressed in protein amounts) and (**B**) SEC chromatograms of bovine serum albumin (BSA) protein standard (in red) and a 1:1 mixture of NVs and exogenously added BSA (in blue) showing the separation efficiency of the SEC.

Fractions 4, 5, and 6 were combined and analyzed by DLS, NTA (Figure 7), SDS-PAGE, and MS-based proteomics (Figure 8). Figure 7A shows the intensity average distribution measured by DLS with a maximum (or mode) at 110 ± 10 nm and 135 ± 10 nm for NVs and MVs, respectively. There was not much difference in the size distribution of NVs before and after SEC purification, but we observed that the size distribution of MVs was slightly changed, as we found nearly 20 nm decrease in the moda of MVs. This may have been due to the entrapment of the larger vesicles into 2% cross-linked agarose gel filtration resins. In addition, the fraction of small-size objects, likely free proteins, detected in the crude dUC-isolated MV and NV samples (Figure 2B) were removed by SEC. The Rayleigh ratios were measured in the 30 SEC fractions of NVs isolated using the small bed volume column (SEC chromatogram is shown in Figure 7B), and they were found to be proportional to the average molecular mass times the mass concentration. Thus, they directly represented the amount of particles purified in each sample. More specifically, in this case, since we were dealing on average with the same nanoparticles, the Rayleigh ratio profile as a function of sample fraction could be taken as a measure of particle concentration.

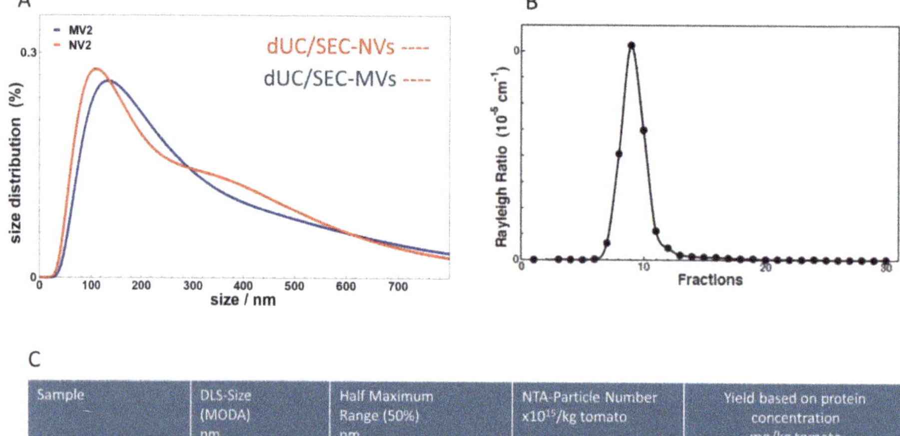

Sample	DLS-Size (MODA) nm	Half Maximum Range (50%) nm	NTA-Particle Number x10¹⁵/kg tomato	Yield based on protein concentration mg/kg tomato
SEC-MVs (blue)	135 ± 10	70 ÷ 320	0.2	13±3
SEC-NVs (Red)	110 ± 10	60 ÷ 320	2.6	7±1.5

Figure 7. Physical characteristics and yields of microvesicles (MVs) and nanovesicles (NVs) isolated using the differential ultracentrifugation (dUC) method and purified by size-exclusion chromatography (SEC). (**A**) Size distribution measured by dynamic light scattering (DLS) in the pooled 4–6 SEC fractions. (**B**) Rayleigh ratio measured in each SEC fraction and (**C**) DLS moda, nanoparticle tracking analysis (NTA) particle size, and protein yields measured in the pooled 4–6 SEC fractions.

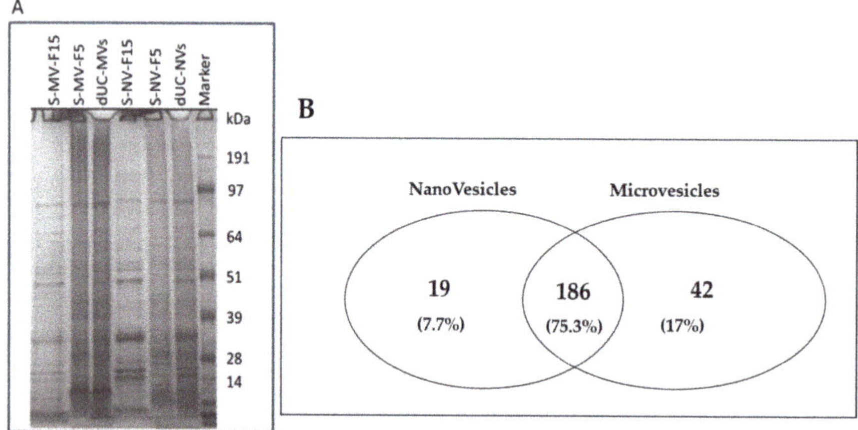

Figure 8. Protein characterization of nanovesicles (NVs) and microvesicles (MVs) isolated by differential ultracentrifugation (dUC) and purified using size-exclusion chromatography (SEC). (**A**) Sodium dodecyl sulphate polyacrylamide gel electrophoresis (SDS-PAGE) image shows the protein profiles of two SEC-separated fractions (F5 and F15) of NVs and MVs and (**B**) Venn diagram of the proteins identified in the proteomics study (refer to Table 2 and Table S1 for the full set of proteins identified and quantified) showing a high level of similarity between the identified protein sets in MVs and NVs.

From NTA data, we found 0.2×10^{15} and 2.6×10^{15} particles for 1 kg of tomato fruit, respectively (Figure 6). Regarding the yields, after SEC purification, we obtained 13.1 (MVs) and 6.9 mg vesicles (NVs) per kilogram of tomato fruit on the basis of the protein content measured. Particle to protein ratios were 1.7×10^{11} vesicles/μg of protein and 4.6×10^{11} vesicles/μg of proteins for MVs and NVs, respectively.

SEC-purified vesicles showed less complex proteins patterns than the bulk NVs and MVs in the SDS-PAGE image (Figure 8A). SEC was proven to be efficient in removing small and medium molecular mass proteins from the vesicle sample, as can be appreciated in the SDS-PAGE image of the late eluting fractions (S-MV-Fraction 15 and S-NV Fraction 15 in Figure 8A). To gain further insights into the composition of the protein biocargo of the SEC-purified vesicles, we performed proteomics analysis (Section 3.2).

3.4. Proteomic Characterization of dUC/SEC-Purified MVs and NVs

Mass spectrometry (MS)-based proteomics was performed to gain insights into the protein cargo of tomato-derived NVs and MVs. MS-based downstream analysis requires highly purified samples; therefore, only SEC-purified MVs and NVs were analyzed. Vesicle samples were lysed and digested by trypsin in-solution prior to nanoHPLC–electrospray ionization (ESI)–MS analysis. A total of 228 proteins were identified in the MV fraction and 205 proteins in the NV fraction (Table S1). Proteomic profile of the SEC-purified NVs and MVs was similar as 75.3% of the identified proteins were detected in both vesicle types (Figure 8B, Venn diagram).

Protein quantification was carried out using label-free proteomics. The 20 most abundant proteins are listed in (Table 2), together with their cellular localization. Of the 20 most abundant proteins, 16 were present in both the MVs and NVs. Most of the top-ranking proteins were enzymes. The most abundant protein in both fractions was lipoxygenase (LOX). Lipoxygenases are a family of enzymes that dioxygenate unsaturated fatty acids and thus participate in oxylipin biosynthesis. LOXs are involved in plant growth and development, ripening, and plant defense mechanisms. Plant LOXs have found some applications in the green food additives' industry as strengthening and bleaching agents [42]. Adenosine triphosphate (ATP) synthase (ATPA and ATPB) and heat shock proteins (HSP70, HSP80 and G5DGD4) have been reported to be highly expressed in other plant-derived vesicle preparations [27]. Amongst the 20 most abundant proteins, several are involved in fruit ripening, e.g., ACCH3_SOLLC, ASR1_SOLLC, and PGLR_SOLLC. Several proteins identified only in the MV sample are localized in subcellular organs such as the mitochondrion (SUCB_SOLLC, B1Q3F8_SOLLC, and Q8GT30_SOLLC). Glucan endo-1,3-beta-glucosidase B, a vacuolar protein detected in the MV sample, participates in defense against pathogens. The NVs also contained proteins that play a role in plant defense such as osmotin-like protein. Our results were compared to a recent article focusing on the proteomics analysis of chromoplast from six crop species including tomato [43]. Tomato fruit is composed of several different tissue types. The pericarp formed from the ovary wall and the placenta comprise the fleshy tissue of the tomato—mainly a pericarp, the locular tissue, and the seeds [44]. During fruit ripening, chloroplasts differentiate into photosynthetically inactive specialized plastids, called chromoplasts, which are present in relatively high amount in tomato fruit. Chromoplasts are where the colored carotenoids are synthetized and accumulate. Chromoplasts are a double lipid bilayer-surrounded structure of about 1 μm in diameter, and they are isolated using the gUC-based method. Interestingly several abundant proteins identified in chromoplasts were also detected in both the MV and NV fractions, suggesting a possible chromoplast origin of some of the vesicles in our preparations. These proteins were alcohol dehydrogenase 2, ATP synthase subunit alpha and beta, and adenosine diphosphate (ADP) adenosine triphosphate (ATP) translocator.

Table 2. List of the 20 top-ranking proteins in the microvesicle (MV) and nanovesicle (NV) samples isolated and purified by differential ultracentrifugation and size-exclusion chromatography (dUC/SEC) from tomato fruit (Figure 1). For the full set of proteins identified, refer to Table S1.

	UniProtKB	Protein Names	Typical Subcellular Location	NV Intesnsity	MV Intensity
1	Q42873_SOLLC	Lipoxygenase	cytoplasm	2.94×10^8	3.62×10^8
2	ADH2_SOLLC	Alcohol dehydrogenase 2	cytoplasm	1.45×10^8	1.47×10^8
3	ACCH3_SOLLC	1-aminocyclopropane-1-carboxylate oxidase homolog		1.23×10^8	1.16×10^8
4	ATPB_SOLLC	ATP synthase subunit beta, chloroplastic	chloroplast	1.14×10^8	6.02×10^7
5	ASR1_SOLLC	Abscisic stress-ripening protein 1	nucleus	1.04×10^8	1.48×10^8
6	PGLR_SOLLC	Polygalacturonase-2 (Pectinase)	cell wall, apoplast	1.04×10^8	2.04×10^8
7	ATPA_SOLLC	ATP synthase subunit alpha, chloroplastic	chloroplast	9.11×10^7	4.44×10^7
8	EF1A_SOLLC	Elongation factor 1-alpha	cytoplasm	8.20×10^7	9.24×10^7
9	P93767_SOLLC	ADP/ATP translocator	membrane	7.28×10^7	5.78×10^7
10	Q40140_SOLLC	Aspartic protease		6.38×10^7	3.58×10^7
11	Q9XEX8_SOLLC	Remorin 1		6.14×10^7	4.06×10^7
12	H1ZXA9_SOLLC	Heat shock protein 70 isoform 3	cytoplasm	6.09×10^7	9.74×10^7
13	HSP80_SOLLC	Heat shock cognate protein 80	cytoplasm	6.06×10^7	7.05×10^7
14	G5DGD4_SOLLC	Class I small heat shock protein		6.01×10^7	4.57×10^7
15	K4CJ46_SOLLC	2-Isopropylmalate synthase	chloroplast, cytoplasm	5.88×10^7	5.40×10^7
16	B0JEU3_SOLLC	Vicilin		5.81×10^7	4.65×10^7
17	Q4W5U7_SOLLC	Calnexin-like protein	ER	5.74×10^7	4.24×10^7
18	Q6IV07_SOLLC	UDP-arabinopyranose mutase	cytosol/Golgi	5.27×10^7	1.50×10^7
19	G8Z279_SOLLC	Hop-interacting protein THI113		4.54×10^7	4.69×10^7
20	Q38JD4_SOLLC	Temperature-induced lipocalin	cytoplasm	4.50×10^7	3.70×10^7
21	FSPM_SOLLC	Fruit-specific protein		6.51×10^6	5.78×10^7
22	O81536_SOLLC	Annexin	cytoplasm	2.92×10^7	5.76×10^7
23	CATA1_SOLLC	Catalase isozyme 1	peroxisome	2.16×10^7	4.42×10^7
24	RS27A_SOLLC	Ubiquitin-40S ribosomal protein S27a	nucleus	3.64×10^7	4.10×10^7

4. Conclusions

Conventional dUC isolates an average 0.5–1.5×10^{13} particles from 1 liter of mammalian cell culture medium or 3–4 µg protein per 10^6 cells [45]. Plant-derived EV-like vesicles can be produced at a considerably higher yield, mainly because they contain a mixture of different populations of intra- and extracellular vesicles. [27] Ginger, for example, is one of the most promising edible resources of NVs. Ginger-derived vesicles show high antinflammatory activity and they can be produced at a high yield [6,9,19,36,40]. Chen et al. isolated 0.5–2×10^{14} particles per kilogram of ginger root using the gUC method [17], which is about 10 times more than that obtainable in current mammalian production systems. Fruit and fruit juices are also valuable sources for the isolation of EV-like vesicles (Table 1).

While grapefruit and lemon have been studied by several authors, tomato fruit has been less exploited thus far (Table 1).

Recently, Wang et al. reported the isolation of tomato-derived vesicles using the dUC/gUC method and obtained 440 ± 20 mg vesicles per kilogram of fruit [25]. However, this yield was based on the measured weight of the obtained NV sample. Different to this, we herein calculated the yield on the basis of commonly used methods for the quantitation of EVs (47) i.e., protein concentration and the number of particles determined by NTA. Isolating the vesicles by the classical dUC method, we obtained exceptionally high yields with high particle to protein ratios: 2.7×10^{16} MVs (35.6 ± 8.6 mg protein) and 3.8×10^{16} (25.8 ± 11 mg protein) NVs particles (Figure 2D). We found that the number of isolated NVs was higher than that of MVs, yet the protein content was lower. This may indicate a higher presence of liposome-like vesicles in the NVs with respect to MVs. Vesicle isolates were analyzed using DLS, NTA, TEM, SDS-PAGE, and TLC. Interestingly, we found no significant differences between the size distribution, morphology, and molecular content of tomato-derived MV and NV isolates. These isolates were further purified and separated using SEC or gUC. Through SEC (Figure 1; Figures 5–7), we purified 27% of the loaded crude MVs and 37% of NVs as pure nano-sized vesicular objects. On the other hand, gUC yielded 25.6% and 17.6% of the loaded crude NVs in the two visible bands, B1 and B2, respectively (Figure 4). When comparing the performance of two methods, gUC was proven to be more useful in the separation of different vesicle populations (on the basis of buoyant density) in comparison with SEC. Instead, SEC was efficient in the removal of the co-purifying proteins and other impurities, and thus improved the quality of NV and MV preparations and enabled them for subsequent downstream analysis, such as omics (Figure 8 and Table 2) and biological characterizations. We should point out that both methods required concentrated samples for the separation. Proteomic and lipidomic analysis revealed the main proteins (lipoxygenase, alcohol dehydrogenase 2, protein E8, ATPases, and abcsistic stress protein 1) and phospholipids (PA, PS, PG, PE, and PC) in these isolates. Since edible plant-derived vesicles can be found at relatively low concentrations in highly diluted samples, their isolation enrichment by dUC, TFF, precipitation, or other means is necessary. We found SEC to be easy to perform.

Summarizing our results, we made a step forward in the purification of MVs and NVs from plant resources by introducing a SEC purification step after the dUC separation with the aim of improving the purity of the vesicle isolates for downstream applications.

Supplementary Materials: The following are available online at http://www.mdpi.com/2304-8158/9/12/1852/s1: Table S1: Details on proteomics data title.

Author Contributions: Conceptualization, G.P.; formal analysis, R.B., I.F., S.R., G.A.; data curation, R.B., M.M., S.R., L.T., S.S., T.C., G.P.; writing—original draft preparation, G.P., R.B., M.M., L.T.; writing—review and editing, G.P.; visualization, A.P.R., T.C.; supervision, G.P.; funding acquisition, G.P., R.B. All authors have read and agreed to the published version of the manuscript.

Funding: This research was funded by the European Union's Horizon 2020 research and innovation program under the Marie Skłodowska-Curie grant acronym nanoTOM, grant agreement no. 798576.

Conflicts of Interest: The authors declare no conflict of interest.

References

1. Ding, Y.; Wang, J.; Wang, J.; Stierhof, Y.-D.; Robinson, D.G.; Jiang, L. Unconventional protein secretion. *Trends Plant Sci.* **2012**, *17*, 606–615. [CrossRef] [PubMed]
2. Rutter, B.D.; Innes, R. Extracellular Vesicles Isolated from the Leaf Apoplast Carry Stress-Response Proteins. *Plant Physiol.* **2017**, *173*, 728–741. [CrossRef] [PubMed]
3. An, Q.; Van Bel, A.J.E.; Hückelhoven, R. Do Plant Cells Secrete Exosomes Derived from Multivesicular Bodies? *Plant Signal. Behav.* **2007**, *2*, 4–7. [CrossRef] [PubMed]
4. Xiao, J.; Feng, S.; Wang, X.; Long, K.; Luo, Y.; Wang, Y.; Ma, J.; Tang, Q.; Jin, L.; Li, X.; et al. Identification of exosome-like nanoparticle-derived microRNAs from 11 edible fruits and vegetables. *PeerJ* **2018**, *6*, e5186. [CrossRef] [PubMed]

5. Baldini, N.; Torreggiani, E.; Roncuzzi, L.; Perut, F.; Zini, N.; Avnet, S. Exosome-like Nanovesicles Isolated from Citrus limon L. Exert Antioxidative Effect. *Curr. Pharm. Biotechnol.* **2018**, *19*, 877–885. [CrossRef] [PubMed]

6. Stanly, C.; Moubarak, M.; Fiume, I.; Turiák, L.; Pocsfalvi, G. Membrane Transporters in Citrus clementina Fruit Juice-Derived Nanovesicles. *Int. J. Mol. Sci.* **2019**, *20*, 6205. [CrossRef]

7. Mu, J.; Zhuang, X.; Wang, Q.; Jiang, H.; Deng, Z.; Wang, B.; Zhang, L.; Kakar, S.; Jun, Y.; Miller, D.; et al. Interspecies communication between plant and mouse gut host cells through edible plant derived exosome-like nanoparticles. *Mol. Nutr. Food Res.* **2014**, *58*, 1561–1573. [CrossRef]

8. Kalarikkal, S.P.; Prasad, D.; Kasiappan, R.; Chaudhari, S.R.; Sundaram, G.M. A cost-effective polyethylene glycol-based method for the isolation of functional edible nanoparticles from ginger rhizomes. *Sci. Rep.* **2020**, *10*, 4456. [CrossRef]

9. Pocsfalvi, G.; Turiák, L.; Ambrosone, A.; Del Gaudio, P.; Puska, G.; Fiume, I.; Silvestre, T.; Vékey, K. Physiochemical and protein datasets related to citrus juice sac cells-derived nanovesicles and microvesicles. *Data Brief* **2019**, *22*, 251–254. [CrossRef]

10. Woith, E.; Melzig, M.F. Extracellular Vesicles from Fresh and Dried Plants—Simultaneous Purification and Visualization Using Gel Electrophoresis. *Int. J. Mol. Sci.* **2019**, *20*, 357. [CrossRef]

11. Fujita, D.; Arai, T.; Komori, H.; Shirasaki, Y.; Wakayama, T.; Nakanishi, T.; Tamai, I. Apple-Derived Nanoparticles Modulate Expression of Organic-Anion-Transporting Polypeptide (OATP) 2B1 in Caco-2 Cells. *Mol. Pharm.* **2018**, *15*, 5772–5780. [CrossRef] [PubMed]

12. Raimondo, S.; Naselli, F.; Fontana, S.; Monteleone, F.; Dico, A.L.; Saieva, L.; Zito, G.; Flugy, A.; Manno, M.; Di Bella, M.A.; et al. Citrus limon-derived nanovesicles inhibit cancer cell proliferation and suppress CML xenograft growth by inducing TRAIL-mediated cell death. *Oncotarget* **2015**, *6*, 19514–19527. [CrossRef] [PubMed]

13. Zhao, Z.; Yu, S.; Li, M.; Gui, X.; Li, P. Isolation of Exosome-Like Nanoparticles and Analysis of MicroRNAs Derived from Coconut Water Based on Small RNA High-Throughput Sequencing. *J. Agric. Food Chem.* **2018**, *66*, 2749–2757. [CrossRef] [PubMed]

14. Jung, J.-S.; Yang, C.; Viennois, E.; Zhang, M.; Merlin, D. Isolation, Purification, and Characterization of Ginger-derived Nanoparticles (GDNPs) from Ginger, Rhizome of Zingiber officinale. *Bio-protocol* **2019**, *9*. [CrossRef] [PubMed]

15. Zhang, M.; Viennois, E.; Xu, C.; Merlin, D. Plant derived edible nanoparticles as a new therapeutic approach against diseases. *Tissue Barriers* **2016**, *4*, e1134415. [CrossRef]

16. Ludwig, N.; Whiteside, T.L.; Reichert, T.E. Challenges in Exosome Isolation and Analysis in Health and Disease. *Int. J. Mol. Sci.* **2019**, *20*, 4684. [CrossRef]

17. Rome, S. Biological properties of plant-derived extracellular vesicles. *Food Funct.* **2019**, *10*, 529–538. [CrossRef]

18. Lee, R.; Ko, H.J.; Kim, K.; Sohn, Y.; Min, S.Y.; Kim, J.A.; Na, D.; Yeon, J.H. Anti-melanogenic effects of extracellular vesicles derived from plant leaves and stems in mouse melanoma cells and human healthy skin. *J. Extracell. Vesicles* **2020**, *9*, 1703480. [CrossRef]

19. Yang, C.; Zhang, M.; Merlin, D. Advances in plant-derived edible nanoparticle-based lipid nano-drug delivery systems as therapeutic nanomedicines. *J. Mater. Chem. B* **2018**, *6*, 1312–1321. [CrossRef]

20. Nielsen, M.E.; Feechan, A.; Böhlenius, H.; Ueda, T.; Thordal-Christensen, H. Arabidopsis ARF-GTP exchange factor, GNOM, mediates transport required for innate immunity and focal accumulation of syntaxin PEN1. *Proc. Natl. Acad. Sci. USA* **2012**, *109*, 11443–11448. [CrossRef]

21. Takov, K.; Yellon, D.M.; Davidson, S.M. Comparison of small extracellular vesicles isolated from plasma by ultracentrifugation or size-exclusion chromatography: Yield, purity and functional potential. *J. Extracell. Vesicles* **2019**, *8*, 1560809. [CrossRef] [PubMed]

22. Iravani, S.; Varma, R.S. Plant-Derived Edible Nanoparticles and miRNAs: Emerging Frontier for Therapeutics and Targeted Drug-Delivery. *ACS Sustain. Chem. Eng.* **2019**, *7*, 8055–8069. [CrossRef]

23. Konoshenko, M.Y.; Lekchnov, E.A.; Vlassov, A.V.; Laktionov, P.P. Isolation of Extracellular Vesicles: General Methodologies and Latest Trends. *BioMed Res. Int.* **2018**, *2018*, 1–27. [CrossRef] [PubMed]

24. Monguió-Tortajada, M.; Gálvez-Montón, C.; Bayes-Genis, A.; Roura, S.; Borràs, F.E. Extracellular vesicle isolation methods: Rising impact of size-exclusion chromatography. *Cell. Mol. Life Sci.* **2019**, *76*, 2369–2382. [CrossRef]

25. Wang, Q.; Zhuang, X.; Mu, J.; Deng, Z.-B.; Jiang, H.; Zhang, L.; Xiang, X.; Wang, B.; Yan, J.; Miller, D.L.; et al. Delivery of therapeutic agents by nanoparticles made of grapefruit-derived lipids. *Nat. Commun.* **2013**, *4*, 1–13. [CrossRef]

26. Akuma, P.; Okagu, O.D.; Udenigwe, C.C. Naturally Occurring Exosome Vesicles as Potential Delivery Vehicle for Bioactive Compounds. *Front. Sustain. Food Syst.* **2019**, *3*. [CrossRef]

27. Teng, Y.; Ren, Y.; Sayed, M.; Hu, X.; Lei, C.; Kumar, A.; Hutchins, E.; Mu, J.; Deng, Z.; Luo, C.; et al. Plant-Derived Exosomal MicroRNAs Shape the Gut Microbiota. *Cell Host Microbe* **2018**, *24*, 637–652.e8. [CrossRef]

28. Pocsfalvi, G.; Turiák, L.; Ambrosone, A.; Del Gaudio, P.; Puska, G.; Fiume, I.; Silvestre, T.; Vékey, K. Protein biocargo of citrus fruit-derived vesicles reveals heterogeneous transport and extracellular vesicle populations. *J. Plant Physiol.* **2018**, *229*, 111–121. [CrossRef]

29. Baldrich, P.; Rutter, B.D.; Zandkarimi, H.; Podicheti, R.; Meyers, B.C.; Innes, R.W. Plant Extracellular Vesicles Contain Diverse Small RNA Species and Are Enriched in 10- to 17-Nucleotide "Tiny" RNAs. *Plant Cell* **2019**, *31*, 315–324. [CrossRef]

30. Regente, M.; Corti-Monzón, G.; Maldonado, A.M.; Pinedo, M.; Jorrín, J.; De La Canal, L. Vesicular fractions of sunflower apoplastic fluids are associated with potential exosome marker proteins. *FEBS Lett.* **2009**, *583*, 3363–3366. [CrossRef]

31. O'Leary, B.M.; Rico, A.; McCraw, S.; Fones, H.N.; Preston, G.M. The Infiltration-centrifugation Technique for Extraction of Apoplastic Fluid from Plant Leaves Using Phaseolus vulgaris as an Example. *J. Vis. Exp.* **2014**, e52113. [CrossRef] [PubMed]

32. Nouchi, I.; Hayashi, K.; Hiradate, S.; Ishikawa, S.; Fukuoka, M.; Chen, C.P.; Kobayashi, K. Overcoming the Difficulties in Collecting Apoplastic Fluid from Rice Leaves by the Infiltration–Centrifugation method. *Plant Cell Physiol.* **2012**, *53*, 1659–1668. [CrossRef] [PubMed]

33. Schmitz, K.S.; Phillies, G.D.J. An Introduction to Dynamic Light Scattering by Macromolecules. *Phys. Today* **1991**, *44*, 66. [CrossRef]

34. Webber, J.; Clayton, A. How pure are your vesicles? *J. Extracell. Vesicles* **2013**, *2*, 19861. [CrossRef] [PubMed]

35. Wei, T.; Hibino, H.; Omura, T. Release of Rice dwarf virusfrom insect vector cells involves secretory exosomes derived from multivesicular bodies. *Commun. Integr. Biol.* **2009**, *2*, 324–326. [CrossRef] [PubMed]

36. Zhang, M.; Viennois, E.; Prasad, M.; Zhang, Y.; Wang, L.; Zhang, Z.; Han, M.K.; Xiao, B.; Xu, C.; Srinivasan, S.; et al. Edible ginger-derived nanoparticles: A novel therapeutic approach for the prevention and treatment of inflammatory bowel disease and colitis-associated cancer. *Biomaterials* **2016**, *101*, 321–340. [CrossRef] [PubMed]

37. Deng, Z.; Rong, Y.; Teng, Y.; Mu, J.; Zhuang, X.; Tseng, M.; Samykutty, A.; Zhang, L.; Yan, J.; Miller, D.; et al. Broccoli-Derived Nanoparticle Inhibits Mouse Colitis by Activating Dendritic Cell AMP-Activated Protein Kinase. *Mol. Ther.* **2017**, *25*, 1641–1654. [CrossRef]

38. Zhang, M.; Merlin, D. Curcuma Longa-Derived Nanoparticles Reduce Colitis and Promote Intestinal Wound Repair by Inactivating the NF-ÎšB Pathway. *Gastroenterology* **2017**, *152*, S567. [CrossRef]

39. Ju, S.; Mu, J.; Dokland, T.; Zhuang, X.; Wang, Q.; Jiang, H.; Xiang, X.; Deng, Z.-B.; Wang, B.; Zhang, L.; et al. Grape Exosome-like Nanoparticles Induce Intestinal Stem Cells and Protect Mice From DSS-Induced Colitis. *Mol. Ther.* **2013**, *21*, 1345–1357. [CrossRef]

40. Chen, X.; Zhou, Y.; Yu, J. Exosome-like Nanoparticles from Ginger Rhizomes Inhibited NLRP3 Inflammasome Activation. *Mol. Pharm.* **2019**, *16*, 2690–2699. [CrossRef]

41. Zhang, M.; Xiao, B.; Wang, H.; Han, M.K.; Zhang, Z.; Viennois, E.; Xu, C.; Merlin, D. Edible Ginger-derived Nano-lipids Loaded with Doxorubicin as a Novel Drug-delivery Approach for Colon Cancer Therapy. *Mol. Ther.* **2016**, *24*, 1783–1796. [CrossRef] [PubMed]

42. Baysal, T.; Demirdöven, A. Lipoxygenase in fruits and vegetables: A review. *Enzym. Microb. Technol.* **2007**, *40*, 491–496. [CrossRef]

43. Wang, Y.-Q.; Yang, Y.; Fei, Z.; Yuan, H.; Fish, T.; Thannhauser, T.W.; Mazourek, M.; Kochian, L.V.; Wang, X.; Li, L. Proteomic analysis of chromoplasts from six crop species reveals insights into chromoplast function and development. *J. Exp. Bot.* **2013**, *64*, 949–961. [CrossRef] [PubMed]

44. Montgomery, A.M.; De Clerck, Y.A.; Langley, K.E.; Reisfeld, R.A.; Mueller, B.M. Melanoma-mediated dissolution of extracellular matrix: Contribution of urokinase-dependent and metalloproteinase-dependent proteolytic pathways. *Cancer Res.* **1993**, *53*, 693–700.
45. Lobb, R.J.; Becker, M.; Wen, S.W.; Wong, C.S.F.; Wiegmans, A.P.; Leimgruber, A.; Möller, A. Optimized exosome isolation protocol for cell culture supernatant and human plasma. *J. Extracell. Vesicles* **2015**, *4*, 27031. [CrossRef] [PubMed]

Publisher's Note: MDPI stays neutral with regard to jurisdictional claims in published maps and institutional affiliations.

 *foods*

Article

Japanese Quince (*Chaenomeles japonica*) as a Potential Source of Phenols: Optimization of the Extraction Parameters and Assessment of Antiradical and Antimicrobial Activities

Ieva Urbanavičiūtė [1,*], Mindaugas Liaudanskas [2,3], Česlovas Bobinas [1], Antanas Šarkinas [4,5], Aistė Rezgienė [4] and Pranas Viskelis [1]

[1] Biochemistry and Technology Laboratory, Institute of Horticulture, Lithuanian Research Centre for Agriculture and Forestry, Kauno st.30, Babtai, LT-54333 Kaunas distr., Lithuania; ceslovas.bobinas@lammc.lt (Č.B.); biochem@lsdi.lt (P.V.)

[2] Laboratory of Pharmaceutical Science, Institute of Pharmaceutical Technologies of the Faculty of Pharmacy of Lithuanian University of Health Sciences, Sukilėlių st.13, LT-50162 Kaunas, Lithuania; mindaugas.liaudanskas@lsmuni.lt

[3] Department of Pharmacognosy, Faculty of Pharmacy, Lithuanian University of Health Sciences, Sukilėlių st.13, LT-50162 Kaunas, Lithuania

[4] Microbiological Research Laboratory, Food Institute of Kaunas University of Technology, Radvilėnų pl. 19, 50292 Kaunas, Lithuania; antanas.sarkinas@ktu.lt (A.Š.); aiste.rezgiene@ktu.lt (A.R.)

[5] Department of Food Science and Technology of Kaunas University of Technology, Radvilėnų pl. 19, LT-50254 Kaunas, Lithuania

* Correspondence: ievaurbanaviciute@yahoo.com; Tel.: +37-0-6753-5112

Received: 28 July 2020; Accepted: 14 August 2020; Published: 17 August 2020

Abstract: The value of fruits is determined by the quantity and variety of biologically active compounds they contain, and their benefits on human health. This work presents the first study of the biochemical composition and antibacterial activity of the new Japanese quince (JQ) cultivars 'Darius', 'Rondo', and 'Rasa' fruits. The total phenolic content (TPC) was determined using the Folin–Ciocalteu method and each compound was identified by HPLC High Performance Liquid Chromatography. The antimicrobial activity against three Gram-positive and three Gram-negative bacteria, and one yeast strain, was evaluated by the agar well diffusion method using three different concentrations. The free radical scavenging activity was determined using DPPH (2,2-diphenyl-1-picrylhydrazyl) and ABTS (2,2′-azino-bis-3-ethylbenzthiazoline-6-sulphonic acid) methods and ranged from 99.1 to 115.9 $\mu mol_{TE}/100$ g, and from 372 to 682 $\mu mol_{TE}/100$ g, respectively. TPC ranged from 3906 to 4550 $mg_{GAE}/100$ g, and five compounds, isoquercitrin, rutin, (+)-catechin, (−)-epicatechin, and chlorogenic acid were identified. All JQ extracts possessed antimicrobial activity against Gram-positive and Gram-negative bacteria, and *Enterococcus faecalis* (ATCC 29212) was the most sensitive strain. These results indicate that JQ fruits are a significant source of bio-compounds, which can enrich the diet with strong antioxidants, and they are very promising as a substitute for chemical preservatives in the food and cosmetic industry.

Keywords: *Chaenomeles japonica* fruit; polyphenols; antioxidants; antimicrobial activity

---

1. Introduction

Japanese quince (*Chaenomeles japonica* (Thunb.) Lindl. ex Spach) is a dwarf shrub that originated in East Asia and was used in Chinese medicine 3000 years ago [1]. Quince of the *Chaenomeles* genus is one of the oldest cultivated plants belonging to the *Rosaceae* family, a subgenus of *Maloideae* [1,2].

Studies of the biological activity of Japanese quince (JQ) fruits have revealed their great potential for human health, including growth promotion of the beneficial intestinal bacteria *Lacticaseibacillus casei* and *Lactiplantibacillus plantarum*, protective effect on the lipid membrane against free radicals, and inhibition of cyclooxygenase involved in the inflammatory reactions [3]. Other researchers have shown that extracts of JQ fruits are promising raw material for cancer treatment and prevention, due to their phenols composition and cytotoxic activity [4–7].

JQ fruit extracts have strong biological activity due to their particular biochemical composition and content of bio compounds. Due and co-authors established 24 phenolic compounds in five *Chaenomeles* species, their quantity and distribution were different only for chlorogenic acid, catechin, procyanidin B1, epicatechin, and procyanidin B2 [8]. Differences in the antioxidant activity of these five species fruits were observed in the same study [8]. Another study identified eleven phenolic compounds, which were dominated by (−)-epicatechin and procyanidin B2 [3]. Besides that, JQ fruits and their juice have a high amount of ascorbic acid (the main biologically active form of vitamin C), which acts as a biological antioxidant and can contribute to chronic disease prevention [1,9,10]. In addition, a number of dietary fibers and pectin were reported [11,12], which are beneficial in bodyweight control, and could prevent the progression of type 2 diabetes and heart diseases [13].

Phenolic compounds are a large and diverse group of molecules, in which the structural characteristics determine their biological activity. The antioxidant activity of phenols depends on the hydroxyl group number, and their configuration in B-ring [14,15]. Structural differences between phenols cause distinct mechanisms of actions against microorganisms, and consequently, their effectiveness [16].

Numerous studies have shown that the phenolic compounds are promising biologically active compounds that may act as a new type of antimicrobial agent [17–19]. Kikowska and co-authors demonstrated the antibacterial activity of JQ leaf and fruit extracts against four bacteria strains *Staphylococcus aureus* (ATCC 25923), *Escherichia coli* (ATCC 25922), *Pseudomonas aeruginosa* (ATCC 27853), and one yeast strain *Candida albicans* (ATCC 10231) [20]. The antibacterial activity of other species such as *Chaenomeles speciosa* essential oil against 10 microorganisms has been studied [21]. However, a limited number of studies have reported the antibacterial activity of *Chaenomeles japonica* species fruits.

The extraction efficiency of phenols depends on many conditions, including the solvent system, extraction time, temperature, ultrasound power, etc. [22–24]. Response surface methodology (RSM) is a convenient tool to estimate several variables and their interaction influence on total phenolic content (TPC), and optimize the extraction conditions [25,26].

Currently, the cultivation of JQ is gaining popularity in northern European countries, especially in the Baltic Sea area [27]. JQ is very diverse in plant and fruit characteristics, and its propagation by the seeds can cause morphological and biochemical heterogeneity. Breeding new cultivars change the genetic context and leads to morphological, physiological, and metabolic variations [28]. Within the project "Japanese Quince—A new European fruit crop for the production of juice, flavor, and fiber" from 1998–2001, the thornless cultivars named 'Darius', 'Rondo', and 'Rasa' were released. The differences of the bio-compounds composition in leaves and seed oils of these cultivars were studied [29–31]. Nevertheless, the biochemical composition and biological activity of their fruits have not yet been investigated.

This study aimed to optimize the phenols extraction conditions, determine the biochemical composition, antiradical, and antibacterial activity of Japanese quince cultivars 'Darius', 'Rondo', and 'Rasa', cultivated in Lithuania.

2. Materials and Methods

2.1. Plant Material

Fresh Japanese quince fruits (cvs. Darius, Rondo, and Rasa) were obtained from the garden of the Institute of Horticulture, Lithuanian Research Center for Agriculture and Forestry, Babtai (55°60′ N,

23°48′ E) Lithuania in 2018. The fruits were cut into slices, and lyophilized with a ZIRBUS sublimator 3 × 4 × 5/20 (ZIRBUS technology, Bad Grund, Germany) at the pressure of 0.01 mbar (condenser temperature, −85 °C). The slices were ground to powder by using a knife mill GRINDOMIX GM 200 (Retsh, Haan, Germany).

2.2. Maceration Extraction Method

First, 0.5 g of the powdered sample with 10 mL solvent in different concentrations (ratio 1:20, *w/v*) were mixed and left in the dark for 24 h at room temperature 22 °C. Then, the mixtures were centrifuged and filtered through a Whatman filter paper. Three different solvents (ethanol, methanol, and acetone) and three concentrations of each solvent (100%, 70%, and 50%) were used for the maceration extraction.

2.3. Ultrasound Extraction Method and Experimental Design

First, 0.5 g of the powdered sample was mixed with 10 mL 50% ethanol. The ultrasound extraction (UE) of phenolic compounds carried out using a Sonorex Digital 10 P ultrasonic bath (Bandelin Electronic GmbH & Co. KG, Berlin, Germany). Response surface methodology (RSM) was used to examine the influence of UE processing variables on phenols extraction. The impact of three factors (ultrasound power, extraction time, and temperature) on the response (TPC) was modeled according to a central composite design. Ultrasonic power ranged from 48 to 480 W and chosen according to the limitations of the ultrasonic device. The selected extraction temperature did not exceed 60 °C to avoid the degradation of compounds. The experimental design of the three-level-three-factor was composed; consisting of twenty experimental runs including six replicates at the center point. Design-Expert 7 (Stat-Ease Inc., Minneapolis, MN, USA) software was used for statistical analysis of the obtained data. The experimental results fit a first-order polynomial model to obtain the regression coefficients by Equation (1):

$$Y = \beta_0 + \beta_1 X_1 + \beta_2 X_2 - \beta_3 X_3, \tag{1}$$

where Y is the predicted response (TPC), X_1, X_2, and X_3 meet the variables namely ultrasonic power, extraction time, and temperature, respectively. The β_0, β_1, β_2, and β_3 values represent their corresponding regression coefficients.

Design-Expert 7 software was used to draw up 3D response surface plots. To estimate the statistical significance of the proposed model, Fisher's test for analysis of variance (ANOVA) was performed. Further optimized terms of the independent variables applied to approve the model and to compare predicted results to the experimental data.

2.4. Determination of Total Phenolic Content

TPC assessed spectrophotometrically using Folin–Ciocalteu reagent [32]. The total phenol content is determined by the equation (y = 10.56X + 0.0189, r^2 = 0.997) of the calibration curve of gallic acid and expressed in mg/100 g, the equivalent of gallic acid for the dry raw material. The absorbance was measured using a Genesys-10 UV/Vis spectrophotometer (Thermo Spectronic, Rochester, NY, USA), at 765 nm wavelength.

2.5. Determination of Total Proanthocyanidins Content

Spectrophotometric measurements were scored using a Genesys-10 UV/Vis spectrophotometer (Thermo Spectronic, Rochester, NY, USA). Total proanthocyanidins were determined by applying the technique described by [33]. Three mL DMCA solution (0.1% 4-dimethylamino cinnamaldehyde in methanol—HCl 9:1 *v/v*) was mixed with 20 µL of the extract. A decrease in absorbance was determined at a wavelength of 640 nm after 5 min. The concentration of condensed tannins in the extract was calculated based on a calibration curve established with catechin as a standard (calibration curve: catechin (mg/100 g) = (y − 0.0066)/3.1312), r^2 = 0.995.

2.6. Antiradical Activity

The DPPH * free radical scavenging activity was determined using the slightly modified spectrophotometric method described by [34]. Two mL DPPH (2,2-diphenyl-1-picrylhydrazyl) solution in 99.0% *v/v* ethanol was mixed with 20 µL of the extract. A decrease in absorbance was determined at a wavelength of 515 nm after storing the samples in the dark for 30 min at a ambient temperature. An ABTS + radical cation decolorization assay was applied according to the methodology described by [35]. Then, 2 mL of ABTS (2,2′-azino-bis (3-ethylbenzthiazoline-6-sulphonic acid)) solution (absorbance 0.800 ± 0.02) was mixed with 20 µL of the extract. A decrease in absorbance was measured at a wavelength of 734 nm after storing the samples in the dark for 30 min. Results were expressed in µmol of Trolox equivalents in 100 g of dry extract and were calculated based on a calibration curve established using Trolox (6-hydroxy-2,5,7,8-tetramethylchromane-2carboxylic acid).

2.7. Determination of Ascorbic Acid (Vitamin C) Content

Ascorbic acid (vitamin C) was measured by AOAC's (Association of Official Analytical Chemists) official titrimetric method (AOAC, 1990) [36].

2.8. Determination of Total Fibre Content

The total fiber content was determined using the enzymatic-gravimetric method, according to AOAC 985.29, 1997 [37].

2.9. High Performance Liquid Chromatography (HPLC) Method for the Determination of Phenolic Compounds

A Waters e2695 chromatograph equipped with a Waters 2998 photodiode array detector (Waters, Milford, MA, USA) was used for the HPLC analysis according to the methodology described by [38]. Chromatographic separations were carried out by using an YMC-Pack ODS-A (5 µm, C18, 250 × 4.6 mm i.d.) column equipped with a YMC-Triart (5 µm, C18, 10 × 3.0 mm i.d.) pre-column (YMC Europe GmbH, Dinslaken, Germany). The column operated at a constant temperature of 25 °C. The injection volume was 10 µL. The flow rate 1 mL/min, and gradient elution was used. The mobile phase consisted of solvent A-2% (*v/v*) acetic acid in water and solvent B-acetonitrile 100% (*v/v*). The following conditions of elution were applied: 0–30 min, 3–15% B; 30–45 min, 15–25% B; 45–50 min, 25–50% B; and 50–55 min, 50–95% B. The identification of the chromatographic peaks was achieved by the retention times and spectral characteristics (λ = 200–400 nm) of the eluting peaks with those of the reference compounds.

2.10. Preparation of Extracts for Antibacterial Testing

Twenty grams of freeze-dried quince fruit powder was mixed with 200 mL of 50% ethanol and extracted at the optimized condition. The extracts were filtered and dried in a rotary vacuum evaporator Büchi R-250, (Büchi Laboratortechnic, Flawil, Switzerland) to remove ethanol and later in a freeze-dryer ILShin FD 85125 (Ilshin Lab., Nam-myun, Yangju-si Gyeonggi-do, Korea) to remove the water. Dry extracts were kept in a freezer in hermetically sealed containers until used. Dry extracts were re-dissolved in 80% methanol to produce 0.5%, 1%, and 5% solutions, which were tested against microorganisms. The bacteria used in this study were stored at Micro-Bank (Pro-Lab Diagnostic, England) at −72 ± 3 °C before the start of the experiments. The bacteria were revitalized in the brain heart infusion broth (BHI, Oxoid, England) for 24 h, at the optimum temperature (30 ± 1 °C or 37 ± 1 °C). *B. subtilis* ATCC 6633 were grown on TSA (Liofilchem, Italy) agar slants for 24 h, at 30 °C. *Enterococcus faecalis* (ATCC 29212), *Staphylococcus aureus* (ATCC 25923), *Escherichia coli* (25922 ATCC), *Pseudomonas aeruginosa* (27853 ATCC), *Salmonella enterica serovar Typhimurium* (ATCC 14028) were grown on TSA agar slants for 24 h at 37 °C. *C. albicans* were grown on Sabouraud dextrose Liofilchem, (LD 610103) agar slants for 24–48 h at 25 °C.

2.11. Antimicrobial Activity Assay

The antimicrobial properties were evaluated by the agar well diffusion method according to the method described by [39]. The bacteria were grown in peptone-soy bouillon (LAB 04, LAB M) for 24 h at 37 °C. After cultivation, culture cells were mixed using a mini shaker MS 1 (Wilmington, NC, USA.) and the cell suspensions were adjusted according to McFarland nr 0.5 standard [40]. The cell suspensions of *C. albicans* were adjusted according to McFarland nr 1.0 standard. Then, 1 mL of the suspension of bacteria cells was introduced into 100 mL dissolved plate count agar *Liofilchem* (LD 610040), medium cooled to 47 °C. Then, 10 mL of the suspension was added into a 90-mm diameter Petri plate, the final concentration of cells in 1 mL was 1.5×10^6. Eight mm diameter wells in agar were filled with 50 µL of extracts. The plates were incubated overnight at 37 °C. *B. subtilis* 30 °C, *C. albicans* 25 °C, in Sabouraud dextrose agar, Liofilchem, (LD 610103). Then, the inhibition zones were measured with calipers to an accuracy of 0.5 mm. As a control in the blank sample, aqueous methanol (80%) was used.

2.12. The Statistical Methods

All the experiments repeated three times and the results were expressed as means ± SD. Data were submitted to the analysis of variance (ANOVA). Tukey's HSD (honest significant difference test) was used to evaluate the significant differences ($p \leq 0.05$) between means (multiple comparison test). The statistical analysis was performed using Statistica 10 software (StatSoft, Inc., Tulsa, OK, USA).

3. Results and Discussion

3.1. Selection of Extraction Parameters

Maceration extraction (ME) with pure acetone provided the lowest content of phenolic compounds (Table 1). The acetone has the lowest dielectric constant from the tested organic solvents, which proves that the lower the relative static permittivity, the extraction efficiency of TPC is weaker [22,23]. The extraction of TPC efficiency significantly improved, when the water concentration in acetone and ethanol increased. It has been reported in previous studies, that dual solvent systems are more efficient for TPC extraction [22,23]. In contrast, the higher water concentration with methanol had a negative impact on extraction, and the highest TPC obtained with pure methanol.

Table 1. The total phenolic compounds in Japanese quince fruit, mg/100 gDW.

Solvent	Solvents Concentration [%]		
	100	**70**	**50**
Ethanol	4409 ± 25 [c]	5104 ± 32 [b]	5256 ± 19 [a]
Methanol	5195 ± 34 [a]	4984 ± 22 [b]	4796 ± 27 [c]
Acetone	3228 ± 61 [b]	5426 ± 83 [a]	5274 ± 52 [a]

The different letters in the same line indicate statistically significant differences between the samples.

There was no significant difference between the highest TPC values, which was obtained with acetone 70% and ethanol 50%, so for all further extractions, the latter was chosen. The central composite design prepared using response surface methodology (RSM) to optimize the extraction condition of TPC (Table 2). Different combinations of parameters had a significant effect on TPC in JQ fruit extracts, ranging from 4522.6 to 6784.9 mg/100 gDW (Table 2).

Table 2. Experimental design of three-level, three-variable central composite design for ultrasound extraction phenols from quince fruit extracts.

Test Set	X_1, Ultrasonic Power (W)	X_2, Extraction Time (min)	X_3, Temperature (°C)	Total Phenols mg/100 g
1	240 (0)	40 (0)	45 (0)	5365.5
2	240 (0)	40 (0)	45 (0)	5219.2
3	480 (+1)	20 (−1)	60 (+1)	4522.6
4	240 (0)	40 (0)	60 (+1)	5047.3
5	48 (−1)	60 (+1)	60 (+1)	4851.1
6	240 (0)	20 (−1)	45 (0)	5579.9
7	48 (−1)	40 (0)	45 (0)	5830.9
8	48 (−1)	20 (−1)	60 (+1)	4729.5
9	240 (0)	40 (0)	45 (0)	5417.3
10	240 (0)	40 (0)	45 (0)	5866.3
11	240 (0)	40 (0)	45 (0)	5840.6
12	240 (0)	60 (+1)	45 (0)	5095.9
13	48 (−1)	60 (+1)	30 (−1)	6061.7
14	240 (0)	40 (0)	30 (−1)	6236.9
15	240 (0)	40 (0)	45 (0)	5007.3
16	480 (+1)	60 (+1)	60 (+1)	4785.2
17	480 (+1)	20 (−1)	30 (−1)	6435.2
18	48 (−1)	20 (−1)	30 (−1)	5515.3
19	480 (+1)	60 (+1)	30 (−1)	6016.7
20	480 (+1)	40 (0)	45 (0)	6784.9

The suitability and significance of design was evaluated using the analysis of the variance (ANOVA), shown in (Table 3). A linear relationship was considered in our analysis, with the selected model (p-value = 0.0022). The results of the analysis also showed that only temperature (X_3) had a significant effect on phenols extraction (p-value = 0.0003).

Table 3. Analysis of variance (ANOVA for response surface linear model) showing the effect of the three independent variables on the extraction efficiency of phenolic compounds from quince fruit.

Source	Sum of Squares	df	Mean Square	F Value	p-Value
Model	4.25×10^6	3	1.42×10^6	7.6519	0.0022
Ultrasonic power	2.42×10^5	1	2.42×10^5	1.30814	0.2696
Extraction time	78.961	1	78.961	4.27×10^{-4}	0.9838
Temperature	4.01×10^6	1	4.01×10^6	21.64713	0.0003
Lack of Fit	2.38×10^6	11	2.16×10^5	1.853435	0.2570
Pure Error	5.83×10^5	5	1.17×10^5		

To determine the most effective values of the variables, the three-dimensional surface plots (Figure 1), were designed according to the final predictive Equation (2), given below:

$$\text{Response (TPC)} = 5510.47 + 155.6X_1 + 2.81X_2 - 633.01X_3. \qquad (2)$$

Figure 1A shows the overall response of extraction time, and ultrasound power to the TPC. The total phenols decreased when the extraction time extended and slightly increased when the ultrasonic power got stronger. This response shows that the long use of strong ultrasound has a negative effect on phenolic compounds.

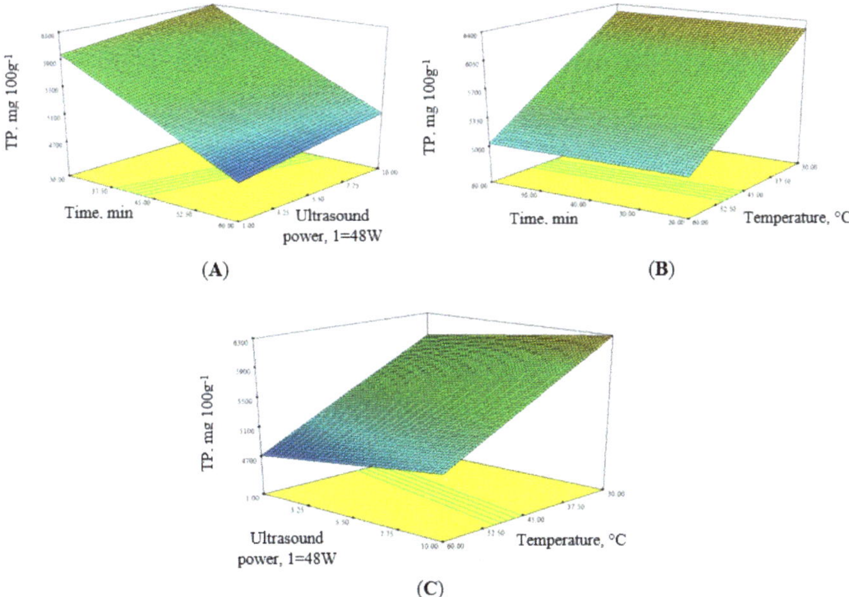

Figure 1. Response surfaces plots for total phenols (TP) in Japanese quince fruit, in the function of ultrasound power, extraction time, and temperature. (**A**)—Extraction time and ultrasound power; (**B**)—extraction time and temperature; (**C**)—extraction temperature, and ultrasound power.

Figure 1B,C confirms the analysis results that the temperature had a significant effect on the extraction efficiency, concerning both extraction time and ultrasonic strength. As the temperature raised, the phenols content decreased, as high temperatures caused degradation of compounds. To verify the model and the predicted amount of phenolic compounds (6296.3 mg/100 g), optimum extraction conditions were as follows: 20 min, at 30 °C with 480 W ultrasound power. The observed value obtained under these conditions was 6147 mg/100 g and the absolute error (AE) value 2.4%. Compared to simple maceration, ultrasonic extraction increased the phenol extraction from quince fruit by 14.5%, and the process time reduced from 24 h to 20 min.

3.2. Biochemical Composition and Antiradical Activity of Japanese Quince Fruit Extracts

The biochemical composition and antiradical activity of JQ fruits are presented in (Table 4). No significant difference was detected between cultivars using Folin–Ciocalteu assay for TPC, ranging from 3906 to 4550 mgGAE/100 gDW. Previous studies found a twice-lower TPC in JQ fruits; these differences may have been due to another extraction method, and that result was expressed in fresh weight [8].

However, the total content of proanthocyanidins (condensed tannins) differed significantly between cultivars, was lowest in 'Rondo' and highest in 'Darius' (Table 4). The total proanthocyanidins content accounts for 22–34% of the total polyphenol content. Proanthocyanidins have several types of bioactivities, e.g., antioxidant, cardio protective, neuroprotective, including antimicrobial activity [41]. The antiradical activity of the extracts with DPPH assay have shown a slight difference between cultivars and had a strong correlation with their TPC and proanthocyanidins content, r = 0.95, and 0.99, respectively. This is in agreement with previously provided studies, which demonstrated a strong correlation between the antiradical activity of berry fruits and their TPC [39,42]. A weaker correlation was found between antiradical activity by ABTS assay with TPC and proanthocyanidins content, r = 0.78, and 0.63, respectively. JQ fruits had a considerable amount of ascorbic acid (vitamin C), and appreciably similar to the previous study of nine JQ genotypes [1]. The total fiber content was not different between cultivars, averaged 30 g/100 g, and it is in agreement with the previously performed study of 12 JQ genotypes [12].

Table 4. The biochemical composition and antiradical activity of Japanese quince fruits.

Properties	Japanese Quince Cultivars		
	'Darius'	'Rondo'	'Rasa'
TPC, mgGAE/100 g	4550 ± 394 [a]	3906 ± 77 [a]	4366 ± 385 [a]
Content of proanthocyanidins, mg/100 g	1550.1 ± 31.4 [a]	879.7 ± 20.1 [c]	1233.4 ± 15.6 [b]
RSA (DPPH), μmol TE/100 gDW	115.9 ± 5.9 [a]	99.1 ± 2.2 [b]	106.5 ± 2.3 [a]
RSA (ABTS), μmol TE/100 gDW	559.7 ± 34.2 [b]	372.0 ± 5.0 [c]	681.6 ± 11.7 [a]
Ascorbic acid (vitamin C), mg/100 g	168 ± 2.1 [a]	169 ± 1.1 [a]	114 ± 2.8 [b]
Total fiber content, g/100 g	28.5 ± 2.5 [a]	28.5 ± 3.0 [a]	31.2 ± 3.2 [a]

The different letters in the same line indicate statistically significant differences between cultivars ($p < 0.05$). Results represent means ± SD ($n = 3$).

Using the HPLC method, five phenolic compounds identified: (−)-epicatechin, (+)-catechin, chlorogenic acid, rutin, and isoquercitrin (Table 5). The total flavan-3-ol (catechin and epicatechin) content accounts for around 94% of the total polyphenol content, indicating that they are the main polyphenol compounds in JQ fruits. Our results, that JQ fruits had more epicatechin, and less catechin and chlorogenic acid, coincided with previous studies [3,8].

Table 5. The quantitative composition of phenolic compounds in quince fruit, μg/g DW.

Compound, μg g^{-1} DW	Japanese Quince Cultivars		
	'Darius'	'Rondo'	'Rasa'
Isoquercitrin	38.8 ± 2.5 [a]	33.3 ± 3.2 [b]	42.4 ± 1.9 [a]
Rutin	37.8 ± 3.4 [c]	57.4 ± 2.2 [b]	66.7 ± 2.2 [a]
(+)-Catechin	131.8 ± 6.1 [c]	182.9 ± 8.3 [a]	157.9 ± 10.2 [b]
(−)-Epicatechin	3535.1 ± 60.2 [a]	3343.1 ± 55.1 [b]	3575.9 ± 50.5 [a]
Chlorogenic acid	152.9 ± 7.1 [a]	98.6 ± 6.3 [c]	113.7 ± 4.4 [b]
Total	3896.3 ± 65.2 [a]	3715.2 ± 58.1 [a]	3956.7 ± 53.7 [a]

The different letters in the same line indicate statistically significant differences between the individual compounds in the samples ($p < 0.05$).

No significant differences in TPC obtained by the HPLC method between cultivars. However, some differences were found between individual phenols, 'Rasa' had the highest amount of rutin, 'Rondo' and 'Darius' had more catechin and chlorogenic acid, respectively. There was a strong correlation between TPC using both Folin–Ciocalteu assay and HPLC method ($r = 0.87$), suggesting that TPC in JQ fruit was stable.

3.3. Antibacterial Activity

Our results show that all JQ cultivar possessed antimicrobial activity against three Gram-positive and three Gram-negative bacteria, and the most sensitive was *Enterococcus faecalis* (ATCC 29212) (Table 6). However, none of the extracts showed antifungal activity against *Candida albicans* yeast. Antibacterial activity of JQ extracts had a concentration-dependent manner, and the strongest inhibition effect was found using a 5% concentration (Table 6). As the concentration of the JQ extracts increased 10-fold, the inhibitory effect doubled. Extracts of 'Rondo' with 0.5% concentration has not inhibited the growth of *Staphylococcus aureus* (ATCC 25923), but the effect was due to the increased concentration. The lowest concentration extracts of all cultivars did not show any antibacterial effect on *Salmonella enterica serovar Typhimurium* (ATCC 14028), but 1% and 5% extracts inhibited this strain quite well.

These results confirm previous studies, that extracts with higher phenols concentrations have stronger antibacterial activity [43,44].

Table 6. Antibacterial activity of Japanese quince fruit extracts.

Microorganism	Extract Concentration, %	Japanese Quince Cultivars		
		'Rasa'	'Darius'	'Rondo'
		Inhibition Zone Size, mm		
Bacillus subtilis (ATCC 6633)	0.5	11.0 ± 0.1	11.0 ± 0.1	11.7 ± 0.5
	1	14.0 ± 0.1	12.0 ± 0.1	12.7 ± 0.4
	5	21.7 ± 0.5	18.3 ± 1.2	18.0 ± 0.1
Enterococcus faecalis (ATCC 29212)	0.5	13.0 ± 0.1	12.0 ± 0.1	17.7 ± 0.4
	1	17.0 ± 0.1	15.0 ± 0.1	20.6 ± 0.5
	5	30.7 ± 0.5	25.3 ± 1.3	27.0 ± 0.2
Staphylococcus aureus (ATCC 25923)	0.5	9.0 ± 0.1	9.0 ± 0.1	0
	1	12.0 ± 0.1	10.0 ± 0.1	13.7 ± 0.5
	5	18.7 ± 0.4	17.3 ± 1.1	18.0 ± 0.1
Escherichia coli (25922 ATCC)	0.5	9.0 ± 0.1	10.0 ± 0.1	9.7 ± 0.5
	1	13.0 ± 0.1	12.0 ± 0.1	12.6 ± 0.5
	5	19.6 ± 0.5	15.3 ± 1.1	17.0 ± 0.1
Pseudomonas aeruginosa (27853 ATCC)	0.5	9.0 ± 0.1	9.0 ± 0.1	9.6 ± 0.4
	1	13.0 ± 0.1	12.0 ± 0.1	11.7 ± 0.5
	5	19.7 ± 0.5	16.3 ± 1.1	16.0 ± 0.1
Salmonella enterica serovar Typhimurium (ATCC 14028)	0.5	0	0	0
	1	12.0 ± 0.1	12.0 ± 0.1	11.7 ± 0.5
	5	19.7 ± 0.4	17.3 ± 1.2	15.0 ± 0.1
Candida albicans (ATCC 10231)	0.5	0	0	0
	1	0	0	0
	5	0	0	0

Results represent means ± SD (*n* = 3).

The extracts of JQ cultivars more inhibited Gram-positive bacteria than Gram-negative, with very few exceptions. These differences probably depend on Gram-negative bacteria cells properties, which have an additional outer membrane with lipopolysaccharide molecules [45]. Besides, there are a number of reports that plant phenolic extracts have a stronger effect against Gram-positive bacteria [46–48]. Cultivar 'Rasa' had the strongest inhibitory activity against both Gram-positive and Gram-negative bacteria (Table 6). Interestingly, the fruit of the 'Rasa' had a significantly higher amount of rutin (Table 5). Other studies concluded that plant extracts, with higher levels of rutin, were more effective against bacteria [49,50]. In addition, rutin showed the ability to enhance the antibacterial activity of other phenols and antibiotics [51,52]. JQ fruit extracts had a stronger inhibitory effect against *E. coli* ATCC 25922 and *P. aeruginosa* ATCC 27853 than *S. aureus* ATCC 25923 [20]. The results of our study showed that only 'Rasa' extracts have similar activity, but 'Darius' and 'Rondo' extracts had the opposite effect, stronger inhibits *S. aureus* (ATCC 25923).

Our results also showed that antimicrobial activity of the extracts against bacteria was in good correlation with their rutin content (r = 0.98, 0.94, 0.92, 0.69, and 0.69 for *S. aureus*, *E. coli*, *E. faecalis*, *B. subtilis*, and *P. aeruginosa*, respectively), and epicatechin content (r = 0.94 for *Salmonella* ssp.). A previous study had reported similar results, that *E. coli* is more sensitive to rutin, and *Salmonella* to epicatechin [53]. It was noticed that the antibacterial potent of each cultivar was significantly different on all tested bacteria, with a few exceptions. The cultivars 'Rasa' and 'Darius' extracts had similar activity on *Staphylococcus aureus* and *Salmonella* ssp., and 'Rasa' and 'Darius' only on the *Salmonella* ssp.

There is no doubt that TPC determines the antibacterial activity, however, no significant differences were detected between cultivars. Differences in efficiency were probably due to the distribution of individual phenols between cultivars. The mechanisms of action of antibacterial activity are unequal for individual phenolic compounds, and their combinations [54–57]. Besides, individual bacteria have different resistance mechanisms, which are based on their biological properties [16]. Nevertheless, all of the bacteria cultures tested were sensitive to JQ fruit extracts.

4. Conclusions

All three JQ cultivars ('Rasa', 'Darius', and 'Rondo') are rich in bio-compounds and showed an important antibacterial activity against all tested bacteria. A considerable amount of phenols, vitamin C, and fiber were determined. The chemical analysis of these cultivars showed the presence of five phenolic compounds, whose main major compound detected was epicatechin. Our study results showed that JQ extracts not only have strong antiradical activity but also could effectively fight against three Gram-positive and three Gram-negative bacteria. The inhibition zone varied between different concentrations of the extracts and between bacterial strains. The food complemented with freeze-dried powder of JQ fruit can enrich the diet with strong antioxidants, which are important in maintaining human health and help to prevent various diseases. In addition, due to their antibacterial activity, they are very promising as a substitute for chemical preservatives in the food and cosmetic industry. These results are significant and quite important, especially due to growing consumer interest for natural products free of chemical additives.

Author Contributions: Conceptualization, I.U., P.V. and A.Š.; methodology, Č.B., I.U., A.R. and A.Š.; software, M.L.; validation, M.L.; investigation, I.U., P.V. and A.R.; writing—original draft preparation, I.U. and A.Š.; writing—review and editing, Č.B and P.V.; supervision, P.V. All authors have read and agreed to the published version of the manuscript.

Funding: This study was financed by the Institute of Horticulture, Lithuanian Research Centre for Agriculture and Forestry. The work is partly attributed to the technological development project through a contract with the Agency for Science, Innovation and Technology Nr.31V-5.

Conflicts of Interest: The authors declare no conflict of interest.

References

1. Bieniasz, M.; Dziedzic, E.; Kaczmarczyk, E. The effect of storage and processing on vitamin C content in Japanese quince fruit. *Folia Hortic.* **2017**, *29*, 83–93. [CrossRef]

2. Hummer, K.E.; Janick, J. Rosaceae: Taxonomy, economic importance, genomics. In *Genetics and Genomics of Rosaceae*; Springer: New York, NY, USA, 2009; pp. 1–17.

3. Strugała, P.; Cyboran-Mikołajczyk, S.; Dudra, A.; Mizgier, P.; Kucharska, A.Z.; Olejniczak, T.; Gabrielska, J. Biological Activity of Japanese quince Extract and Its Interactions with Lipids, Erythrocyte Membrane, and Human Albumin. *J. Membr. Biol.* **2016**, *249*, 393–410. [CrossRef] [PubMed]

4. Gorlach, S.; Wagner, W.; Podsędek, A.; Szewczyk, K.; Koziołkiewicz, M.; Dastych, J. Procyanidins from Japanese quince (*Chaenomeles japonica*) fruit induce apoptosis in human colon cancer Caco-2 cells in a degree of polymerization-dependent manner. *Nutr. Cancer* **2011**, *63*, 1348–1360. [CrossRef] [PubMed]

5. Lewandowska, U.; Szewczyk, K.; Owczarek, K.; Hrabec, Z.; Podsędek, A.; Koziołkiewicz, M.; Hrabec, E. Flavanols from Japanese quince (*Chaenomeles japonica*) fruit inhibit human prostate and breast cancer cell line invasiveness and cause favorable changes in Bax/Bcl-2 mRNA ratio. *Nutr. Cancer* **2013**, *65*, 273–285. [CrossRef] [PubMed]

6. Owczarek, K.; Hrabec, E.; Fichna, J.; Sosnowska, D.; Koziołkiewicz, M.; Szymański, J.; Lewandowska, U. Flavanols from Japanese quince (*Chaenomeles japonica*) fruit suppress expression of cyclooxygenase-2, metalloproteinase-9, and nuclear factor-kappaB in human colon cancer cells. *Acta Biochim. Pol.* **2017**, *64*, 567–576. [CrossRef]

7. Strek, M.; Gorlach, S.; Podsedek, A.; Sosnowska, D.; Koziolkiewicz, M.; Hrabec, Z.; Hrabec, E. Procyanidin oligomers from Japanese quince (*Chaenomeles japonica*) fruit inhibit activity of MMP-2 and MMP-9 metalloproteinases. *J. Agric. Food Chem.* **2007**, *55*, 6447–6452. [CrossRef]

8. Du, H.; Wu, J.; Li, H.; Zhong, P.-X.; Xu, Y.-J.; Li, C.-H.; Wang, L.-S. Polyphenols and triterpenes from *Chaenomeles* fruits: Chemical analysis and antioxidant activities assessment. *Food Chem.* **2013**, *141*, 4260–4268. [CrossRef]

9. Ros, J.; Laencina, J.; Hellin, P.; Jordan, M.; Vila, R.; Rumpunen, K. Characterization of juice in fruits of different *Chaenomeles* species. *Lebensm. Wiss. Technol.* **2004**, *37*, 301–307. [CrossRef]

10. Frei, B.; Birlouez-Aragon, I.; Lykkesfeldt, J. Authors' perspective: What is the optimum intake of vitamin C in humans? *Crit. Rev. Food Sci. Nutr.* **2012**, *52*, 815–829. [CrossRef]

11. Thomas, M.; Guillemin, F.; Guillon, F.; Thibault, J.F. Pectins in the fruits of Japanese quince (*Chaenomeles japonica*). *Carbohydr. Polym.* **2003**, *53*, 361–372. [CrossRef]

12. Thomas, M.; Crépeau, M.J.; Rumpunen, K.; Thibault, J.F. Dietary fibre and cell wall polysaccharides in the fruits of Japanese quince (*Chaenomeles japonica*). *LWT Food Sci. Technol.* **2000**, *33*, 124–131. [CrossRef]

13. Kendall, C.W.; Esfahani, A.; Jenkins, D.J. The link between dietary fibre and human health. *Food Hydrocoll.* **2010**, *24*, 42–48. [CrossRef]

14. Burda, S.; Oleszek, W. Antioxidant and antiradical activities of flavonoids. *J. Agric. Food Chem.* **2001**, *49*, 2774–2779. [CrossRef] [PubMed]

15. Pannala, A.S.; Chan, T.S.; O'Brien, P.J.; Rice-Evans, C.A. Flavonoid B-ring chemistry and antioxidant activity: Fast reaction kinetics. *Biochem. Biophys. Res. Commun.* **2001**, *282*, 1161–1168. [CrossRef] [PubMed]

16. Farhadi, F.; Khameneh, B.; Iranshahi, M.; Iranshahy, M. Antibacterial activity of flavonoids and their structure–activity relationship: An update review. *Phytother. Res.* **2019**, *33*, 13–40. [CrossRef]

17. Česonienė, L.; Daubaras, R.; Viškelis, P.; Šarkinas, A. Determination of the total phenolic and anthocyanin contents and antimicrobial activity of *Viburnum opulus* fruit juice. *Plant. Foods Hum. Nutr.* **2012**, *67*, 256–261. [CrossRef]

18. Viskelis, P.; Rubinskienė, M.; Jasutienė, I.; Šarkinas, A.; Daubaras, R.; Česonienė, L. Anthocyanins, antioxidative, and antimicrobial properties of American cranberry (*Vaccinium macrocarpon* Ait.) and their press cakes. *J. Food Sci.* **2009**, *74*, C157–C161. [CrossRef]

19. Puupponen-Pimiä, R.; Nohynek, L.; Alakomi, H.L.; Oksman-Caldentey, K.M. Bioactive berry compounds—Novel tools against human pathogens. *Appl. Microbiol. Biotechnol.* **2005**, *67*, 8–18. [CrossRef]

20. Kikowska, M.; Włodarczyk, A.; Rewers, M.; Sliwinska, E.; Studzińska-Sroka, E.; Witkowska-Banaszczak, E.; Thiem, B. Micropropagation of *Chaenomeles japonica*: A Step towards Production of Polyphenol-rich Extracts Showing Antioxidant and Antimicrobial Activities. *Molecules* **2019**, *24*, 1314. [CrossRef]

21. Xianfei, X.; Xiaoqiang, C.; Shunying, Z.; Guolin, Z. Chemical composition and antimicrobial activity of essential oils of *Chaenomeles speciosa* from China. *Food Chem.* **2007**, *100*, 1312–1315. [CrossRef]

22. Dent, M.; Dragović-Uzelac, V.; Penić, M. The effect of extraction solvents, temperature and time on the composition and mass fraction of polyphenols in Dalmatian wild sage (*Salvia officinalis* L.) extracts. *Food Technol. Biotechnol.* **2013**, *51*, 84–91.

23. Tan, M.C.; Tan, C.P.; Ho, C.W. Effects of extraction solvent system, time and temperature on total phenolic content of henna (*Lawsonia inermis*) stems. *Int. Food Res. J.* **2013**, *20*, 3117.

24. Aybastıer, Ö.; Işık, E.; Şahin, S.; Demir, C. Optimization of ultrasonic-assisted extraction of antioxidant compounds from blackberry leaves using response surface methodology. *Ind. Crop. Prod.* **2013**, *44*, 558–565. [CrossRef]

25. Irakli, M.; Chatzopoulou, P.; Ekateriniadou, L. Optimization of ultrasound-assisted extraction of phenolic compounds: Oleuropein, phenolic acids, phenolic alcohols and flavonoids from olive leaves and evaluation of its antioxidant activities. *Ind. Crop. Prod.* **2018**, *124*, 382–388. [CrossRef]

26. Pandey, A.; Belwal, T.; Sekar, K.C.; Bhatt, I.D.; Rawal, R.S. Optimization of ultrasonic-assisted extraction (UAE) of phenolics and antioxidant compounds from rhizomes of Rheum moorcroftianum using response surface methodology (RSM). *Ind. Crop. Prod.* **2018**, *119*, 218–225. [CrossRef]

27. Rumpunen, K. *Chaenomeles*: Potential new fruit crop for northern Europe. In *Trends in New Crops and New Uses*; ASHA Press: Alexandria, VA, USA, 2002; pp. 385–392.

28. Allam, M.; Ordás, B.; Djemel, A.; Tracy, W.F.; Revilla, P. Linkage disequilibrium between fitness QTLs and the sugary1 allele of maize. *Mol. Breed.* **2019**, *39*, 3. [CrossRef]

29. Urbanaviciute, I.; Liaudanskas, M.; Seglina, D.; Viskelis, P. Japanese Quince *Chaenomeles Japonica* (Thunb.) Lindl. ex Spach Leaves a New Source of Antioxidants for Food. *Int. J. Food Prop.* **2019**, *22*, 795–803. [CrossRef]

30. Mišina, I.; Sipeniece, E.; Rudzińska, M.; Grygier, A.; Radzimirska-Graczyk, M.; Kaufmane, E.; Górnaś, P. Associations between Oil Yield and Profile of Fatty Acids, Sterols, Squalene, Carotenoids, and Tocopherols in Seed Oil of Selected Japanese Quince Genotypes during Fruit Development. *Eur. J. Lipid Sci. Technol.* **2020**, *122*, 1900386. [CrossRef]

31. Radziejewska-Kubzdela, E.; Górnaś, P. Impact of Genotype on Carotenoids Profile in Japanese quince (*Chaenomeles japonica*) Seed Oil. *J. Am. Oil Chem. Soc.* **2020**. [CrossRef]

32. Slinkard, K.; Singleton, V.L. Total phenol analysis: Automation and comparison with manual methods. *Am. J. Enol. Vitic.* **1977**, *28*, 49–55.

33. Heil, M.; Baumann, B.; Andary, C.; Linsenmair, E.K.; McKey, D. Extraction and quantification of "condensed tannins" as a measure of plant anti-herbivore defence? Revisiting an old problem. *Naturwissenschaften* **2002**, *89*, 519–524. [CrossRef] [PubMed]

34. Brand-Williams, W.; Cuvelier, M.E.; Berset, C. Use of a free radical method to evaluate antioxidant activity. *LWT-Food Sci. Technol.* **1995**, *28*, 25–30. [CrossRef]

35. Re, R.; Pellegrini, N.; Proteggente, A.; Pannala, A.; Yang, M.; Rice-Evans, C. Antioxidant activity applying an improved ABTS radical cation decolorization assay. *Free Radic. Biol. Med.* **1999**, *26*, 1231–1237. [CrossRef]

36. AOAC. *Official Methods of Analysis of the Association of Official Analytical Chemists*, 15th ed.; AOAC: Arlington, VA, USA, 1990; pp. 1058–1059.

37. Cunniff, P. Official methods of analysis of aoac international. *J. AOAC Int.* **1997**, *80*, 127A.

38. Liaudanskas, M.; Viškelis, P.; Raudonis, R.; Kviklys, D.; Uselis, N.; Janulis, V. Phenolic composition and antioxidant activity of *Malus domestica* leaves. *Sci. World J.* **2014**. [CrossRef] [PubMed]

39. Bobinaitė, R.; Viškelis, P.; Šarkinas, A.; Venskutonis, P.R. Phytochemical composition, antioxidant and antimicrobial properties of raspberry fruit, pulp, and marc extracts. *CyTA J. Food* **2013**, *11*, 334–342. [CrossRef]

40. Hood, J.R.; Wilkinson, J.M.; Cavanagh, H.M. Evaluation of common antibacterial screening methods utilized in essential oil research. *J. Essent. Oil Res.* **2003**, *15*, 428–433. [CrossRef]

41. Tao, W.; Zhang, Y.; Shen, X.; Cao, Y.; Shi, J.; Ye, X.; Chen, S. Rethinking the mechanism of the health benefits of proanthocyanidins: Absorption, metabolism, and interaction with gut microbiota. *Compr. Rev. Food Sci. Food Saf.* **2019**, *18*, 971–985. [CrossRef]

42. Viškelis, P.; Rubinskienė, M.; Bobinaitė, R.; Dambrauskienė, E. Bioactive compounds and antioxidant activity of small fruits in Lithuania. *J. Food Agric. Environ.* **2010**, *8*, 259–263.

43. Vaquero, M.R.; Alberto, M.R.; de Nadra, M.M. Antibacterial effect of phenolic compounds from different wines. *Food Control.* **2007**, *18*, 93–101. [CrossRef]

44. Boulekbache-Makhlouf, L.; Slimani, S.; Madani, K. Total phenolic content, antioxidant and antibacterial activities of fruits of *Eucalyptus globulus* cultivated in Algeria. *Ind. Crop. Prod.* **2013**, *41*, 85–89. [CrossRef]

45. Vaara, M. Agents that increase the permeability of the outer membrane. *Microbiol. Rev.* **1992**. [CrossRef]

46. Wendakoon, C.; Calderon, P.; Gagnon, D. Evaluation of selected medicinal plants extracted in different ethanol concentrations for antibacterial activity against human pathogens. *J. Med. Act. Plants* **2012**, *1*, 60–68.

47. Tajkarimi, M.M.; Ibrahim, S.A.; Cliver, D.O. Antimicrobial herb and spice compounds in food. *Food Control.* **2010**, *21*, 1199–1218. [CrossRef]

48. Biswas, B.; Rogers, K.; McLaughlin, F.; Daniels, D.; Yadav, A. Antimicrobial activities of leaf extracts of guava (*Psidium guajava* L.) on two gram-negative and gram-positive bacteria. *Int. J. Microbiol.* **2013**. [CrossRef]

49. Basile, A.; Sorbo, S.; Giordano, S.; Ricciardi, L.; Ferrara, S.; Montesano, D.; Ferrara, L. Antibacterial and allelopathic activity of extract from *Castanea sativa* leaves. *Fitoterapia* **2000**, *71*, S110–S116. [CrossRef]

50. Radovanović, B.C.; Anđelković, S.M.; Radovanović, A.B.; Anđelković, M.Z. Antioxidant and antimicrobial activity of polyphenol extracts from wild berry fruits grown in southeast Serbia. *Trop. J. Pharm. Res.* **2013**, *12*, 813–819. [CrossRef]

51. Arima, H.; Ashida, H.; Danno, G.I. Rutin-enhanced antibacterial activities of flavonoids against *Bacillus cereus* and *Salmonella enteritidis*. *Biosci. Biotechnol. Biochem.* **2002**, *66*, 1009–1014. [CrossRef]

52. Amin, M.U.; Khurram, M.; Khattak, B.; Khan, J. Antibiotic additive and synergistic action of rutin, morin and quercetin against methicillin resistant *Staphylococcus aureus*. *BMC Complement. Altern. Med.* **2015**, *15*, 59. [CrossRef]

53. Cetin-Karaca, H.; Newman, M.C. Antimicrobial efficacy of plant phenolic compounds against *Salmonella* and *Escherichia coli*. *Food Biosci.* **2015**, *11*, 8–16. [CrossRef]

54. Nohynek, L.J.; Alakomi, H.L.; Kähkönen, M.P.; Heinonen, M.; Helander, I.M.; Oksman-Caldentey, K.M.; Puupponen-Pimiä, R.H. Berry phenolics: Antimicrobial properties and mechanisms of action against severe human pathogens. *Nutr. Cancer* **2006**, *54*, 18–32. [CrossRef] [PubMed]

55. Katalinić, V.; Možina, S.S.; Skroza, D.; Generalić, I.; Abramovič, H.; Miloš, M.; Boban, M. Polyphenolic profile, antioxidant properties and antimicrobial activity of grape skin extracts of 14 *Vitis vinifera* varieties grown in Dalmatia (Croatia). *Food Chem.* **2010**, *119*, 715–723. [CrossRef]

56. Omojate Godstime, C.; Enwa Felix, O.; Jewo Augustina, O.; Eze Christopher, O. Mechanisms of antimicrobial actions of phytochemicals against enteric pathogens—A review. *J. Pharm. Chem. Biol. Sci.* **2014**, *2*, 77–85.

57. Caillet, S.; Côté, J.; Sylvain, J.-F.; Lacroix, M. Antimicrobial effects of fractions from cranberry products on the growth of seven pathogenic bacteria. *Food Control.* **2012**, *23*, 419–428. [CrossRef]

Article

TOCOSH FLOUR (*Solanum tuberosum* L.): A Toxicological Assessment of Traditional Peruvian Fermented Potatoes

Jonas Roberto Velasco-Chong [1], Oscar Herrera-Calderón [1,*], Juan Pedro Rojas-Armas [2], Renán Dilton Hañari-Quispe [3], Linder Figueroa-Salvador [4], Gilmar Peña-Rojas [5], Vidalina Andía-Ayme [6], Ricardo Ángel Yuli-Posadas [7], Andres F. Yepes-Perez [8] and Cristian Aguilar [9]

[1] Academic Department of Pharmacology, Bromatology and Toxicology, Faculty of Pharmacy and Biochemistry, Universidad Nacional Mayor de San Marcos, Jr Puno 1002, Lima 15001, Peru; jvelasco9490@gmail.com

[2] Department of Dynamic Sciences, Faculty of Medicine, Universidad Nacional Mayor de San Marcos, Av. Miguel Grau 755, Cercado de Lima 15001, Peru; jprojasarmas@yahoo.com

[3] Clinical Pathology Laboratory, Faculty of Veterinary Medicine and Zootechnics, Universidad Nacional del Altiplano, Av Floral 1153, Puno 21001, Peru; rhanari@unap.edu.pe

[4] School of Medicine, Faculty of Health Sciences, Universidad Peruana de Ciencias Aplicadas, Prolongación Primavera 2390, Lima 15023, Peru; pcmelfig@upc.edu.pe

[5] Laboratory of Cellular and Molecular Biology, Biological Sciences Faculty, Universidad Nacional de San Cristóbal de Huamanga, Portal Independencia 57, Ayacucho 05003, Peru; gilmar.pena@unsch.edu.pe

[6] Food Microbiology Laboratory, Biological Sciences Faculty, Universidad Nacional de San Cristóbal de Huamanga, Portal Independencia 57, Ayacucho 05003, Peru; vidalina.andia@unsch.edu.pe

[7] Universidad Continental, Av San Carlos 1980, Huancayo 12000, Peru; ryuli@continental.edu.pe

[8] Chemistry of Colombian Plants, Institute of Chemistry, Faculty of Exact and Natural Sciences, University of Antioquia-UdeA, Calle 70 52–21, A.A 1226, Medellin 050010, Colombia; andresf.yepes@udea.edu.co

[9] Laboratory of Pathology, Instituto Nacional Cardiovascular, Jirón Coronel Zegarra 417, Jesús María 15072, Peru; a.crisaguilar@gmail.com

* Correspondence: oherreraca@unmsm.edu.pe; Tel.: +51-956-550-510

Received: 20 April 2020; Accepted: 29 May 2020; Published: 2 June 2020

Abstract: Potato tocosh is a naturally processed potato for nutritional and curative purposes from traditional Peruvian medicine. The aim of this study was to investigate the acute and sub-acute toxicity of tocosh flour (TF). For sub-acute toxicity, TF was administered orally to rats daily once a day for 28 days at doses of 1000 mg/kg body weight (BW). Animals were observed for general behaviors, mortality, body weight variations, and histological analysis. At the end of treatment, relative organ weights, histopathology, hematological and biochemical parameters were analyzed. For acute toxicity, TF was administered orally to mice at doses of 2000 and 5000 mg/kg BW at a single dose in both sexes. Body weight, mortality, and clinical signs were observed for 14 days after treatment. The results of acute toxicity showed that the median lethal dose (LD_{50}) value of TF is higher than 2000 g/kg BW but less than 5000 mg/Kg BW in mice. Death and toxicological symptoms were not found during the treatment. For sub-acute toxicity, we found that no-observed-adverse-effect levels (NOAEL) of TF in rats up to 1000 g/kg BW. There were statistically significant differences in body weight, and relative organ weight in the stomach and brain. No differences in hematological and biochemical parameters were observed when compared with the control group. For sub-acute toxicity, histopathological studies revealed minor abnormalities in liver and kidney tissues at doses of 5000 mg/Kg. Based on these results, TF is a traditional Peruvian medicine with high safety at up to 1000 mg/kg BW for 28 days in rats.

Keywords: *Solanum tuberosum* L.; oral toxicity; tocosh; fermented foods; traditional medicine

1. Introduction

Solanum tuberosum L. (Family: Solanaceae) is one of the most important Andean crops, cultivated along the Andean mountain range of South America and spread to other regions worldwide [1]. Over time, Andean farmers have developed frost and drought-resistant crops, which can be planted at heights greater than 3800 m above sea level (m.a.s.l.). In Peru, there are around 3800 varieties of potato and it is one of the main contributors to the world. The potato was domesticated just under 10,000 years ago; staple food crops of the ancient Peruvians not only used fresh potato but also consumed the product in the fermentation state, named tocosh [2].

Tocosh is a naturally processed potato for curative and nutritional purposes, which consists of leaving the potato in pools protected with straw or mesh near a stream for an average of six months, then it is extracted for consumption [3]. At the end of the process, the potato is reduced in size, except for its peel and it gets a very peculiar unpleasant smell. Since Inca and Pre-Incas times, the inhabitants of Ancash, Huánuco, and Junín regions have used tocosh as a medicine, in the form of flour or in its natural state to prepare *mazamorra* (Api in the Quechua language), is the best-known form of consumption (see Figure 1: Elaboration of tocosh). Tocosh flour (TF) is attributed to some beneficial properties such as gastritis, ulcers, gastro-esophagi reflux, and gastric cancer. People consume it by dissolving one teaspoonful per 100 mL in water before food as an alternative treatment. Although tocosh flour consumption is invariable, the normal dose known in traditional medicine is between 500 and 1000 mg/Kg daily (this information was taken according to an interview at the place where potato tocosh was collected).

Figure 1. Extraction processing of "potato tocosh" in Yaca district, Huanuco region, Peru.

Potato flour (PF) is characterized by its unpleasant smell, which is the first thing to be perceived, a peculiarity that does not limit its consumption or commercialization, affirming by empirical knowledge that it contains natural penicillin [4] and that among its innumerable benefits it is able to protect the gastric mucosa from damage or inflammation, according to popular customs, this product is used in postpartum, colds, pneumonia, in wound healing, as an antibacterial, healing of hemorrhoids and gastric ulcers, to avoid gastrointestinal infections and mountain sickness [5,6].

The potato species (*Solanum tuberosum* L.) are specific products consumed massively, these species present steroidal alkaloids, and when it is not well stored, can cause symptoms of poisoning such as respiratory distress, nausea, vomiting, and diarrhea related to acetylcholinesterase inhibition [7,8]. The primary steroidal glycoalkaloids in potato tubers are R-solanine and R-chaconine, being glycosylated forms of the steroidal alkaloid solanidine, these often improve the flavor of the potato [9]. The concentration of steroidal glycoalkaloids increases in response to several factors, such as injury, fungal attack, poor growing conditions, weather, and unsuitable storage conditions [10]. Nowadays, tocosh flour and derivates are sold as natural products in Peruvian markets but toxicity studies have not been reported to assess its consumption over a long period of time, which could induce any organ damage or death when there is no available data about the correct doses of administration. The main objective of this research was to evaluate the toxicological effect of tocosh flour following the guidelines for subacute and acute oral toxicity in rodents.

2. Material and Methods

2.1. Collection of Plant Material

Tocosh was collected in December 2019 from the Amarilis district, crossing the border of Yaca and Panao Pampa hamlets until reaching the Chicchuy village, Huanuco province (10°00′13″ S, 76°12′17″ W), Peru (see, Figure 1). Identification and authentication of the potato variety "walash" was used for elaborating tocosh and was carried out at the Natural History Museum of the Universidad Nacional Mayor de San Marcos (UNMSM), and a sample specimen was deposited with Ref. No. 038-USM-2020.

2.2. Preparation of the Tocosh Flour

The collected tocosh was washed in order to remove foreign matter and dust, then allowed to dry for three weeks under a shade in the place where it was obtained (−4 °C–18 °C). The dried tocosh was then pulverized using a grinder. The obtained product was named tocosh flour (TF) and stored until further use.

2.3. Phytochemical Analysis

A solution of tocosh dissolved and filtered was used to determine some phytochemicals such as alkaloids, phenols, terpenes, steroids, flavonoids, tannins, sugar, and saponins following the methodology of Herrera et al. [11]. The reaction to identify these components was done by using specific reagents for each chemical group showing any change of color or precipitation.

2.4. Experimental Animals

Balb/C albino mice and Holtzman rats of both sex were obtained from the bioterium of the National Institute of Health (Lima, Peru) with Sanitary Certificate No. 230-2019. Adult healthy male and female Holtzman albino rats (age, 12 weeks: body weight (males), 160–180 g; body weight (females), 150–170 g) were used to evaluate the sub-acute toxicity. Male and female rats were housed separately, and the selected female rats were nulliparous and non-pregnant. For acute toxicity, adult healthy male and female albino mice (age, 8 weeks: body weight (males), 30–32 g; body weight (females), 25–30 g) were used during the evaluation. These were kept in cages for approximately 15 days before the start of the study. The acclimatization of the experimental animals was carried out under environmental control conditions (12-h light/dark cycle) and temperature (22 ± 3 °C). Animals were given sterilized pellet food (National Center of Biological Products, NIH, Peru) and purified (reverse osmotic) water via a water bottle, ad libitum. All procedures were performed in reference to the institutional parameters and the guide for the care and use of laboratory animals. The protocol was presented and approved by the Ethics Committee of the Research Unit of the Faculty of Pharmacy and Biochemistry, UNMSM, (Document No 0198/FFB-UDI-2019. 11OCT2019. is certified with the REGISTRY No 010-CE-UDI-FFB-2).

2.5. Sub-Acute Toxicity at Repeated Dose for 28 Days

Sub-acute toxicity was performed in reference to the OECD 407 test guidelines [12]. Twenty rats were used and distributed in four groups. Group I (n = 5 males) and Group II (n = 5 females) named control groups, received only distilled water at repeated doses of 10 mL/Kg. Group III (n = 5 males) and IV (n = 5 females): rats received a limit dose of 1000 mg/Kg BW respectively, which according to animal safety criteria and empirical information for the consumption of tocosh by the population, which was established correspondingly to the repeated dose of 1000 mg/kg body weight, for 28 days. Each animal received a dose of tocosh flour suspension of 10 mL/kg body weight/day. The weights of each animal were recorded weekly during the treatment. After dosages of the product, the rats' body weights were measured and recorded during the test every 7 days until completion at 28 days.

On the 29th day, blood samples were collected by intracardiac puncture, under anesthesia with ethyl ether, biochemical and hematological parameters were evaluated. At the same time, all the animals were sacrificed with sodium pentobarbital (100 mg/kg) by subcutaneous route. The organs were fixed in 10% formalin for histopathology examinations.

Organs such as the heart, lungs, liver, spleen, stomach, brain, kidneys, and testes or uterus were removed immediately after sacrifice, washed with 0.9% sodium chloride, dried on filter paper, and weighed calculating the relative weights of the organs (ratio of organ weight and animal body weight (at the end of the experiment) × 100). Organs were examined for gross and/or microscopic pathology.

The biochemistry exams were performed with the Liquid Kinetic Chemistry method by using a clinical chemical analyzer brand MEDICA—EasyRA (5 Oak Park Dr, Bedford, MA 01730, USA), according to the manufacturer's specifications. The levels of aspartate aminotransferase (AST), alanine aminotransferase (ALT), alkaline phosphatase, total protein, bilirubin, total cholesterol, triglycerides, high-density lipoprotein (HDL), low-density lipoprotein (LDL), albumin, glucose, serum urea, and serum creatinine were determined.

Hematology examinations were performed by the flow cytometry method using a ZYBIO Brand Hematology Analyzer, Model Z31(Building J No. 70-1, 70-2 of Keyuan 4th Street Jiulongpo District, Chongqing Municipality 400039, China). The automated blood count (white blood cells, red blood cells, hematocrit, hemoglobin concentration, and platelet count) was evaluated.

2.6. Acute Oral Toxicity—Fixed-Dose Study Procedure

The acute oral toxicity of a fixed-dose procedure was evaluated according to the guidance of the Organization for Economic Cooperation and Development (OECD) method 420, with slight modifications in the animal selection, sex, and fasting [13]. This method grouped animals of both sexes dosing in a fixed-dose procedure using the highest doses of 2000 mg/kg and 5000 mg/kg (justified by criteria in animal welfare and related to the protection of human health based on the reference in the knowledge of empirical observation of the inhabitants of the area according to sample collection, being a product of frequent consumption in the area of the province of Huánuco).

Mice were kept with water ad libitum and were fasted for a fixed period such as 4–6 h before the administration of samples. Next, the animals were weighed and the test substance was administered. The toxicological evaluation was performed in 4 groups (n = 20). Group I (n = 5 males) and Group II (n = 5 females) received a single dose of 2000 mg/Kg body weight respectively, Group III (n = 5 males) and IV (n = 5 females): mice received a dose of 5000 mg/Kg BW, respectively. Each animal received a single dose of tocosh flour suspension at 10 mL/kg body weight. Mice were observed separately for 30 min, daily for 24 h, with rigorous observation in the first 4 h, and daily for 14 days. The individual weights of each mouse were determined before the administration of the test product and were re-calculated at the end of the 14 days.

The animal registry during the specific time of treatment was based on signs and symptoms of toxicity. The observations were recorded according to the duration of the treatment, including the specific external changes of minimal toxicity. At the end of the study, mice were sacrificed by sodium pentobarbital (100 mg/kg), immediately followed by necropsy. Organs were examined grossly for abnormal lesions.

2.7. Histopathological Analysis

Brain, heart, lung, liver, spleen. stomach, kidney, testes, and uterus of both studies (sub-acute and acute toxicity) were preserved in 10% formalin and fixed for a minimum of 24 h for a maximum of three days, dehydrated with alcohol of 70%, 96%, and absolute alcohol, the tissues were rinsed with xylol, impregnation with Paraffin, all these procedures were performed with a minimum of 1 h each. Inclusion in paraffin (formation of the paraffin block with the tissue) was performed, the cut was in a microtome, and at the last, they were sectioned at 3 μm depending on the tissue, and stained with hematoxylin and eosin (H&E). The sheets were assembled with coverslips and Entellan® (Sigma Chemical Co, St. Louis, MO, USA), which is a rapid non-aqueous mounting medium. Finally, the organ slides were examined microscopically and photographed with an optical microscope Nikon Eclipse E200 (Shinagawa Intercity Tower C, 2-15-3, Konan, Minato-ku, Tokyo 108-6290, Japan) at 40× and 200× magnification [14].

2.8. Prediction of Drug-Likeness Properties for Steroidal Glycoalkaloids: α-Solanin, α-Chaconine and Solanidine

Drug-likeness prediction along with further ADME properties presents wide of opportunities to evaluate a rapid prediction of chemical compounds with possible toxicological effects. The drug-like and ADME properties for the most active components of tocosh flour (constituents chemicals from *Solanum tuberosum* L., Figure 2) were screened using open-access cheminformatics platforms such as Molinspiration (for molecular weight-MW, rotatable bonds, and polar surface area-PSA descriptors), ALOGPS 2.1 (for Log Po/w descriptor) and the Pre-ADMET 2.0 to predicted four pharmaceutical relevant properties such as intestinal permeability (App. Caco-2), albumin-binding proteins (KHSA), Madin-Darby Canine Kidney (MDCK Line) cells permeation and intestinal absorption (%HIA). These parameters establish movement, permeability, absorption, and action of potential drugs [15]. The interpretation of both MDCK and Caco-2 permeability using PreADMET is as follows:

1. Permeability lower than 25: low permeability.
2. Permeability between 25 and 500: medium permeability.
3. Permeability higher 500: high permeability.

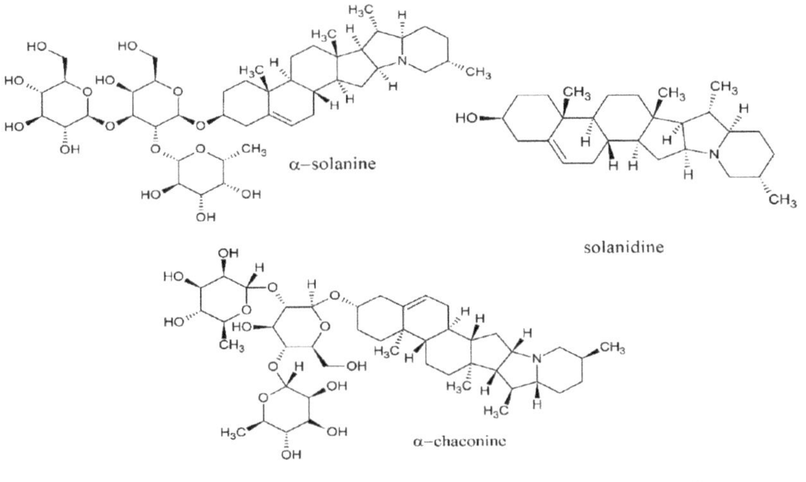

Figure 2. 2D-structures for the main alkaloids found in potato tubers [10].

2.9. Statistical Analysis

Data were recorded indicating a mean ± SD of five animals in each group and were analyzed by Student *t*-test in Graph Pad Prism software v6. The results were significant when the *p* value is less than 0.05.

3. Results

3.1. Phytochemical Analysis

The phytochemicals present in the solution of tocosh flour were sugars, phenols, alkaloids, saponins, and steroids.

3.2. Repeated Dose Toxicity Study for 28 Days

All female and male rats that received tocosh flour at doses of 1000 mg/kg/day for 28 days survived and no signs of toxicity were observed. In relative organ weights, livers of both sexes had a significant increase ($p < 0.05$), which was compared with the control group. The other organs evaluated did not have a significant difference ($p > 0.05$) compared to the control group (Table 1).

Table 1. Effect of tocosh flour solution on the relative organ weights in rats treated for 28 days.

Organs	Control Group	Tocosh Flour Dose: 1000 mg/kg
Male rats	$n = 5$	$n = 5$
Brain	0.70 ± 0.03	0.65 ± 0.01
Heart	0.35 ± 0.00	0.37 ± 0.05
Lung	0.48 ± 0.06	0.50 ± 0.15
Liver	2.96 ± 0.34	3.48 ± 0.33 *
Spleen	0.18 ± 0.04	0.26 ± 0.05
Stomach	0.76 ± 0.03	0.73 ± 0.06
Kidney	0.60 ± 0.07	0.76 ± 0.05
Testes	1.73 ± 0.05	1.77 ± 0.28
Female rats	$n = 5$	$n = 5$
Brain	0.78 ± 0.10	0.69 ± 0.05
Heart	0.40 ± 0.01	0.41 ± 0.03
Lung	0.47 ± 0.12	0.62 ± 0.04
Liver	2.83 ± 0.02	3.28 ± 0.25 *
Spleen	0.24 ± 0.04	0.26 ± 0.04
Stomach	0.62 ± 0.01	0.54 ± 0.04
Kidney	0.66 ± 0.03	0.68 ± 0.01
Uterus	0.47 ± 0.00	0.45 ± 0.06

Values are expressed as mean ± SD, compared with the control is significant when * ($p < 0.05$). Data were analyzed using the Student *t*-test. No historical control values are available.

There were no cases of death for 28 days. Microscopic examination of each organ in female and male rats showed no abnormalities due to toxicity in any of the organs, such as liver, kidney, heart, lung, stomach, brain, spleen, testes, and uterus, compared to the control group (Figure S1).

At the end of 28 days, liver enzyme levels were maintained relative to the control group in male and female rats, there was no variation in the respective biochemical parameters, only a slight variation in the significant increase in LDL cholesterol was recorded in rats and decreased triglycerides, in male rats. (Table 2). In hematological parameter, there was no variation except for a slight significant variation of the percentage of monocytes in male rats, the other results of the hematological parameters did not have significant differences with the control group (Table 3).

Table 2. Biochemical parameters of rats after administration of repeated oral doses of tocosh flour at doses of 1000 mg/kg for 28 days.

Parameters	Control Group	Tocosh Flour Dose: 1000 mg/kg
Male rats	$n = 5$	$n = 5$
AST (IU/L)	113.00 ± 4.00	108.50 ± 1.50
ALT (IU/L)	69.50 ± 3.50	70.00 ± 1.50
Alkaline Phosphatase (IU/L)	231.00 ± 9.00	238.50 ± 0.50
Total Bilirubin (mg/dL)	0.10 ± 0.00	0.10 ± 0.00
Total Protein (g/dL)	6.95 ± 0.05	7.01 ± 0.05
Albumin (g/dL)	3.80 ± 0.20	3.95 ± 0.05
Globulin (g/dL)	3.15 ± 0.25	3.00 ± 0.10
Total cholesterol (mg/dL)	59.50 ± 0.50	60.00 ± 0.00
Triglycerides (mg/dL)	134.50 ± 8.50	107.00 ± 13.00 *
HDL (mg/dL)	12.50 ± 0.50	10.60 ± 0.40
LDL (mg/dL)	20.00 ± 3.00	28.00 ± 3.00 *
Glucose (mg/dL)	100.50 ± 0.50	108.50 ± 1.50
Serum urea (mg/dL)	37.50 ± 1.50	40.50 ± 2.50
Serum creatinine (mg/dL)	0.44 ± 0.01	0.44 ± 0.03
Female rats	$n = 5$	$n = 5$
AST (IU/L)	116.50 ± 0.50	106.00 ± 0.50
ALT (IU/L)	65.50 ± 0.50	63.00 ± 1.00
Alkaline Phosphatase (IU/L)	235.50 ± 5.50	223.00 ± 17.00
Total Bilirubin (mg/dL)	0.10 ± 0.00	0.10 ± 0.00
Total Protein (g/dL)	6.90 ± 0.30	6.90 ± 0.20
Albumin (g/dL)	3.80 ± 0.20	4.10 ± 0.00
Globulin (g/dL)	3.05 ± 0.05	2.80 ± 0.20
Total cholesterol (mg/dL)	57.00 ± 3.00	59.50 ± 0.50
Triglycerides (mg/dL)	139.00 ± 1.00	130.50 ± 10.50
HDL (mg/dL)	9.55 ± 0.25	8.25 ± 0.25
LDL (mg/dL)	18.00 ± 2.00	25.00 ± 2.00 *
Glucose (mg/dL)	110.50 ± 3.50	113.00 ± 2.00
Serum urea (mg/dL)	35.50 ± 2.50	34.00 ± 0.00
Serum creatinine (mg/dL)	0.40 ± 0.02	0.46 ± 0.01

Values are expressed as mean ± SD, compared with the control is significant when * ($p < 0.05$). Data were analyzed using the Student test. AST (Aspartate aminotransferase); ALT (Alanine Aminotransferase); HDL (high density lipoprotein); LDL (low density lipoprotein). No historical control values are available.

Table 3. Hematological evaluation of rats after administration of repeated oral doses of tocosh flour at doses of 1000 mg/kg for 28 days.

Parameters	Control Group	Tocosh Flour Dose: 1000 mg/kg
Male rats	$n = 5$	$n = 5$
RBC ($\times 10^6$/mm^3)	7.09 ± 0.30	7.30 ± 0.08
White blood cells ($\times 10^3$/mm^3)	4.14 ± 0.13	3.91 ± 0.17
Hemoglobin (g/dL)	14.70 ± 0.60	14.65 ± 0.15
Hematocrit (%)	41.80 ± 1.80	43.55 ± 1.45
Eosinophils (%)	1.50 ± 0.50	1.00 ± 0.00
Basophil (%)	0.00 ± 0.00	0.00 ± 0.00
Monocytes (%)	1.00 ± 0.00	2.00 ± 0.00 *
Segmented (%)	20.50 ± 8.50	17.00 ± 4.00
Lymphocytes (%)	77.00 ± 9.00	80.50 ± 4.50
Platelets ($\times 10^3$/mm^3)	7.5 ± 0.26	7.22 ± 0.22
Female rats	$n = 5$	$n = 5$
RBC ($\times 10^6$/mm^3)	7.00 ± 0.10	6.93 ± 0.22
White blood cells ($\times 10^3$/mm^3)	4.03 ± 0.18	4.20 ± 0.30
Hemoglobin (g/dL)	14.65 ± 0.25	14.60 ± 0.10
Hematocrit (%)	42.00 ± 0.00	41.50 ± 0.50
Eosinophils (%)	0.00 ± 0.00	0.00 ± 0.00

Table 3. *Cont.*

Parameters	Control Group	Tocosh Flour Dose: 1000 mg/kg
Basophil (%)	0.00 ± 0.00	0.00 ± 0.00
Monocytes (%)	2.50 ± 0.50	1.50 ± 0.50 *
Segmented (%)	24.00 ± 1.00	17.00 ± 1.00
Lymphocytes (%)	79.00 ± 1.00	80.50 ± 0.50
Platelets ($\times 10^3/\text{mm}^3$)	7.3 ± 0.20	7.6 ± 0.35

Values are expressed as mean ± SD, compared with the control is significant when * ($p < 0.05$). Data were analyzed using the Student test. No historical control values are available.

The body weight of the male rats administered with potato flour at a dose of 1000 mg/kg/day increased from the first week of 181.8 ± 9.28 g until the end of the fourth week, 261.4 ± 15.67 g, in contrast to the control group. In the control group of male rats, it was from 170.2 ± 9.20 g to 228.4 ± 20.24 g on day-28, which was significant ($p < 0.05$). Similarly, female rats administered at a dose of 1000 mg/kg had a weight gain that started in the first week at 172.8 ± 7.14 g to 228.4 ± 7.66 g corresponding to the end of the last week, compared to the control group of female rats which was significant ($p < 0.001$, see Figure 3).

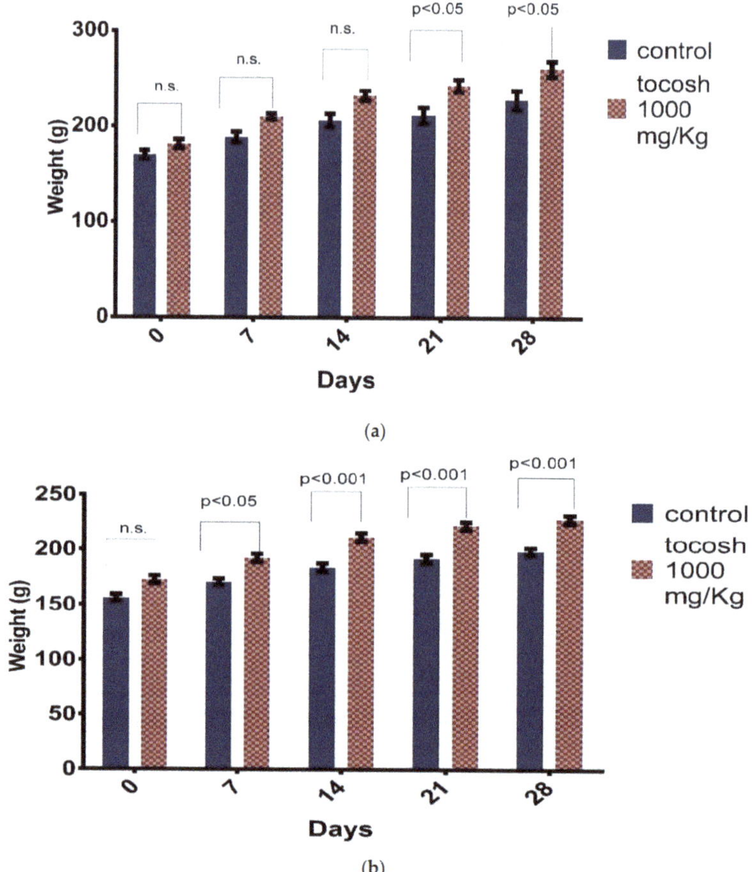

Figure 3. Body weights of rats treated with repeated oral doses of tocosh flour (1000 mg/kg) for 28 days. (**a**) male rats and (**b**) female rats. * $p < 0.05$, ** $p < 0.001$ compared to control, by using the Student *t*-test.

3.3. Acute Oral Toxicity—Fixed Dose Procedure Study

The individual weights of the mice to which the tocosh flour was administered at doses of 2000 and 5000 mg/kg were determined before the administration of the test product and were calculated and re-recorded at the end of the 14 days, with weight gain at the end of the experiment.

Mice administered at doses of 2000 mg/Kg and 5000 mg/kg body weight exhibited minor organ damage in the liver (parenchymal lymphocyte) and kidney (lymphocyte) according to Figures 4 and 5. However, no external aspects of toxicity were observed during the study linked to liver and kidney damage.

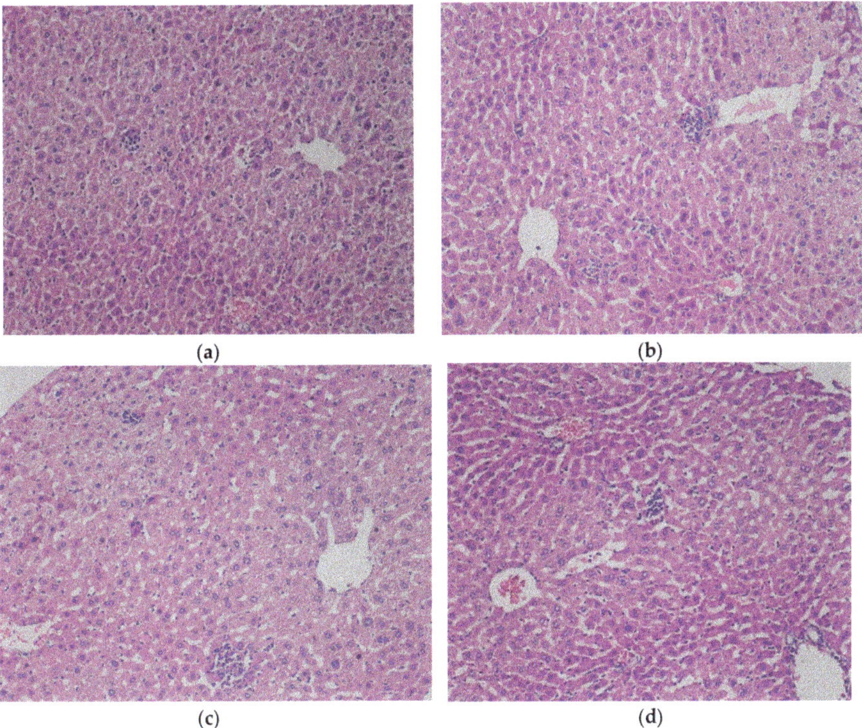

Figure 4. Liver tissue images of mice receiving a fixed dose of 2000 mg/kg and 5000 mg/kg of tocosh flour. (Stained with H&E, 200×). (**a**) Male group at the dose of 2000 mg/kg. Liver parenchyma with few lymphocytes around the central vein. (**b**) Female group at the 2000 mg/Kg dose. Liver parenchyma with isolated lobular lymphocytes (mild lobular inflammation). (**c**) Male group at the dose of 5000 mg/Kg. Liver parenchyma with isolated lobular lymphocytes (mild lobular inflammation). (**d**) Female group at the dose of 5000 mg/kg. Liver parenchyma with isolated lobular lymphocytes (mild lobular inflammation).

All this evaluation was followed in accordance with the OECD guideline 420, classifying the product as category B toxic (evident toxicity and/or $\leq$ 1 death), using the highest dose for safety and animal welfare to limit the number of animals, therefore the use of lower doses was restricted. In other organs, such as the brain, spleen, stomach, heart, and testes, no toxic damage was observed (Figures S2 and S3).

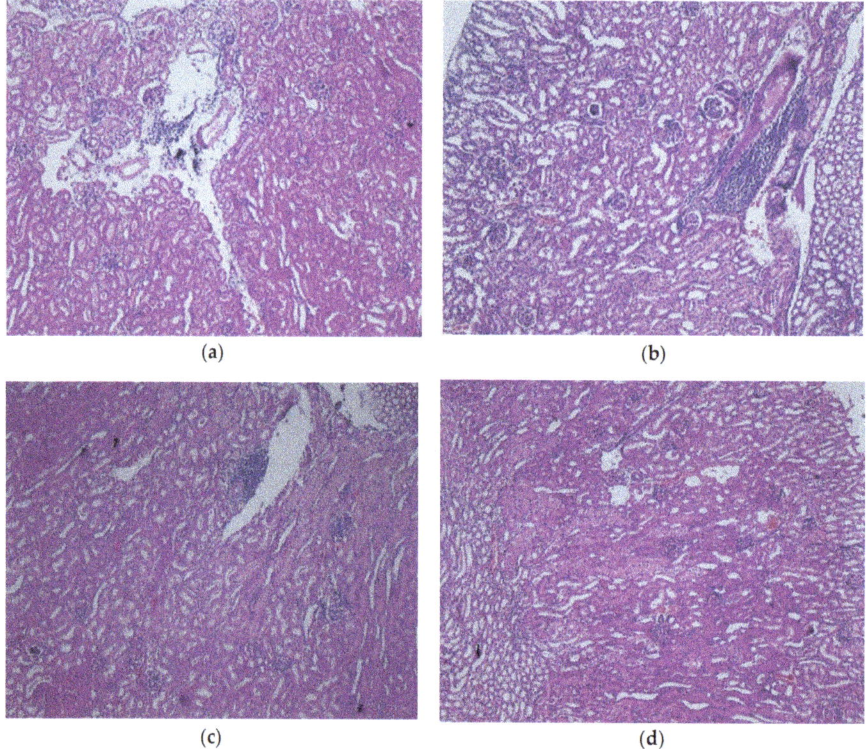

Figure 5. Renal tissue images of male and female mice that received a fixed dose of 2000 mg/kg BW and 5000 mg/kg BW of tocosh flour. (Stained with H&E, 200×). (**a**) Male group at the dose of 2000 mg/kg. Renal parenchyma with isolated chronic interstitial inflammatory infiltrate. (**b**) Female group at the dose of 2000 mg/kg. Renal parenchyma showing mild chronic interstitial inflammatory infiltrate with a tendency to lymphoid accumulation. (**c**) Male group at the dose of 5000 mg/kg. Renal parenchyma showing mild chronic interstitial inflammatory infiltrate with a tendency to the formation of lymphoid accumulation. (**d**) Female group at the dose of 5000 mg/kg. Renal parenchyma with isolated chronic interstitial inflammatory infiltrate.

3.4. Prediction of Drug-Likeness Properties for Steroidal Glycoalkaloids: α-Solanin, α-Chaconine and Solanidine

We cannot show these results as toxicological finds due to these are only drug-likeness parameters. The properties showed the steroidal alkaloids as α-solanine, α-chaconine, and solanidine could be considered as pharmacokinetic behavior and were included because they are the main alkaloids found in *Solanum tuberosum* as is reported in other literature (Table 4).

Table 4. Calculated drug-likeness properties for glycoalkaloids α-solanine, α-chaconine, and solanidine.

Compound	M.W. [a]	PSA [b]	n-Rot Bond (0–10)	n-ON (<10) [c]	n-OHNH [d]	Log $P_{o/w}$ [e]	LogK$_{HSA}$ [f]	Caco-2 [g] (nm/s)	App.MDCK (nm/s) [h]	% HIA [i]	% HOA [j]	Lipinski Rule of Five
α-chaconine	852.070	199.560	15	15	8	−0.091	−0.875	<25 poor	<25 poor	<25poor	Low	3
α-solanine	868.069	220.534	17	16	9	−0.860	−1.157	<25 poor	<25 poor	<25poor	Low	3
Solanidine	397.643	22.288	1	2	1	5.007	0.554	1049	576	100	High	0

[a] Molecular weight of the hybrid (150–500). [b] Polar surface area (PSA) (7.0–200 Å^2). [c] n-ON number of hydrogen bond acceptors ≤10. [d] n-OHNH number of hydrogens bonds donors ≤5. [e] Octanol water partition coefficient (log $P_{o/w}$) (−2.0 to 6.5). [f] Binding-serum albumin (K$_{HSA}$) (−1.5 to 1.5). [g] Human intestinal permeation (<25 poor, >500 great). [h] Madin-Darby canine kidney (MDCK) cells permeation. [i] Human intestinal absorption (% HIA) (>80% is high, <25% is poor). [j] Model for Human Oral Absorption.

4. Discussion

The glycoalkaloids from potatoes such as α-solanine, α-chaconine, and solanidine have a special function as a natural defense against plagues and its consumption may result in different symptoms that may include, diarrhea, fever, nausea, and even death [16,17]. The main studied mechanism of this kind of alkaloids is related to an inhibitory effect on the enzymes acetylcholinesterase (AChE) and butyrylcholinesterase (BuChE) [7,18]. Furthermore, it has been reported that glycoalkaloids interferes with ion-transport in cell membranes. The European Commission based on toxicological studies decided that the total glycoalkaloid content must not exceed the limit of 150 mg/Kg in potato protein powder for food applications [19]. However, the toxic effects produced by the inhibition of AchE and BuChE during the administration of tocosh flour were not observed in both subacute and acute toxicity treatments. Currently, there are no data on serious poisonings in the population, who consume tocosh flour as an alternative traditional treatment for digestive and respiratory diseases.

Tocosh comes from a fermentation process (Andean technique), which is suitable for distribution and consumption in the different markets of Peru as flour or in its raw form. On the other hand, the amount of glycoalkaloids are related to the cultivation method, storage and temperature, depending overall on the Andean techniques destined for its production and can be distributed in different rates in *Solanum tuberosum* tubers, they have been found in the tuber (smaller quantity), leaves and peel (greater quantity), and some analysis showed quantities such as 300–600 mg/kg in peel, 2000–4000 mg/kg in buds, and 3000–5000 mg/kg in flowers [20]. In our work, we did not evaluate the identification of constituents in tocosh flour by chromatography, but based on several studies of *Solanum tuberosum*, its glycoalkaloids are known by its fingerprint (α-solanine, α-chaconine and solanidine). Furthermore, probably other phytochemicals, lactobacillus and other could have been generated during the fermentation process and should be studied depending on the variety of potato and climatic factors. Otherwise, potato flour is an excellent food additive that can be applied in cakes, puffed food, breakfast food, baby food, condiments and soup and some functional factors such as anthocyanin, rosterone, and mucus proteins have been found, so its effect on health care is significant [21].

In regard to the repeated dose toxicity study using the limit dose of 1000 mg/Kg BW in rats, they had an increase in body weight. The increase of body weights differs from other studies where mice fed with 2130 mg/kg and 2170 mg/Kg of potato alkaloids such as α-chaconine and α-solanine for 7 days showed a decrease in body weight and in organs such as the liver, similarly in the administration for 14 days of the alkaloids: α-solasodine and solanidine [9,22]. This could be explained due to the animals being fed with tocosh flour which might contain a high content of carbohydrate and proteins, increasing the body weight in rats.

Significant increases in liver weight are commonly associated with adaptive changes such as accumulation of lipid, glycogen or other substances or a result of cell damage, congestion, hepatocellular hypertrophy or hyperplasia [23,24]. This variation does not always correlate with the amount of hepatic enzyme induction in rats as AST and ALT [25]. In our findings, we did not evidence any damage in liver tissues analyzed by using microscopy and any alteration in hepatic enzyme. This research revealed that presence of any phytochemical groups found in tocosh flour did not alter the histology in rats at the repeated dose of 1000 mg/kg BW for 28 days.

In hematological examination, there were no significant differences between rats administered with tocosh flour and the control group. The biochemical analysis LDL showed a significant increase in female and male rats. Similarly, there was a significant decrease of Tryglicerides in male rats but not in female rats. These findings might be altered due to potato starch, which is more phosphorylated than other starches of cereals [26], and indigestive polysaccharides promote excretion of bile acids, producing a reduction of Tryglicerides levels. However, an increase in LDL levels could be linked to carbohydrate consumption [27], thereby LDL production. No historical control data on the clinical chemistry or hematology values from the animal supplier were available for comparison; therefore, it is possible that the statistically changed parameters noted were within the normal range of average parameter values.

In acute toxicity according to OCED 420, minor changes in liver and kidney were observed with doses of 2000 and 5000 mg/Kg BW of tocosh administered in a single dose, probably attributed to its alkaloidal content and other components not determined in this study. *Solanum tuberosum* is known to have glycoalkaloids such as α-chaconin and α-solanine, mainly in tubers in almost 95%, also β-solanine, β-chaconin, γ-solanine, γ-chaconin, α-solamargin and β-solamargin but in less quantity [9,28]. However, these findings are unclear and not necessarily due to the consumption of tocosh flour.

The joint committee for food additives of FAO and WHO (JECFA) considers amounts of glycoalkaloids between 20 and 100 mg/kg as safe [29]. The toxic dose in the population corresponds to levels higher than 100 mg of total glycoalkaloids/kg of potato but this value could be influenced by environmental and storage conditions [17,30,31]. On the other hand, in the administration of 1.25 mg total glycoalkaloids/kg BW considered the highest dose in humans, gastrointestinal signs as vomiting appeared within 4 h [32,33]. Potato alkaloids at a dose of 75 mg/Kg of α-chaconin and α-solanin were lethally fatal within 4 to 5 days in Syrian golden hamster [34].

In an in vitro study, porcine oocytes were exposed to α-solanine (10 μM) and negatively affected early porcine embryo development by suppressing blastocyst formation and reducing embryo quality [35]. Another study found that the toxic and cytotoxic effects of α-Solanin in potatoes altered the proliferation and function of testicular cells in mice by regulating Sertoli and Leydig cells, affecting the testes and the reproductive function of male mice [36]. Likewise, α-solanine and α-chaconin at micromolar concentrations cause a cytotoxic effect on C6 rat glioma cells at the plasma membrane [37].

Currently, toxicological studies of the tocosh flour is not documented, according to the results obtained from in silico models, α-chaconine and α-solanine do not present Lipinski's rules, this also results in the permeability values calculated using the Caco-2 and MDCK models being so poor (<25 nm/s). Additionally, the oral and intestinal absorption values are very low (<25%). However, it is important to highlight that the prediction values obtained for its ability to bind to plasma transporter proteins are within the range of 95%, so they would have a high toxicity when it is given by intravenous administration. Nevertheless, solanidine showed results for a quick absorption and might be due to its small molecular weight and chemical structure not glycosylated. In animal studies, it has been shown that orally administered glycoalkaloids are less toxic than intraperitoneal administration (I.P.) due to poor absorption in the gut. In mice, LD_{50} by I.P. administration have been reported to be 23 mg/Kg for α-chaconine, 34 mg/Kg for α-solanine, 500 mg/Kg for solanidine, and greater than 1000 mg/Kg for α-solanine by oral administration [38].

The NOAEL in rats of tocosh flour was 1000 mg/Kg, the estimate of the human equivalent dose (HED) [39] corresponds to a dose of 1000 mg as an initial dose in humans with 60 Kg of body weight. According to the traditional consumption of tocosh, this dose is less than the dose consumed by the population. Although, it seems to be safe compared with the limit dose of glycoalkaloids found in potato tubers.

In the present study, we could not identify and quantify each chemical constituent of tocosh potato. It is known that tocosh could contain other components such as lactobacillus and antibiotics produced during its fermentative process. However, tocosh can be consumed in established doses up to 1000 mg daily. Future studies of genotoxicity and chronic toxicity are needed as well to standardize its dose for consumption in foods.

5. Conclusions

The tocosh flour did not present toxicity at the repeated dose for 28 days in the highest dose corresponding to 1000 mg/kg BW. There were no deaths at up to 5000 mg/kg BW, therefore, the oral LD50 was greater than 5000 mg/kg.

Supplementary Materials: The following are available online at http://www.mdpi.com/2304-8158/9/6/719/s1, Figure S1: Images of some tissues of male and female mice without specific alterations that received fixed doses of 2000 mg/kg and 5000 mg/kg of Tocosh Flour stained with H&E and 40X, Figure S2: Images of the organ sections: Stomach (A), liver (B), Kidney (C) and heart (D), of male rats who were administered Tocosh flour for 28 days. Subscript C and 1 refer to the control group and the group administered at dose of 1000 mg/kg/day, respectively. H&E staining, 40X, Figure S3: Images of the organ sections such as: Stomach (A), liver (B), Kidney (C) and heart (D), of female rats that were administered Tocosh flour for 28 days. Subscript C and 1 refer to the control and group administered at dose of 1000 mg/kg/day, respectively. H&E staining, 40X.

Author Contributions: Conceptualization, O.H.-C. and J.R.V.-C.; methodology, J.R.V.-C., A.F.Y.-P. and J.P.R.-A.; software, C.A.; validation, R.D.H.-Q. and L.F.-S.; formal analysis, G.P.-R. and V.A.-A.; investigation, O.H.-C. and J.R.V.-C.; writing and editing-original draft preparation, O.H.-C.; writing—review and editing, O.H.-C., R.Á.Y.-P. and L.F.-S.; visualization, J.P.R.-A.; supervision, A.F.Y.-P. All authors have read and agree to the published version of the manuscript.

Funding: This research received no external funding.

Data Availability: The data sets used and/or analyzed during the current study are available from the corresponding author upon reasonable request.

Conflicts of Interest: The authors declare that there are no conflict of interest regarding the publication of this paper.

References

1. Hardigan, M.A.; Laimbeer, F.P.E.; Newton, L.; Crisovan, E.; Hamilton, J.P.; Vaillancourt, B.; Wiegert-Rininger, K.; Wood, J.C.; Douches, D.S.; Farré, E.M.; et al. Genome diversity of tuber-bearing solanum uncovers complex evolutionary history and targets of domestication in the cultivated potato. *Proc. Natl. Acad. Sci. USA* **2017**. [CrossRef] [PubMed]

2. Berdugo-Cely, J.; Valbuena, R.I.; Sánchez-Betancourt, E.; Barrero, L.S.; Yockteng, R. Genetic diversity and association mapping in the colombian central collection of solanum tuberosum L. Andigenum group using SNPs markers. *PLoS ONE* **2017**. [CrossRef] [PubMed]

3. Velásquez-Milla, D.; Casas, A.; Torres-Guevara, J.; Cruz-Soriano, A. Ecological and socio-cultural factors influencing in situ conservation of crop diversity by traditional andean households in Peru. *J. Ethnobiol. Ethnomed.* **2011**. [CrossRef] [PubMed]

4. Mayta-Tovalino, F.; Sedano-Balbin, G.; Romero-Tapia, P.; Alvítez-Temoche, D.; álvarez-Paucar, M.; Gálvez-Calla, L.; Sacsaquispe-Contreras, S. Development of new experimental dentifrice of peruvian solanum tuberosum (Tocosh) fermented by water stress: Antibacterial and cytotoxic activity. *J. Contemp. Dent. Pract.* **2019**. [CrossRef]

5. Mosso, A.L.; Jimenez, M.E.; Vignolo, G.; LeBlanc, J.G.; Samman, N.C. Increasing the folate content of tuber based foods using potentially probiotic lactic acid bacteria. *Food Res. Int.* **2018**. [CrossRef] [PubMed]

6. LeBlanc, J.G.; Vignolo, G.; Todorov, S.D.; de Giori, G.S. Indigenous fermented foods and beverages produced in Latin America. In *Food Intake: Regulation, Assessing and Controlling*; Nova Science Publishers, Inc.: Hauppauge, NY, USA, 2013.

7. Langkilde, S.; Mandimika, T.; Schrøder, M.; Meyer, O.; Slob, W.; Peijnenburg, A.; Poulsen, M. A 28-day repeat dose toxicity study of steroidal glycoalkaloids, α-solanine and α-chaconine in the Syrian Golden hamster. *Food Chem. Toxicol.* **2009**. [CrossRef]

8. Barceloux, D.G. Potatoes, Tomatoes, and Solanine Toxicity (Solanum tuberosum L., Solanum lycopersicum L.). *Dis. Mon.* **2009**, *55*, 391–402. [CrossRef]

9. Friedman, M. Potato glycoalkaloids and metabolites: Roles in the plant and in the diet. *J. Agric. Food Chem.* **2006**, *54*, 8655–8681. [CrossRef]

10. Chen, Z.; Miller, A.R. Steroidal alkaloids in solanaceous vegetable crops. *Hortic. Rev.* **2010**, 171–196. [CrossRef]

11. Herrera-Calderon, O.; Arroyo-Acevedo, J.L.; Rojas-Armas, J.; Chumpitaz-Cerrate, V.; Figueroa-Salvador, L.; Enciso-Roca, E.; Tinco-Jayo, J.A. Phytochemical screening, total phenolic content, antioxidant and cytotoxic activity of chromolaena laevigata on human tumor cell lines. *Annu. Res. Rev. Biol.* **2017**. [CrossRef]

12. OECD/OECDE. *Test No. 407: Repeated Dose 28-Day Oral Toxicity Study in Rodents*; OECD Publishing: Paris, France, 2008.

13. OECD. Test No. 420: Acute oral toxicity-fixed dose procedure. *Oecd Guidel. Test. Chem.* **2002**. [CrossRef]

14. Enciso-Roca, E.; Aguilar-Felices, E.J.; Tinco-Jayo, J.A.; Arroyo-Acevedo, J.L.; Herrera-Calderon, O.; Aguilar-Carranza, C.; Justil, H.G. Effects of acute and sub-acute oral toxicity studies of ethanol extract of tanacetum parthenium (L) Sch. Bip. Aerial parts in mice and rats. *Annu. Res. Rev. Biol.* **2017**. [CrossRef]

15. Butina, D.; Segall, M.D.; Frankcombe, K. Predicting ADME properties in silico: Methods and models. *Drug Discov. Today* **2002**. [CrossRef]

16. Morris, S.C.; Lee, T.H. The toxicity and teratogenicity of Solanaceae glycoalkaloids, particularly those of the potato (Solanum tuberosum): A review. *Food Technol. Aust.* **1984**, *36*, 195–202.

17. Romanucci, V.; Pisanti, A.; Di Fabio, G.; Davinelli, S.; Scapagnini, G.; Guaragna, A.; Zarrelli, A. Toxin levels in different variety of potatoes: Alarming contents of α-chaconine. *Phytochem. Lett.* **2016**. [CrossRef]

18. Crawford, L.; Myhr, B. A preliminary assessment of the toxic and mutagenic potential of steroidal alkaloids in transgenic mice. *Food Chem. Toxicol.* **1995**. [CrossRef]

19. Abduh, S.B.M.; Leong, S.Y.; Agyei, D.; Oey, I. Understanding the properties of starch in potatoes (Solanum tuberosum var. Agria) after being treated with pulsed electric field processing. *Foods* **2009**. [CrossRef]

20. Nielsen, S.D.; Schmidt, J.M.; Kristiansen, G.H.; Dalsgaard, T.K.; Larsen, L.B. Liquid chromatography mass spectrometry quantification of α-solanine, α-chaconine, and solanidine in potato protein isolates. *Foods* **2020**. [CrossRef]

21. Cui, L.; Tian, Y.; Tian, S.; Wang, Y.; Gao, F. Preparation of potato whole flour and its effects on quality of flour products: A review. *Grain Oil Sci. Technol.* **2018**. [CrossRef]

22. Patil, B.C.; Sharma, R.P.; Salunkhe, D.K.; Salunkhe, K. Evaluation of solanine toxicity. *Food Cosmet. Toxicol.* **1972**. [CrossRef]

23. Lau, J.K.C.; Zhang, X.; Yu, J. Animal models of non-alcoholic fatty liver disease: Current perspectives and recent advances. *J. Pathol.* **2016**. [CrossRef] [PubMed]

24. Brunt, E.M.; Tiniakos, D.G. Histopathology of nonalcoholic fatty liver disease. *World J. Gastroenterol.* **2010**. [CrossRef] [PubMed]

25. Greaves, P. *Histopathology of Preclinical Toxicity Studies: Interpretation and Relevance in Drug Safety Evaluation*, 4th ed.; Academic Press: Amsterdam, The Netherlands, 2011.

26. Hashimoto, N.; Ito, Y.; Han, K.H.; Shimada, K.I.; Sekikawa, M.; Topping, D.L.; Bird, A.R.; Noda, T.; Chiji, H.; Fukushima, M. Potato pulps lowered the serum cholesterol and triglyceride levels in rats. *J. Nutr. Sci. Vitaminol.* **2006**. [CrossRef] [PubMed]

27. Ma, Y.; Chiriboga, D.E.; Olendzki, B.C.; Li, W.; Leung, K.; Hafner, A.R.; Li, Y.; Ockene, I.S.; Hebert, J.R. Association between Carbohydrate Intake and Serum Lipids. *J. Am. Coll. Nutr.* **2006**. [CrossRef] [PubMed]

28. Wang, S.; Panter, K.E.; Gaffield, W.; Evans, R.C.; Bunch, T.D. Effects of steroidal glycoalkaloids from potatoes (Solanum tuberosum) on in vitro bovine embryo development. *Anim. Reprod. Sci.* **2005**. [CrossRef] [PubMed]

29. Zhou, X.; Gao, Q.; Praticò, G.; Chen, J.; Dragsted, L.O. Biomarkers of tuber intake. *Genes Nutr.* **2019**. [CrossRef]

30. Friedman, M. Analysis of biologically active compounds in potatoes (Solanum tuberosum), tomatoes (Lycopersicon esculentum), and jimson weed (Datura stramonium) seeds. *J. Chromatogr. A* **2004**. [CrossRef]

31. Friedman, M.; Rayburn, J.R.; Bantle, J.A. Structural Relationships and Developmental Toxicity of Solanum Alkaloids in the Frog Embryo Teratogenesis Assay-Xenopus. *J. Agric. Food Chem.* **1992**. [CrossRef]

32. Connors, N.J.; Glover, R.L.; Stefan, C.; Patterson, D.; Wong, E.; Milstein, M.; Swerdlow, M.; Hoffman, R.S.; Nelson, L.S.; Smith, S.W. Biological and botanical confirmation of solanaceous glycoalkaloid poisoning by susumber berries (Solanum torvum). *Clin. Toxicol.* **2014**. [CrossRef]

33. Nigg, H.N.; Ramos, L.E.; Graham, E.M.; Sterling, J.; Brown, S.; Cornell, J.A. Inhibition of human plasma and serum butyrylcholinesterase (EC 3.1.1.8) by α-chaconine and α-solanine. *Fundam. Appl. Toxicol.* **1996**. [CrossRef]

34. Gaffield, W.; Keeler, R.F. Induction of terata in hamsters by solanidane alkaloids derived from Solanum tuberosum. *Chem. Res. Toxicol.* **1996**. [CrossRef] [PubMed]

35. Lin, T.; Oqani, R.K.; Lee, J.E.; Kang, J.W.; Kim, S.Y.; Cho, E.S.; Jeong, Y.D.; Baek, J.J.; Jin, D. Il α-Solanine impairs oocyte maturation and quality by inducing autophagy and apoptosis and changing histone modifications in a pig model. *Reprod. Toxicol.* **2018**. [CrossRef] [PubMed]

36. Bell, D.P.; Gibson, J.G.; McCarroll, A.M.; McClean, G.A. Embryotoxicity of solanine and aspirin in mice. *J. Reprod. Fertil.* **1976**. [CrossRef] [PubMed]

37. Yamashoji, S.; Matsuda, T. Synergistic cytotoxicity induced by α-solanine and α-chaconine. *Food Chem.* **2013**. [CrossRef]

38. Al Sinani, S.S.S.; Eltayeb, E.A. The steroidal glycoalkaloids solamargine and solasonine in solanum plants. *South. Afr. J. Bot.* **2017**. [CrossRef]

39. Nair, A.; Jacob, S. A simple practice guide for dose conversion between animals and human. *J. Basic Clin. Pharm.* **2016**. [CrossRef]

 *foods*

Review

The Pharmacological Activity, Biochemical Properties, and Pharmacokinetics of the Major Natural Polyphenolic Flavonoid: Quercetin

Gaber El-Saber Batiha [1,2,*,†] , Amany Magdy Beshbishy [1,†], Muhammad Ikram [3], Zohair S. Mulla [4], Mohamed E. Abd El-Hack [5] , Ayman E. Taha [6] , Abdelazeem M. Algammal [7] and Yaser Hosny Ali Elewa [8,9]

[1] National Research Center for Protozoan Diseases, Obihiro University of Agriculture and Veterinary Medicine, Nishi 2-13, Inada-cho, Obihiro 080-8555, Japan; amanimagdi2008@gmail.com

[2] Department of Pharmacology and Therapeutics, Faculty of Veterinary Medicine, Damanhour University, Damanhour 22511, Egypt

[3] Department of Chemistry, Abdul Wali Khan University Mardan, Mardan 23200, Pakistan; ikrambiochem2014@gmail.com

[4] Department of Public Health, College of Veterinary Medicine, King Faisal University, Hofuf 31982, Saudi Arabia; drzomu@gmail.com

[5] Department of Poultry, Faculty of Agriculture, Zagazig University, Zagazig 44511, Egypt; dr.mohamed.e.abdalhaq@gmail.com

[6] Department of Animal Husbandry and Animal Wealth Development, Faculty of Veterinary Medicine, Alexandria University, Edfina 22758, Egypt; Ayman.Taha@alexu.edu.eg

[7] Department of Bacteriology, Immunology and Mycology, Faculty of Veterinary Medicine, Suez Canal University, Ismailia 41522, Egypt; abdelazeem.algammal@vet.suez.edu.eg or abdelazeem.algammal@gmail.com

[8] Department of Histology and Cytology, Faculty of Veterinary Medicine, Zagazig University, Zagazig 44519, Egypt; y-elewa@vetmed.hokudai.ac.jp

[9] Laboratory of Anatomy, Department of Biomedical Sciences, Graduate School of Veterinary Medicine, Hokkaido University, Sapporo 060-0818, Japan

* Correspondence: dr_gaber_batiha@vetmed.dmu.edu.eg or gaberbatiha@gmail.com; Tel.: +20-45-2716024; Fax: +20-45-2716024

† These authors contributed equally.

Received: 21 February 2020; Accepted: 20 March 2020; Published: 23 March 2020

Abstract: Flavonoids are a class of natural substances present in plants, fruits, vegetables, wine, bulbs, bark, stems, roots, and tea. Several attempts are being made to isolate such natural products, which are popular for their health benefits. Flavonoids are now seen as an essential component in a number of cosmetic, pharmaceutical, and medicinal formulations. Quercetin is the major polyphenolic flavonoid found in food products, including berries, apples, cauliflower, tea, cabbage, nuts, and onions that have traditionally been treated as anticancer and antiviral, and used for the treatment of allergic, metabolic, and inflammatory disorders, eye and cardiovascular diseases, and arthritis. Pharmacologically, quercetin has been examined against various microorganisms and parasites, including pathogenic bacteria, viruses, and *Plasmodium*, *Babesia*, and *Theileria* parasites. Additionally, it has shown beneficial effects against Alzheimer's disease (AD), and this activity is due to its inhibitory effect against acetylcholinesterase. It has also been documented to possess antioxidant, antifungal, anti-carcinogenic, hepatoprotective, and cytotoxic activity. Quercetin has been documented to accumulate in the lungs, liver, kidneys, and small intestines, with lower levels seen in the brain, heart, and spleen, and it is extracted through the renal, fecal, and respiratory systems. The current review examines the pharmacokinetics, as well as the toxic and biological activities of quercetin.

Keywords: quercetin; herbal remedies; pharmacological activities; pharmacokinetics; Alzheimer's disease

1. Introduction

Plants have been used since ancient times to cure certain infectious diseases, some of which are now standard treatments for several diseases [1,2]. Over the last decade, there has been a huge increase in acceptance and public interest in natural therapy in both developing and developed countries, and these herbal medicines are now available, not only in drug stores but also in supermarkets and food stores. Approximately 80 percent of people in Africa and other developing nations still depend on traditional herbal remedies to treat ailments due to their easy availability and lower cost compared to synthetic compounds [3,4]. They also demonstrate a number of promising activities against various health problems (e.g., respiratory and gastrointestinal disorders) and show anti-inflammatory, spasmolytic, antioxidant, sedative, antimicrobial, antiviral, antiseptic, anti-diabetic, immunostimulant, and hepatoprotective activities [5–7]. In addition, numerous phytoconstituents and plenty of chemical compounds with different biological and pharmacological activities have been isolated and identified from medicinal plants [8–10]. For instance, Batiha et al. [8], as well as Beshbishy et al. [9], reported the antiprotozoal activity of chalcones and ellagic acid, the naturally derived phytoconstituents isolated from herbal extracts against *Plasmodium*, *Leishmania*, *Trypanosoma*, *Babesia*, and *Theileria* parasites. These phytochemical compounds have been shown to be lead compounds for the development of new synthetic compounds, with higher efficacy and lower toxic side effects [11].

Quercetin (Figure 1: (2-(3,4-dihydroxyphenyl)-3,5,7-trihydroxy-4-Hchromen-4-one)) is classified as a flavonol, which is one of the six subcategories of flavonoid compounds and is the major polyphenolic flavonoid found in various vegetables and fruits, such as berries, lovage, capers, cilantro, dill, apples, and onions [12]. It is yellow in color and completely soluble in lipids and alcohol, insoluble in cold water, while sparingly soluble in hot water. Quercetin's name derives from the Latin word *"Quercetum"*, which means Oak Forest, and also belongs to the flavonol category, which is not produced in the human body [13]. The name of the International Union of Pure and Applied Chemistry (IUPAC) and the chemical formula of quercitin are as follows: 2-(3,4-dihydroxyphenyl)-3,5,7-trihydroxychromen-4-one and $C_{15}H_{10}O_7$, respectively. Quercetin is one of the most important plant molecules that has shown many pharmacological activities, such as being anticancer, antiviral, and treating allergic, metabolic, and inflammatory disorders, eye and cardiovascular diseases, and arthritis [14]. It has also shown a wide range of anticancer properties, and several reports indicate its efficacy as a cancer-preventing agent. Quercetin also has psychostimulant properties and has been documented to prevent platelet aggregation, capillary permeability, and lipid peroxidation, and enhance mitochondrial biogenesis [15]. The current review aims to further understand quercetin's beneficial and pharmacological effects, as well as its clinical application and concerns around safety.

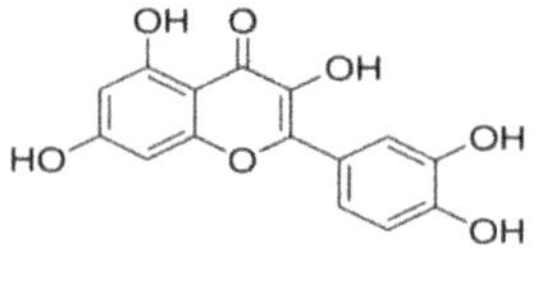

Figure 1. Quercetin's chemical structure.

2. Bioavailability and Pharmacokinetics of Quercetin

Previous animal and human research have reported poor oral bioavailability of quercetin after a single oral dose due to macronutrient absorption [16]. For instance, quercetin is ingested in the form of glycosides, and glycosyl groups are released during chewing, digestion, and absorption. Afterward, quercetin glycosides are converted into aglycone in the intestine before they are absorbed into enterocytes by the action of β-glycosidases enzymes. According to Walle et al. [17], previous studies have reported that intestinal and oral bacteria are involved in this enzymatic hydrolysis. Quercetin is a lipophilic compound, so it is assumed that it can cross the intestinal membranes by simple diffusion, and theoretically, this absorption is better than its glycoside forms which reach the intestines without degradation [18]. To date, a number of human studies have been conducted on the bioavailability of quercetin glycosides extracted from different species. For example, quercetin glycosides from onions were absorbed in patients with ileostomy at a higher percentage than pure aglycone, which has been reported by Hollman et al. [19]. On the other hand, Scholz and Williamson [20] documented the existence of significant amounts of aglycone in ileostomy fluid taken from patients with ileostomy who had eaten a meal with onions. They also reported the presence of a high quantity of quercetin glycosides and a small amount of quercetin aglycone, but quercetin glycosides were not observed in the fluid. One possible explanation is that the hydrolysis of quercetin glycosides takes place as it is converted by β-glycosidases enzymes to aglycone. These enzymes are found in the small intestine, and most of them are then absorbed. Ferry et al. [21] studied the pharmacokinetic properties of intravenous quercetin injection in cancer patients at dose levels of 60–2000 mg/m^2. They determined that 945 mg/m^2 was a safe dose of quercetin, while its toxic dose was reported to cause emesis, hypertension, nephrotoxicity, and decreased serum potassium. The distribution and elimination half-life of intravenous quercetin were found to be 0.7–7.8 min and 3.8–86 min, respectively, whereas its clearance and distribution volume were 0.23–0.84 L/min/m^2 and 3.7 L/m^2, respectively. Erlund et al. [22] examined the pharmacokinetic properties of 8, 20, and 500 mg quercetin aglycone orally in healthy participants. Graefe et al. [23] also studied the pharmacokinetic properties of quercetin and maintained a dose level of up to 200 mg and demonstrated that quercetin C_{max} and T_{max} were 2.3 ± 1.5 µg/mL and 0.7 ± 0.3 h, respectively.

3. Sources of Quercetin and Its Pharmacological Activity

Quercetin is one of the most significant bioflavonoid compounds found in vegetables, grains, and fruits for more than 20 plant species—namely, *Foeniculum vulgare*, *Curcuma domestica valeton*, *Santalum album*, *Cuscuta reflexa*, *Withania somnifera*, *Emblica officinalis*, *Mangifera indica*, *Daucus carota*, *Momordica charantia*, *Ocimum sanctum*, *Psoralea corylifolia*, *Swertia chirayita*, *Solanum nigrum*, and *Glycyrrhiza glabra*, *Morua alba*, *Camellia sinensis* [3], *Allium fistulosum*, *A. cepa*, *Calamus scipionum*, *Moringa oleifera*, *Centella asiatica*, *Hypericum hircinum*, *H. perforatum*, *Apium graveolens*, *Brassica oleracea* var. italica, *B. oleracea* var. sabellica, *Coriandrum sativum*, *Lactuca sativa*, *Nasturtium officinale*, *Asparagus officinalis*, *Capparis spinosa*, *Prunus domestica*, *P. avium*, *Malus domestica*, *Vaccinium oxycoccus*, and *Solanum Lycopersicum* [12]. It pharmacologically possesses antiobesity, anti-inflammatory, and vasodilator effects, and antioxidant, immunostimulant, anti-diabetic, antihypertensive, antiatherosclerosis, and antihypercholesterolemic activities (Figure 2) [24]. It is available as a food supplement in capsule and powder form.

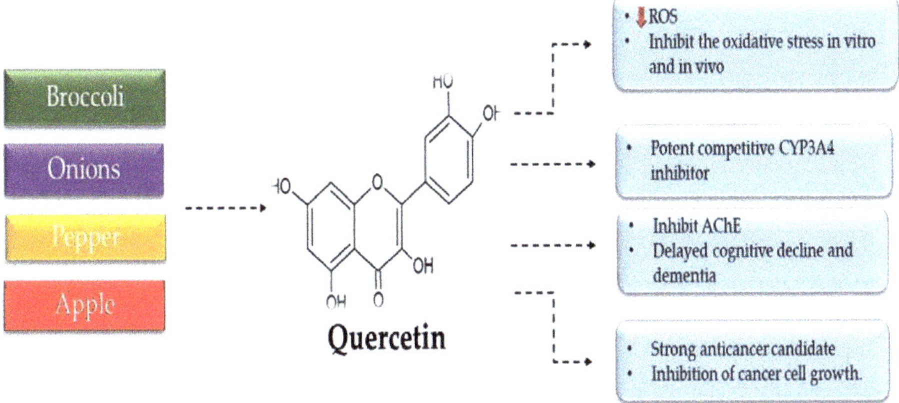

Figure 2. The pharmacological activity of quercetin.

3.1. General Pharmacological/Biochemical Properties of Quercetin

Some of the sources and pharmacological activity of quercetin are shown below in Table 1.

Table 1. Sources of quercetin and its traditional uses.

Plant Name	Family	Pharmacological Activity
Apium graveolens	Apiaceae	Lowers blood pressure and glucose, anti-inflammatory, antibacterial
Allium fistulosum	Amaryllidaceae	Spring onions as food ingredients
Allium cepa (red onions)	Amaryllidaceae	Immunostimulant, cardioprotective, antioxidant
Calamus scipionum	Arecaceae	Source of cane
Moringa oleifera	Moringaceae	Multipurpose medicinal use anti-inflammatory, antihypertensive, antibacterial
Centella asiatica	Apiaceae	Wound healing
Hypericum hircinum	Hypericaceae	Antioxidant
Hypericum perforatum	Hypericaceae	Major depressive disorders, Neurological effects
Brassica oleracea var. sabellica (Kale)	Brassicaceae	Reduce the risk of stroke, reduces blood glucose, neuropathy
Brassica oleracea var. italica (broccoli)	Brassicaceae	Edible plant prevents fluid retention and cancer
Solanum lycopersicum	Solanaceae	Food supplement and salads
Coriandrum Sativum	Apiaceae	Reduce blood pressure, cholesterol, and dyspepsia
Morua alba	Moraceae	Diet
Nasturtium officinale	Brassicaceae	Reduces the risk of cancers
Asparagus officinalis	Asparagaceae	Antineoplastic, antiulcer, antitussive
Lactuca sativa	Asteraceae	Iron deficiency anemia, osteoporosis
Prunus domestica	Rosaceae	Laxative
Malus domestica	Rosaceae	Decrease the risk of cardiovascular disease and cancer
Capparis spinosa	Capparaceae	Vermifuges, disinfectants, antiatherosclerotic agent
Vaccinium oxycoccus	Ericaceae	Urinary tract infections
Prunus avium	Rosaceae	Tonic, astringent, diuretic
Camellia sinensis	Theaceae	Antiviral, antispasmodic, analgesic, antidiabetic, bronchodilator

3.2. Antioxidant Activity

Interestingly, the beneficial effects of quercetin have been attributed to its antioxidant activity. Quercetin is a large class of flavonoids, consisting of five classes of hydroxyl groups, 3,5,7,3', and 4' of the basic flavonol skeleton. Some of these classes of hydroxyls are glycosylated to different quercetin glycosides and form the major quercetin derivatives. It is noteworthy that several studies have shown the relationship between the structural activities of quercetin and its derivatives on antioxidant and anti-inflammatory activities [25]. They found that the modification of quercetin reduces its antioxidant activity, and the total activity was found to be as follows: quercetin > tamarixetin = isorhamnetin > quercetin-3-O-glucuronide > isorhamnetin-3-O-glucoside > quercetin-3,5,7,3',4'-pentamethylether > quercetin-3,4'-di-glucoside, indicating that the 3-hydroxyl quercetin group plays a significant role in antioxidant activity [26]. Moreover, Lesjak et al. [25] reported that methylated quercetin metabolites (e.g., isorhamnetin and tamarixetin) showed higher antioxidant activity than quercetin by inhibiting lipid peroxidation. The antioxidant activity of quercetin has been documented because it can scavenge reactive oxygen species [27]. Quercetin is used to prevent cancer by modulating oxidative stress factors and antioxidant enzymes to prevent the spread of various cancers, such as lung, prostate, liver, breast, colon, and cervical cancers. The in vivo study examined the antioxidant activity of quercetin compared to carcinogen and testosterone by measuring histology and oxidative stress markers, such as reduced glutathione (GSH), lipid peroxidation (LPO), and hydrogen peroxide (H_2O_2) in rats. They found that rats treated with carcinogen and testosterone had higher levels of LPO and H_2O_2 and lower levels of GSH compared to quercetin-treated rats [28]. Sharmila et al. [29] reported that quercetin increased the levels of apoptosis proteins and antioxidant enzymes in animals infected with prostate cancer. Moreover, they documented that quercetin regulated the expression of androgen receptors (AR), protein kinase B (AKT), insulin-like growth factor receptor 1 (IGFIR), and cell proliferation and anti-apoptotic proteins that are increased in cancer. In addition to that, quercetin has been documented to lower malondialdehyde (MDA) content while increasing catalase and superoxide dismutase (SOD) activity to control the anti-inflammatory and anti-apoptosis processes to effectively protect the heart from secondary cardiac dysfunction due to oxidative stress and inflammation [30]. Quercetin also reduces the overproduction of ROS, damage caused by trauma, improves TNF-α, and prevents myocardial cell injury caused by Ca^{2+} overload. Quercetin can thus effectively prevent injury caused by oxidative stress [31].

Interestingly, the antioxidant efficacy of quercetin has been documented in earlier reports to reduce and inhibit oxidative stress and damage, both in vivo and in vitro [32,33]. For illustration, Moretti et al. [34] demonstrated the efficacy of quercetin in the prevention of lipid peroxidation caused by *tert*-Butyl hydroperoxide in human sperm cells in vivo. An additional report in rats revealed that quercetin administered at dose levels of 25–50 mg/kg showed antioxidant action against oxidative stress, which results in streptozotocin-induced diabetes mellitus [35]. Moreover, it has been reported that quercetin acts as a stabilizer in the polyethylene when it is administered at a dose level of 250 µg/mL in addition to its antioxidant activity, and therefore the polymer's residual stability is increased for a long time [36]. Furthermore, the use of quercetin as a chelating agent in chelation therapy for the removal of toxic metallic ions such as cadmium as quercetin-cadmium complexes has been shown to have a high stability constant (Kf) value [37]. Quercetin reduces oxidative stress by controlling the oxidant–antioxidant balance. Several studies have reported that quercetin inhibits oxidative damage caused by acrylamide, brain damage caused by radiation in rats, neurodegenerative disorders, oxidative stress induced by cadmium fluoride, and nerve damage in diabetic rats' retinas. Quercetin protects the nerves, brain, or other body cells from oxidation-induced damage by regulating the antioxidant levels [31]. Quercetin prevents free radicals and enhances the body's antioxidant defense systems and therefore reduces oxidative stress, including the production of nicotine-induced ROS for the treatment of diseases such as nicotine addiction [38]. In vivo studies have shown that quercetin has antioxidant and hepatoprotective activity against acute hepatic injury caused by tertiary butyl hydrogen peroxide. Quercetin effectively protects cells from genetic toxicity and radiation-induced damage by scavenging free radicals and increasing the levels of endogenous antioxidants [39].

3.3. Antiviral Activity

Quercetin has shown antiviral activity towards a wide range of viruses. For instance, quercetin has been documented for its efficacy against the human T-lymphotropic virus 1, as well as the Japanese encephalitis virus (JEV) caused by Japanese encephalitis, the mosquito-borne disease [40,41]. Furthermore, quercetin has been reported to suppress the dengue virus type-2 and hepatitis C virus by suppressing the nonstructural protein 3 protease activity [42,43]. Other Quercetin formulations, such as quercetin-3-O-β-D-glucuronide, quercetin-enriched lecithin formulations, and quercetin 7-rhamnoside have been reported for their efficacy against the porcine epidemic diarrhea virus and influenza-A virus, respectively [44–46].

3.4. Antimicrobial Activity

Quercetin has exhibited potent bacteriostatic activity against different strains of bacteria, such as *Salmonella enterica* serotype Typhimurium, *Pseudomonas aeruginosa*, *P. fluorescens*, *Helicobacter pylori*, *Staphylococcus epidermidis*, *S. aureus*, *Yersinia enterocolitica*, *Micrococcus luteus*, *Campylobacter jejuni*, and *Escherichia coli*, which have been more effective against Gram-positive than Gram-negative bacteria [47]. Jaisinghani. [48] also reported its efficacy against *Shigella flexeneri* NCIM5265 and *Lactobacillus casei var Shirota*. Strikingly, Osonga et al. [49] documented that quercetin derivatives (e.g., quercetin 4',5-diphosphate (QDP), quercetin 3',4',3,5,7-pentaphosphate (QPP), quercetin 5'-sulfonic acid (QSA)) resulted in highly biocompatible, soluble, and potent antibacterial activity with 100% inhibition of *Listeria monocytogenes*, *Pseudomonas aeruginosa*, and *Aeromonas hydrophila*. Moreover, quercetin revealed the strongest antifungal activities against *Candida albicans*, *Cryptococcus neoformans*, and *Aspergillus niger* [50].

3.5. Antiprotozoal Activity

Several reports have demonstrated the growth inhibitory effects of quercetin against various protozoan parasites, namely *Toxoplasma, Babesia, Theileria, Trypanosoma,* and *Leishmania.* Interestingly, quercetin is well-known for its growth inhibitory efficacy against *Trypanosoma brucei rhodesiense, T. brucei brucei, T. cruzi,* and *Leishmania donovani* parasites in vitro and in vivo [51]. It resulted in potent leishmanicidal and trypanocidal activity in vitro, with an IC_{50} of 1.0 µg/mL and 8.3 µg/mL, respectively, while in an in vivo experiment, among six tested flavonoids, only quercetin showed in vivo activity by inhibiting the multiplication of *L. donovani.* Moreover, Weiss et al. [52] documented the remarkable inhibitory effects of quercetin against *Toxoplasma gondii* by preventing the heat shock protein 90 (hsp90), hsp70, and hsp27 synthesis, and thus suppressing the induction of bradyzoite development. Lehane and Saliba. [53] described the antiplasmodial activity of quercetin against a chloroquine-sensitive (3D7) and chloroquine-resistant (7G8) strain of *Plasmodium falciparum.*

3.6. Anti-Inflammatory Effects of Quercetin

Quercetin has been shown to be a long-lasting anti-inflammatory agent with good anti-inflammatory activity. Several in vitro studies have shown that quercetin prevents the development of lipopolysaccharide (LPS)-mediated tumor necrosis factor-α (TNF-α) in macrophages and the development of IL-8 induced LPS in lung A549 cells [54]. In addition, quercetin can inhibit TNF-α and Interleukin (IL)-1α levels of LPS-induced mRNA, which results in reduced apoptotic neuronal cell death caused by microglial activation [55]. Quercetin suppresses the production of inflammatory enzymes (e.g., lipoxygenase (LOX) and cyclooxygenase (COX)). It regulates inflammation caused by LPS by inhibiting Src- and Syk-mediated phosphatidylinositol-3-Kinase (PI3K)-(p85) tyrosine phosphorylation and subsequent complex formation of Toll-like Receptor 4 (TLR4)/MyD88/PI3 K, which restricts downstream signaling pathway activation in RAW 264.7 cells [56]. It may also inhibit the release of pro-inflammatory cytokines, tryptase, and histamine from human umbilical cord blood-derived mast cells; this inhibition is likely to involve the inhibition of calcium influx and Phospho-protein kinase C (PKC) [54]. Quercetin substantially stimulates the gene expression and the development of interferon-γ (IFN-γ) derived from T helper cell-1 (Th-1) and down-regulates IL-4 derived from Th-2 by normal peripheral blood mononuclear cells (PBMC). Quercetin is also known to have inhibitory activity against COX-2, nuclear factor-kappa B (NF-κB), activator protein 1 (AP-1), mitogen-activated protein kinase (MAPK), reactive nitric oxide synthase, (NOS) and reactive C-protein (CRP) expression that causes inflammation [57]. Due to its weak absorption through the surface of the skin, quercetin and its glycoside derivatives have been reported to be ineffective against topical inflammation, while pentamethyl ether, which is a quercetin derivative, has shown potent anti-inflammatory activity with higher absorption through the skin's surface in the rat [58]. Several reports have been documented that quercetin prevents the secretion of iNOS, IL-1β, and TNF-α caused by bacterial LPS in macrophages, TNF-α secretion in RAW2647 cells, and cytokine-stimulated vascular cell adhesion molecules (VCAM-1) and intracellular cell adhesion molecule (ICAM-1) expression, and E-selection in human umbilical vein endothelial cells. Notably, quercetin and its glycoside rutin have shown a reduction in the inflammatory markers TNF-α and IL-6 in NASH mice [59].

3.7. Efficacy in Diseases

3.7.1. Anticancer Activity of Quercetin

Quercetin has been documented to possess anticancer activity both in vitro and in vivo. In in vitro experiments, the anticancer efficacy of quercetin against different cell lines was determined by the prevention of angiogenesis in tamoxifen-resistant cancer, while its in vivo efficacy was attributed to its antioxidant activity [60–62]. According to Gibellini et al. [63], quercetin is considered to be a strong anticancer candidate due to its chemoprotective activity through metastasis and apoptosis against tumor cell lines. Moreover, Du et al. [64] demonstrated the potent efficacy of the quercetin-doxorubicin combined treatment in persistent T-cell tumor-specific responses, resulting in improved the immune responses against breast tumor growth [64]. It is worth noting that quercetin prevents the proliferation of several types of cancers (e.g., breast, lung, prostate, cervical, liver, and colon cancer) and it acts by a various mechanism of actions, including cellular signaling, binding to cellular receptors and proteins, and inhibiting enzymes responsible for carcinogen activation [65]. Recently, quercetin has been reported to increase the chemosensitivity of breast cancer cells to doxorubicin by preventing cell propagation and invasion that promote cell apoptosis. Furthermore, quercetin demonstrated an inhibitory effect on MCF-7 and MDA-MB-231 human breast cancer cell lines by regulating miR-146a expression, cell apoptosis induction, Caspase 3 activation, and mitochondrial pathways [66]. Quercetin also exhibits anti-colon cancer effects with the TLR4- and NF-κB-mediated signaling pathway, and it was found that quercetin showed significant inhibition of human colon cancer proliferation in CACO-2 and SW-620 cells by preventing the NF-κB pathway, down-regulation of B-cell lymphoma 2, and up-regulation of Bcl-2-associated X protein [67].

3.7.2. Quercitin Hepatoprotective and Antihypertensive Activities

Recently, an in vivo study found that quercetin increased heme oxygenase 1 activity in D-galactosamine- and LPS-treated rats by lowering plasma concentrations of alanine aminotransferase and stimulating its hepatotoxic and hepatoprotective activity [68]. Moreover, Liu et al. [69] revealed the ability of quercetin to treat ethanol-induced oxidative damage in rat hepatocytes, suggesting that quercetin may be an appropriate hepatoprotective natural product. Duarte et al. [70] reported that quercetin had antihypertensive activity in spontaneously hypertensive rats, and noted that quercetin had induced a dose-dependent, advanced, and potential reduction in pressure of the blood when given chronically to several hypertensive rat models.

3.7.3. The Important Role of Quercetin in the Treatment of Alzheimer's Disease

Alzheimer's disease (AD) is considered to be the most prevalent cause of dementia, a chronic neurodegenerative disorder characterized by memory loss and mental deficits, such as apraxia, aphasia, and agnosia, and is associated with neuroinflammatory processes in the central nervous system [71,72]. The memory contains several types: visual, olfactory, episodic, and vocal. These are classified into two categories: explicit (active or passive recall of facts) and implicit (nonverbal habitual memory) [73].

Oxidative stress is caused by a free radical imbalance in the body and is included in the establishment of neurodegenerative disorders involving AD. Flavonoids like quercetin have different activities in the vascular system, leading to several modifications in cerebrovascular blood flow, which can alter the neuronal morphology that causes neurogenesis and angiogenesis. In addition to that, it can also protect neurons from neurotoxin-induced injury. Rich food consumption of flavonoids limits neurodegeneration and inverts the age-related injury to cognitive performance [74]. Moreover, quercetin and ascorbic acid combined treatment have been shown to reduce the prevalence of oxidative injury to human lymphocytes and neurovascular structures in the skin and thus prevent neuron injury, which particularly protects the brain cells from oxidative stress that leads to AD and other neurological conditions [13].

Quercetin's beneficial effects against AD are ascribed due to its inhibitory efficacy against acetylcholinesterase (AChE) [72]. Recently, in vivo experiments have documented the ability of quercetin to reduce the oxidative stress caused by 6-hydroxydopamine in the neurons of rats [75]. Another study conducted on healthy P19 neurons revealed that neuron survival is not affected by quercetin, while it depletes the glutathione content that may affect the functioning of the nervous system [76]. Furthermore, recent findings have shown that quercetin improves the pathology of AD and related cognitive deficits in triple-transgenic, aged AD mice [77]. Additionally, combined oral ingestion of quercetin with fish oil improved neuroprotection in 3-nitropropionic acid-treated rats or chronic rotenone-treated rats [78,79].

In the AD, quercetin acts by the following mechanism of action: α-tocopherol (vitamin E), a type of antioxidant that enhances quercetin penetration through the blood–brain barrier (BBB), which leads to significant improvement in quercetin concentration and thus reduces the prevalence of oxidative damage in the brain. Moreover, quercetin acts by activating the NF-E2-related factor 2- antioxidant responsive element (Nrf2-ARE) that offers a neuroprotective effect against oxidative injury and cell death. Recently, previous studies have shown that the formation and deterioration of undisciplined protein aggregates in various neurodegenerative diseases, such as Huntington's diseases, Alzheimer's, Parkinson's, and amyotrophic lateral sclerosis may be altered by the Nrf2-ARE pathway [80,81].

4. Combination Therapy of Quercetin with Other Drugs

The combined effect of quercetin with other antioxidants (e.g., ascorbic acid), decreases the prevalence of oxidative damage in human lymphocytes and neurovascular structures in the skin and inhibits the neuron injury. Moreover, it has been reported to possess a potent effect against AD by protecting the brain cells from oxidative stress that induces tissue damage, resulting in AD and other neurological conditions [13]. Notably, quercetin has been documented to possess neuroprotective and neurotoxic activity, and its combined effect with fish oil has shown neuroprotective efficacy in rat brains and has subsequently shown beneficial effects against neurodegenerative diseases [79]. Quercetin is well-known to influence the pharmacokinetics of different drugs, such as curcumin and resveratrol by controlling their transfer and metabolism, as well as some of its significant activities, including CYP3A4, P-gp efflux pump, and phenol sulfotransferase (SULT 1A1) inhibition. These combined treatments have resulted in an increase in curcumin and resveratrol permeability and acute bioavailability compared to single treatments [82]. Moreover, Sahyon et al. [83] investigated the combination effect of sulfamethoxazole with quercetin against *S. aureus*, and quercetin has been shown to reduce the side effects of sulfamethoxazole while improving its bactericidal efficacy, indicating the importance of this combination therapy for the treatment of human clinical cases. Also, Qu et al. [84] revealed the synergetic effect of quercetin-tetracycline combination treatment against multi-drug resistant (MDR) *E. coli* by disrupting the bacterial cell envelope, thus improving its permeability and cell lysis. Quercetin has been documented to improve the antifungal efficacy of amphotericin B against *Candida* sp and *Cryptococcus neoformans* strains by reducing its toxic effect [85]. Another study demonstrated the potent synergistic efficacy of quercetin against fluconazole-resistant strains of *Candida tropicalis* by enhancing mitochondrial membrane alterations that affect the mitochondrial respiratory function and inhibiting rhodamine-123 accumulation in the mitochondria [86]. Recently, quercetin has been documented to possess synergistic effects when combined with chemotherapeutic drugs (e.g., cisplatin) [65].

5. Dose Use

Typical dietary quercetin intake based on fruit and vegetable consumption is estimated to range from 5 to 100 mg per day. Heavy consumption of foods rich in quercetin, such as apples or onions, could lead to a daily intake of up to 500 mg [87–89]. The effective dose is increased when taken with a fatty meal or in the presence of apple pectin, oligosaccharides, and lecithin [87,90]. Most clinical studies use quercetin at 500 to 1000 mg per day in divided doses [91,92]. As a supplementary food, 2 weeks of quercetin 50 mg achieved a 178% increase in serum levels, while quercetin 100 mg had a 359% increase in the serum levels, and quercetin 500 mg had a 570% increase in the serum levels, although with wide individual variation [93]. Based on animal studies, quercetin accumulates in the lungs, liver, kidneys, and small intestines, with lower levels seen in the brain, heart, and spleen. It is eliminated through the renal, fecal, and respiratory systems [89,94].

6. Metabolism and Excretion of Quercetin

After quercetin administration and absorption, it is transferred to the liver where the first and second phases of metabolism take place, resulting in metabolic products entering the bloodstream for distribution in the body's tissues [95]. Mullen et al. [96] examined the main metabolites of quercetin in the urine and plasma of healthy people after the ingestion of onions. Three major metabolites were identified in the plasma—namely, quercetin-30-sulfate, quercetin-3-glucuronide, and quercetin-3-sulfate with the highest concentrations at 0.8 and 0.6 h, while quercetin-30-glucuronide, quercetin-diglucuronide, isorhamnetin–glucuronide sulfate, isorhamnetin-methyl quercetin, and diglucuronide isorhamnetin-glucuronide were the major urinary metabolites that reached their highest concentrations after 4 h. Notably, quercetin had a short half-life and rapid clearance in the blood, and its metabolites appeared in the plasma 30 min after ingestion, but considerable amounts were excreted over 24 h [97]. Moon et al. [94] identified the aggregation of quercetin conjugates in human plasma following multiple administrations of quercetin-rich foods. The highest concentration of quercetin metabolites was identified following the uptake of onions, and sulfate and glucuronide metabolites were significantly ($p < 0.05$) elevated from 0.04 to 0.63 µM in the plasma of fasting participants.

The use of quercetin in the pharmaceutical industry is limited due to its poor bioavailability, poor aqueous solubility, poor permeability, and instability. Therefore, several studies have been conducted to modify its structure to increase its water solubility and bioavailability and thereby enhance its antioxidant and antimicrobial activity [98]. Recently, new quercetin preparations have appeared, including quercetin-loaded gel, quercetin-loaded mucoadhesive nanoemulsion, quercetin-loaded nanoparticles, and quercetin-loaded polymeric micelle, which may provide new drug formulations for research and development (Figure 3) [31]. Moreover, quercetin bioavailability has also been improved by structural modification with glucoside–sulfate conjugates and the preparation of some complex ionic complexes, such as quercetin–germanium nanoparticles, calcium phosphate–quercetin nanocomplex (CPQN), and glucan–quercetin conjugate that showed higher antioxidant activity than free quercetin [99]. Quercetin also exhibits excellent antioxidant activity and scavenging capacity when combined with metal ions, such as cadmium, vanadium, calcium, magnesium, copper, cobalt, iron, and ruthenium [31].

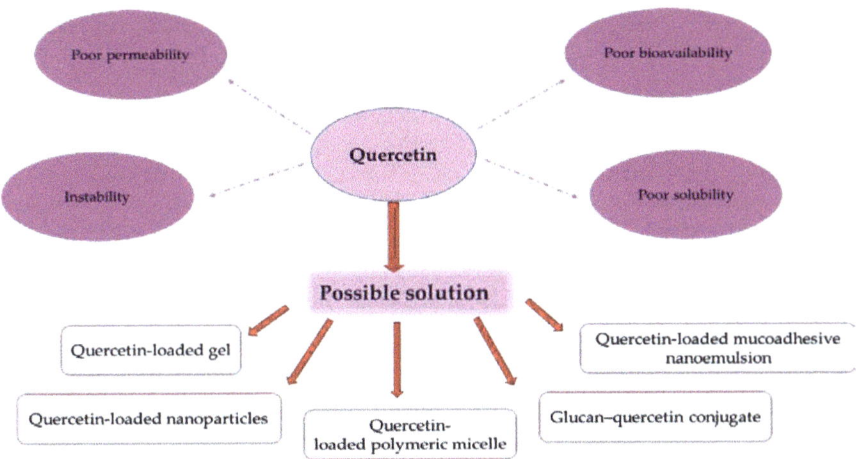

Figure 3. Quercetin formulations for improving its bioavailability.

7. Toxic Side Effects of Quercetin

Quercetin is known to be a mutagenic agent based on the Ames test; however, most in vivo animal studies have shown that quercetin is a safe compound without any carcinogenic effects. It is worth noting that in 1999, the International Agency for Research on Cancer (IARC) stated that quercetin should not be listed as a human carcinogen compound [100,101]. There is no definite proof of quercetin teratogenic activity on embryonic growth; however, in vitro studies suggest that quercetin can have a mild negative impact on fetal growth and demonstrate protective efficacy against toxic agents [102]. In vivo experiments have shown that quercetin resulted in a small increase in the prevalence of malignant tumors to the young offspring of mice lacking DNA repair [103]. An in vivo experiment performed on a four-week rat showed that the ratio of liver and kidney weights increased remarkably in rats fed greater than 314 mg and 157 mg quercetin/kg body weight/day, respectively. Moreover, a pro-oxidant efficacy was observed at doses higher than 157 mg quercetin/kg body weight/day [104]. Quercetin was usually well-tolerated in human clinical studies. Notably, quercetin administration for several months at a concentration higher than 1000 mg/day did not show any side effects on serum electrolytes, kidney, and liver function blood parameters, or hematology. At present, co-administration of high quercetin doses with digoxin is known to be the greatest cause of toxicity; thus, the use of quercetin in digoxin-treated patients should be restricted before more information on appropriate dosage levels is available [105]. Quercetin shows mutagenicity in vitro in the Ames test, and reports of mutagenicity in the 1970s have led to concerns about its safety [87]. Under certain circumstances, quercetin exhibits both radical scavenging and pro-oxidant activity [88].

The majority of in vivo experiments have shown that quercetin is not a carcinogen and may be protective against Geno toxicants. Dietary quercetin, faced with the first-pass metabolism in the intestine and liver, is almost completely metabolized, reducing the potential for toxicity. At oral supplemental doses higher than 1000 mg per day taken for up to three months, no evidence of toxicity has been found; however, data on long-term safety at high doses are lacking [87]. Nephrotoxicity has been reported with the use of high-dose IV quercetin in patients with compromised health [89].

8. Quercetin-Drug Interaction

Quercetin has been reported to competitively bind to bacterial DNA gyrase and is, therefore, contraindicated to be administered with fluoroquinolone antibiotics [106]. Moreover, quercetin is a potent competitive inhibitor of CYP3A4 (the enzyme responsible for drug degradation in the body) and was thus predicted to increase the serum concentrations of drugs (e.g., diltiazem) that are metabolized by this enzyme [107]. Therefore, research towards the optimum mechanism of action of natural compounds that prevents the adverse effects of plants and the development of new molecules with new pharmacological effects will continue [108–110].

9. Conclusions

This review examined the therapeutic and toxic activities of quercetin. Quercetin is the major polyphenolic flavonoid present in several food products that have shown many pharmacological activities, such as anticancer, antiviral, antiprotozoal, and antimicrobial effects, treatment of allergic, metabolic, and inflammatory disorders, eye and cardiovascular diseases, and arthritis. Previous studies documented the poor oral bioavailability of quercetin after a single oral dose, as its absorption was impaired by the macronutrients. It is a famous AChE inhibitor and has been used in treating neurodegenerative diseases, including AD. Quercetin has been documented to have both neurotoxic and neuroprotective activities, and its combined effect with fish oil and ascorbic acid has demonstrated beneficial effects against neurodegenerative diseases. The finding that quercetin can combine with other drugs is a property that can be explored in the chemotherapies' development against AD. However, due to the presence of adverse side effects, its therapeutic use as a treatment has been banned. Moreover, quercetin has been confirmed to be competitively bound to bacterial DNA gyrase and is, therefore, contraindicated to be administered with fluoroquinolone antibiotics.

Author Contributions: A.M.B., G.E.-S.B., Z.S.M., M.I., M.E.A.E.-H., A.E.T., A.M.A., and Y.H.A.E. wrote the paper. A.M.B. and G.E.B. revised the paper. All authors have read and agreed to the published version of the manuscript.

Funding: This research received no external funding.

Conflicts of Interest: The authors declare no conflict of interest.

Abbreviations

AD	Alzheimer's disease
IUPAC	International Union of Pure and Applied Chemistry
Kf	stability constant value
JEV	Japanese encephalitis virus
QDP	quercetin 4′,5-diphosphate
QPP	quercetin 3′,4′,3,5,7-pentaphosphate
QSA	quercetin 5′-sulfonic acid
hsp90	heat shock protein 90
COX	cyclooxygenase
LOX	lipoxygenase
TLR4	Toll-like Receptor 4
PKC	Phospho-protein kinase C
Th-1	T helper cell-1
NF-κB	nuclear factor-kappa B
AP-1	activator protein 1
MAPK	mitogen-activated protein kinase
NOS	nitric oxide synthase
CRP	reactive C-protein
IL-1β	interleukin-1β
TNF-α	tumor necrosis factor-α
LPS	lipopolysaccharide
VCAM-1	vascular cell adhesion molecules
ICAM-1	intracellular cell adhesion molecules
CPQN	calcium phosphate–quercetin nanocomplex
BBB	blood-brain barrier
Nrf2-ARE	NF-E2-related factor 2-antioxidant responsive element
SULT 1A1	phenol sulfotransferase
MDR	multi-drug resistant
IARC	International Agency for Research on Cancer
AChE	acetylcholinesterase

References

1. Batiha, G.-S.; Beshbishy, A.M.; Adeyemi, O.S.; Nadwa, E.H.; Rashwan, E.M.; Alkazmi, L.M.; Elkelish, A.A.; Igarashi, I. Phytochemical screening and antiprotozoal effects of the methanolic *Berberis vulgaris* and acetonic *Rhus coriaria* extracts. *Molecules* **2020**, *25*, 550. [CrossRef]

2. Batiha, G.-S.; Alkazmi, L.M.; Wasef, L.G.; Beshbishy, A.M.; Nadwa, E.H.; Rashwan, E.K. *Syzygium aromaticum* L. (Myrtaceae): Traditional uses, bioactive chemical constituents, pharmacological and toxicological activities. *Biomolecules* **2020**, *10*, 202. [CrossRef] [PubMed]

3. Shakya, A.K. Medicinal plants: Future source of new drugs. *Int. J. Herb. Med. IJHM* **2016**, *59*, 59–64.

4. Bakkali, F.; Averbeck, S.; Averbeck, D.; Idaomar, M. Biological effects of essential oils—A review. *Food Chem. Toxicol.* **2008**, *46*, 446–475. [CrossRef]

5. Ríos, J.L.; Recio, M.C. Medicinal plants and antimicrobial activity. *J. Ethnopharmacol.* **2005**, *100*, 80–84. [CrossRef]

6. Lemma, M.T.; Ahmed, A.M.; Elhady, M.T.; Ngo, H.T.; Vu, T.L.-H.; Sang, T.K.; Campos-Alberto, E.; Sayed, A.; Mizukami, S.; Na-Bangchang, K.; et al. Medicinal plants for in vitro antiplasmodial activities: A systematic review of literature. *Parasitol. Int.* **2017**, *66*, 713–720. [CrossRef]

7. Naghibi, F.; Esmaeili, S.; Abdullah, N.R.; Nateghpour, M.; Taghvai, M.; Kamkar, S.; Mosaddegh, M. In vitro and in vivo antimalarial evaluations of myrtle extract, a plant traditionally used for treatment of parasitic disorders. *Biomed. Res. Int.* **2013**, *2013*, 1–5. [CrossRef]

8. Batiha, G.E.S.; Beshbishy, A.M.; Tayebwa, D.S.; Adeyemi, O.S.; Shaheen, H.; Yokoyama, N.; Igarashi, I. The effects of *trans*-chalcone and chalcone 4 hydrate on the growth of *Babesia* and *Theileria*. *PLoS Negl. Trop. Dis.* **2019**, *13*, e0007030. [CrossRef]

9. Beshbishy, A.M.; Batiha, G.E.; Yokoyama, N.; Igarashi, I. Ellagic acid microspheres restrict the growth of *Babesia* and *Theileria* in vitro and *Babesia microti* in vivo. *Parasites Vectors* **2019**, *12*, 269. [CrossRef]

10. Sulaiman, F.A.; Nafiu, M.O.; Yusuf, B.O.; Muritala, H.F.; Adeyemi, S.B.; Omar, S.A.; Dosumu, K.A.; Adeoti, Z.J.; Adegbesan, O.A.; Busari, B.O.; et al. The GC-MS fingerprints of *Nicotiana tabacum* L. extract and propensity for renal impairment and modulation of serum triglycerides in Wistar rats. *J. Pharm. Pharmacogn. Res.* **2020**, *8*, 191–200.

11. Batiha, G.-S.; Alkazmi, L.M.; Nadwa, E.H.; Rashwan, E.K.; Beshbishy, A.M. Physostigmine: A plant alkaloid isolated from Physostigma venenosum: A review on pharmacokinetics, pharmacological and toxicological activities. *J. Drug Deliv. Therap.* **2020**, *10*. [CrossRef]

12. Anand David, A.V.; Arulmoli, R.; Parasuraman, S. Overviews of Biological Importance of Quercetin: A Bioactive Flavonoid. *Pharmacogn. Rev.* **2016**, *10*, 84–89. [PubMed]

13. Lakhanpal, P.; Rai, D.K. Quercetin: A versatile flavonoid. *Int. J. Med.* **2007**, *2*, 22–37. [CrossRef]

14. Dabeek, W.M.; Marra, M.V. Dietary quercetin and kaempferol: Bioavailability and potential cardiovascular-related bioactivity in humans. *Nutrients* **2019**, *11*, 2288. [CrossRef] [PubMed]

15. Aguirre, L.; Arias, N.; Macarulla, M.T.; Gracia, A.; Portillo, M.P. Beneficial effects of quercetin on obesity and diabetes. *Open Nutraceuticals J.* **2011**, *4*, 189–198.

16. Rich, G.T.; Buchweitz, M.; Winterbone, M.S.; Kroon, P.A.; Wilde, P.J. Towards an understanding of the low bioavailability of quercetin: A study of its interaction with intestinal lipids. *Nutrients* **2017**, *9*, 111. [CrossRef]

17. Walle, T.; Browning, A.M.; Steed, L.L.; Reed, S.G.; Walle, U.K. Flavonoid glucosides are hydrolyzed and thus activated in the oral cavity in humans. *J. Nutr.* **2005**, *135*, 48–52. [CrossRef]

18. Nemeth, K.; Piskula, M.K. Food content, processing, absorption and metabolism of onion flavonoids. *Crit. Rev. Food Sci. Nutr.* **2007**, *47*, 397–409. [CrossRef]

19. Hollman, P.C.; de Vries, J.H.; van Leeuwen, S.D.; Mengelers, M.J.; Katan, M.B. Absorption of dietary quercetin glycosides and quercetin in healthy ileostomy volunteers. *Am. J. Clin. Nutr.* **1995**, *62*, 1276–1282. [CrossRef]

20. Scholz, S.; Williamson, G. Interactions affecting the bioavailability of dietary polyphenols in vivo. *Int. J. Vitam. Nutr. Res.* **2007**, *77*, 224–235. [CrossRef]

21. Ferry, D.R.; Smith, A.; Malkhandi, J.; Fyfe, D.W.; deTakats, P.G.; Anderson, D.; Baker, J.; Kerr, D.J. Phase I clinical trial of the flavonoid quercetin: Pharmacokinetics and evidence for in vivo tyrosine kinase inhibition. *Clin. Cancer Res.* **1996**, *2*, 59–68.

22. Erlund, I.; Kosonen, T.; Alfthan, G.; Mäenpää, J.; Perttunen, K.; Kenraali, J.; Parantainen, J.; Aro, A. Pharmacokinetics of quercetin from quercetin aglycone and rutin in healthy volunteers. *Eur. J. Clin. Pharmacol.* **2000**, *56*, 545–553. [CrossRef]

23. Graefe, E.U.; Wittig, J.; Mueller, S.; Riethling, A.K.; Uehleke, B.; Drewelow, B.; Pforte, H.; Jacobasch, G.; Derendorf, H.; Veit, M. Pharmacokinetics and bioavailability of quercetin glycosides in humans. *J. Clin. Pharmacol.* **2001**, *41*, 492–499. [CrossRef] [PubMed]

24. Salvamani, S.; Gunasekaran, B.; Shaharuddin, N.A.; Ahmad, S.A.; Shukor, M.Y. Antiartherosclerotic effects of plant flavonoids. *Biomed. Res. Int.* **2014**, *2014*. [CrossRef] [PubMed]

25. Lesjak, M.; Beara, I.; Simin, I.; Pintać, D.; Majkić, T.; Bekvalac, K.; Orčić, D.; Mimica-Dukić, N. Antioxidant and anti-inflammatory activities of quercetin and its derivatives. *J. Funct. Foods.* **2018**, *40*, 68–75. [CrossRef]

26. Magar, R.T.; Sohng, J.K. A review on structure, modifications and structure-activity relation of quercetin and its derivatives. *J. Microbiol. Biotechnol.* **2020**, *30*, 11–20. [CrossRef]

27. Boots, A.W.; Haenen, G.R.; Bast, A. Health effects of quercetin: From antioxidant to nutraceutical. *Eur. J. Pharmacol.* **2008**, *585*, 325–337. [CrossRef]

28. Sharmila, G.; Athirai, T.; Kiruthiga, B.; Senthilkumar, K.; Elumalai, P.; Arunkumar, R.; Arunakaran, J. Chemopreventive effect of quercetin in MNU and testosterone-induced prostate cancer of Sprague-Dawley rats. *Nutr. Cancer.* **2014**, *66*, 38–46. [CrossRef]

29. Sharmila, G.; Bhat, F.A.; Arunkumar, R.; Elumalai, P.; Raja Singh, P.; Senthilkumar, K.; Arunakaran, J. Chemopreventive effect of quercetin, a natural dietary flavonoid on prostate cancer in in vivo model. *Clin. Nutr.* **2014**, *33*, 718–726. [CrossRef]

30. Li, B.; Yang, M.; Liu, J.W.; Yin, G.T. Protective mechanism of quercetin on acute myocardial infarction in rats. *Genet. Mol. Res.* **2016**, *15*, 15017117. [CrossRef]

31. Xu, D.; Hu, M.J.; Wang, Y.Q.; Cui, Y.L. Antioxidant activities of quercetin and its complexes for medicinal application. *Molecules* **2019**, *24*, 1123. [CrossRef]

32. Stefek, M.; Karasu, C. Eye lens in aging and diabetes: Effect of quercetin. *Rejuvenation Res.* **2011**, *14*, 525–534. [CrossRef]
33. Coballase-Urrutia, E.; Pedraza-Chaverri, J.; Cardenas Rodriguez, N.; Huerta-Gertrudis, B.; Garcia-Cruz, M.E.; Montesinos-Correa, H.; Sanchez-Gonzalez, D.J.; Camacho-Carranza, R.; Espinosa-Aguirre, J.J. Acetonic and methanolic extracts of Heterotheca inuloides, and quercetin, decrease CCl (4)-oxidative stress in several rat tissues. *Evid. Based Complement. Alternat. Med* **2013**, *2013*, 659165. [CrossRef]
34. Moretti, E.; Mazzi, L.; Terzuoli, G.; Bonechi, C.; Iacoponi, F.; Martini, S.; Rossi, C.; Collodel, G. Effect of quercetin, rutin, naringenin and epicatechin on lipid peroxidation induced in human sperm. *Reprod. Toxicol.* **2012**, *34*, 651–657. [CrossRef]
35. Maciel, R.M.; Costa, M.M.; Martins, D.B.; Franca, R.T.; Schmatz, R.; Graca, D.L.; Duarte, M.M.; Danesi, C.C.; Mazzanti, C.M.; Schetinger, M.R.; et al. Antioxidant and anti-inflammatory effects of quercetin in functional and morphological alterations in streptozotocin-induced diabetic rats. *Res. Vet. Sci.* **2013**, *95*, 389–397. [CrossRef]
36. Tátraaljai, D.; Földes, E.; Pukánszky, B. Efficient melt stabilization of polyethylene with quercetin, a flavonoid type natural antioxidant. *Polym. Degrad. Stab.* **2014**, *102*, 41–48. [CrossRef]
37. Ravichandran, R.; Rajendran, M.; Devapiriam, D. Antioxidant study of quercetin and their metal complex and determination of stability constant by spectrophotometry method. *Food Chem.* **2014**, *146*, 472–478. [CrossRef]
38. Yarahmadi, A.; Zal, F.; Bolouki, A. Protective effects of quercetin on nicotine-induced oxidative stress in 'HepG2 cells'. *Toxicol. Mech. Methods.* **2017**, *27*, 609–614. [CrossRef]
39. Kalantari, H.; Foruozandeh, H.; Khodayar, M.J.; Siahpoosh, A.; Saki, N.; Kheradmand, P. Antioxidant and hepatoprotective effects of *Capparis spinosa* L. fractions and quercetin on tert-butyl hydroperoxide- induced acute liver damage in mice. *J. Tradit. Complement. Med.* **2018**, *81*, 120–127. [CrossRef]
40. Coelho-Dos-Reis, J.G.; Gomes, O.A.; Bortolini, D.E.; Martins, M.L.; Almeida, M.R.; Martins, C.S.; Carvalho, L.D.; Souza, J.G.; Vilela, J.M.; Andrade, M.S.; et al. Evaluation of the effects of quercetin and kaempherol on the surface of MT-2 cells visualized by atomic force microscopy. *J. Virol. Methods* **2011**, *174*, 47–52. [CrossRef]
41. Johari, J.; Kianmehr, A.; Mustafa, M.; Abubakar, S.; Zandi, K. Antiviral activity of Baicalein and quercetin against the Japanese Encephalitis Virus. *Int. J. Mol. Sci.* **2012**, *13*, 16785–16795. [CrossRef]
42. Zandi, K.; Teoh, B.T.; Sam, S.S.; Wong, P.F.; Mustafa, M.R.; Abubakar, S. Antiviral activity of four types of bioflavonoid against dengue virus type-2. *Virol. J.* **2011**, *8*, 560. [CrossRef]
43. Bachmetov, L.; Gal-Tanamy, M.; Shapira, A.; Vorobeychik, M.; Giterman-Galam, T.; Sathiyamoorthy, P.; Golan Goldhirsh, A.; Benhar, I.; Tur-Kaspa, R.; Zemel, R. Suppression of hepatitis C virus by the flavonoid quercetin is mediated by inhibition of NS3 protease activity. *J. Viral. Hepat.* **2012**, *19*, 81–88. [CrossRef]
44. Ramadan, M.F.; Selim, A.M.M. Antimicrobial and antiviral impact of novel quercetin-enriched lecithin. *J. Food Biochem.* **2009**, *33*, 557–571. [CrossRef]
45. Fan, D.; Zhou, X.; Zhao, C.; Chen, H.; Zhao, Y.; Gong, X. Antiinflammatory, antiviral and quantitative study of quercetin-3-O-beta-D-glucuronide in *Polygonum perfoliatum* L. *Fitoterapia* **2011**, *82*, 805–810. [CrossRef]
46. Song, J.H.; Shim, J.K.; Choi, H.J. Quercetin 7-rhamnoside reduces porcine epidemic diarrhea virus replication via independent pathway of viral-induced reactive oxygen species. *Virol. J.* **2011**, *8*, 460. [CrossRef]
47. Wang, S.; Yao, J.; Zhou, B.; Yang, J.; Chaudry, M.T.; Wang, M.; Xiao, F.; Li, Y.; Yin, W. Bacteriostatic effect of quercetin as an antibiotic alternative in vivo and its antibacterial mechanism in vitro. *J. Food Prot.* **2018**, *81*, 68–78. [CrossRef]
48. Jaisinghani, R. Antibacterial properties of quercetin. *Microb. Res.* **2017**, *8*. [CrossRef]
49. Osonga, F.J.; Akgul, A.; Miller, R.M.; Eshun, G.B.; Yazgan, I.; Akgul, A.; Sadik, O.A. Antimicrobial activity of a new class of phosphorylated and modified flavonoids. *ACS. Omega* **2019**, *4*, 12865–12871. [CrossRef]
50. Abd-Allah, W.E.; Awad, H.M.; Abdel Mohsen, M.M. HPLC analysis of quercetin and antimicrobial activity of comparative methanol extracts of *Shinus molle* L. *Int. J. Curr. Microbiol. Appl. Sci.* **2015**, *4*, 550–558.
51. Tasdemir, D.; Kaiser, M.; Brun, R.; Yardley, V.; Schmidt, T.J.; Tosun, F.; Rüedi, P. Antitrypanosomal and antileishmanial activities of flavonoids and their analogues: In vitro, in vivo, structure-activity relationship, and quantitative structure-activity relationship studies. *Antimicrob. Agents Chemother.* **2006**, *50*, 1352–1364. [CrossRef]

52. Weiss, L.M.; Ma, Y.F.; Takvorian, P.M.; Tanowitz, H.B.; Wittner, M. Bradyzoite development in *Toxoplasma gondii* and the hsp70 stress response. *Infect. Immun.* **1998**, *66*, 3295–3302. [CrossRef]
53. Lehane, A.M.; Saliba, K.J. Common dietary flavonoids inhibit the growth of the intraerythrocytic malaria parasite. *BMC Res. Notes* **2008**, *1*, 26. [CrossRef]
54. Li, Y.; Yao, J.; Han, C.; Yang, J.; Chaudhry, M.T.; Wang, S.; Liu, H.; Yin, Y. Quercetin, inflammation and immunity. *Nutrients* **2016**, *8*, 167. [CrossRef]
55. Bureau, G.; Longpre, F.; Martinoli, M.G. Resveratrol and quercetin, two natural polyphenols, reduce apoptotic neuronal cell death induced by neuroinflammation. *J. Neurosci. Res.* **2008**, *86*, 403–410. [CrossRef]
56. Boots, A.W.; Wilms, L.C.; Swennen, E.L.; Kleinjans, J.C.; Bast, A.; Haenen, G.R. In vitro and ex vivo anti-inflammatory activity of quercetin in healthy volunteers. *Nutrition* **2008**, *24*, 703–710. [CrossRef]
57. Garcia-Mediavilla, V.; Crespo, I.; Collado, P.S.; Esteller, A.; Sanchez-Campos, S.; Tunon, M.J.; Gonzalez-Gallego, J. The anti-inflammatory flavones quercetin and kaempferol cause inhibition of inducible nitric oxide synthase, cyclooxygenase-2 and reactive C-protein, and down-regulation of the nuclear factor kappaB pathway in Chang Liver cells. *Eur. J. Pharmacol.* **2007**, *557*, 221–229. [CrossRef]
58. Lin, C.F.; Leu, Y.L.; Al-Suwayeh, S.A.; Ku, M.C.; Hwang, T.L.; Fang, J.Y. Anti-inflammatory activity and percutaneous absorption of quercetin and its polymethoxylated compound and glycosides: The relationships to chemical structures. *Eur. J. Pharm. Sci.* **2012**, *47*, 857–864. [CrossRef]
59. Ginwala, R.; Bhavsar, R.; Chigbu, D.I.; Jain, P.; Khan, Z.K. Potential role of flavonoids in treating chronic inflammatory diseases with a special focus on the anti-Inflammatory activity of apigenin. *Antioxidants* **2019**, *8*, 35. [CrossRef]
60. Oh, S.J.; Kim, O.; Lee, J.S.; Kim, J.A.; Kim, M.R.; Choi, H.S.; Shim, J.H.; Kang, K.W.; Kim, Y.C. Inhibition of angiogenesis by quercetin in tamoxifen-resistant breast cancer cells. *Food Chem. Toxicol.* **2010**, *48*, 3227–3234. [CrossRef]
61. Dajas, F. Life or death: Neuroprotective and anticancer effects of quercetin. *J. Ethno. Pharmacol.* **2012**, *143*, 383–396. [CrossRef] [PubMed]
62. Baghel, S.S.; Shrivastava, N.; Baghel, R.S.; Agrawal, P.; Rajput, S. A review of quercetin: Antioxidant and anticancer properties. *World J. Pharm. Pharm. Sci.* **2012**, *1*, 146–160.
63. Gibellini, L.; Pinti, M.; Nasi, M.; Montagna, J.P.; De Biasi, S.; Roat, E.; Bertoncelli, L.; Cooper, E.L.; Cossarizza, A. Quercetin and cancer chemoprevention. *Evid. Based Complement. Alternat. Med.* **2011**, *2011*, 591356. [CrossRef]
64. Du, G.; Lin, H.; Yang, Y.; Zhang, S.; Wu, X.; Wang, M.; Ji, L.; Lu, L.; Yu, L.; Han, G. Dietary quercetin combining intratumoral doxorubicin injection synergistically induces rejection of established breast cancer in mice. *Int. Immunopharmacol.* **2010**, *10*, 819–826. [CrossRef] [PubMed]
65. Rauf, A.; Imran, M.; Khan, I.A.; ur-Rehman, M.; Gilani, S.A.; Mehmood, Z.; Mubarak, M.S. Anticancer potential of quercetin: A comprehensive review. *Phytother. Res.* **2018**, *32*, 2109–2130. [CrossRef] [PubMed]
66. Tao, S.F.; He, H.F.; Chen, Q. Quercetin inhibits proliferation and invasion acts by up-regulating miR-146a in human breast cancer cells. *Mole. Cell. Biochem.* **2015**, *402*, 93–100. [CrossRef]
67. Zhang, X.A.; Zhang, S.; Yin, Q.; Zhang, J. Quercetin induces human colon cancer cells apoptosis by inhibiting the nuclear factor-kappa B Pathway. *Pharm. Mag.* **2015**, *11*, 404–409. [CrossRef]
68. Lekic, N.; Canova, N.K.; Horinek, A.; Farghali, H. The involvement of heme oxygenase 1 but not nitric oxide synthase 2 in a hepatoprotective action of quercetin in lipopolysaccharide-induced hepatotoxicity of D-galactosamine sensitized rats. *Fitoterapia* **2013**, *87*, 20–26. [CrossRef]
69. Liu, S.; Hou, W.; Yao, P.; Li, N.; Zhang, B.; Hao, L.; Nussler, A.K.; Liu, L. Heme oxygenase-1 mediates the protective role of quercetin against ethanol-induced rat hepatocytes oxidative damage. *Toxicol. In Vitro* **2012**, *26*, 74–80. [CrossRef]
70. Duarte, J.; Pérez-Palencia, R.; Vargas, F.; Ocete, M.A.; Pérez-Vizcaino, F.; Zarzuelo, A.; Tamargo, J. Antihypertensive effects of the flavonoid quercetin in spontaneously hypertensive rats. *Br. J. Pharm.* **2001**, *133*, 117–124. [CrossRef]
71. Brown, R.C.; Lockwood, A.H.; Sonawane, B.R. Neurodegenerative diseases: An overview of environmental risk factors. *Environ. Health Pers.* **2005**, *113*, 1250–1256. [CrossRef]
72. Choi, G.N.; Kim, J.H.; Kwak, J.H.; Jeong, C.H.; Jeong, H.R.; Lee, U.; Heo, H.J. Effect of quercetin on learning and memory performance in ICR mice under neurotoxic trimethyltin exposure. *Food Chem.* **2012**, *132*, 1019–1024. [CrossRef]

73. Arlt, S. Non-Alzheimer's disease-related memory impairment and dementia. *Dialog. Clinic. Neurosc.* **2013**, *15*, 465–473.

74. Vauzour, D.; Vafeiadou, K.; Rodriguez-Mateos, A.; Rendeiro, C.; Spencer, J.P. The neuroprotective potential of flavonoids: A multiplicity of effects. *Genes Nutr.* **2008**, *3*, 115–126. [CrossRef] [PubMed]

75. Haleagrahara, N.; Siew, C.J.; Ponnusamy, K. Effect of quercetin and desferrioxamine on 6hydroxydopamine (6-OHDA) induced neurotoxicity in striatum of rats. *J. Toxicol. Sci.* **2013**, *38*, 25–33. [CrossRef] [PubMed]

76. Jazvinscak, J.M.; Cipak, G.A.; Vukovic, L.; Vlainic, J.; Zarkovic, N.; Orsolic, N. Quercetin supplementation: Insight into the potentially harmful outcomes of neurodegenerative prevention. *Naunyn-Schmiedeberg Arch. Pharmacol.* **2012**, *385*, 1185–1197. [CrossRef]

77. Sabogal-Guáqueta, A.M.; Muñoz-Manco, J.I.; Ramírez-Pineda, J.R.; Lamprea-Rodriguez, M.; Osorio, E.; Cardona-Gómez, G.P. The flavonoid quercetin ameliorates Alzheimer's disease pathology and protects cognitive and emotional function in aged triple transgenic Alzheimer's disease model mice. *Neuropharm* **2015**, *93*, 134–145. [CrossRef]

78. Denny Joseph, K.M. Enhanced neuroprotective effect of fish oil in combination with quercetin against 3-nitropropionic acid-induced oxidative stress in rat brain. *Prog. Neuro-Psychopharmacol. Biol. Psychiatry* **2013**, *40*, 83–92. [CrossRef]

79. Denny Joseph, K.M. Combined oral supplementation of fish oil and quercetin enhance neuroprotection in a chronic rotenone rat model: relevance to Parkinson's disease. *Neurochem. Res.* **2015**, *40*, 894–905. [CrossRef] [PubMed]

80. Gan, L.; Johnson, J.A. Oxidative damage and the Nrf2-ARE pathway in neurodegenerative diseases. *Biochem. Biophys. Acta.* **2014**, *842*, 1208–1218. [CrossRef]

81. Ferri, P.; Angelino, D.; Gennari, L.; Benedetti, S.; Amrogini, P.; Del Grande, P.; Ninfali, P. Enhancement of flavonoid ability to cross the blood-brain barrier of rats by co-administration with α-tocopherol. *Food Funct.* **2015**, *6*, 394–400. [CrossRef] [PubMed]

82. Lund, K.C.; Pantuso, T. Combination effects of quercetin, resveratrol and curcumin on in vitro intestinal absorption. *J. Rest. Med.* **2014**, *3*, 112–120. [CrossRef]

83. Sahyon, H.A.; Ramadan, E.N.M.; Mashaly, M.M.A. Synergistic effect of quercetin in combination with sulfamethoxazole as new antibacterial agent: In vitro and in vivo study. *Pharm. Chem. J.* **2019**, *53*, 803–813. [CrossRef]

84. Qu, S.; Dai, C.; Shen, Z.; Tang, Q.; Wang, H.; Zhai, B.; Zhyao, L.; Hao, Z. Mechanism of synergy between tetracycline and quercetin against antibiotic resistant *Escherichia coli*. *Front. Microbiol.* **2019**, *10*, 2536. [CrossRef] [PubMed]

85. Oliveira, V.M.; Carraro, E.; Auler, M.E.; Khalil, N.M. Quercetin and rutin as potential agents antifungal against *Cryptococcus* spp. *Brazil. J. Biol.* **2016**, *76*, 1029–1034. [CrossRef] [PubMed]

86. Da Silva, C.R.; de Andrade Neto, J.B.; de Sousa Campos, R.; Figueiredo, N.S.; Sampaio, L.S.; Magalhães, H.I.; Cavalcanti, B.C.; Gaspar, D.M.; de Andrade, G.M.; Lima, I.S.; et al. Synergistic effect of the flavonoid catechin, quercetin, or epigallocatechin gallate with fluconazole induces apoptosis in *Candida tropicalis* resistant to fluconazole. *Antimicrob. Agents Chemother.* **2014**, *58*, 1468–1478. [CrossRef]

87. Harwood, M.; Danielewska-Nikiel, B.; Borzelleca, J.F.; Flamm, G.W.; Williams, G.M.; Lines, T.C. A critical review of the data related to the safety of quercetin and lack of evidence of in vivo toxicity, including lack of genotoxic/carcinogenic properties. *Food Chem. Toxicol.* **2007**, *45*, 2179–2205. [CrossRef]

88. Bischoff, S.C. Quercetin: potentials in the prevention and therapy of disease. *Curr. Opin. Clin. Nutr. Metab. Care* **2008**, *11*, 733–740. [CrossRef]

89. Russo, M.; Spagnuolo, C.; Tedesco, I.; Bilotto, S.; Russo, G.L. The flavonoid quercetin in disease prevention and therapy: Facts and fancies. *Biochem. Pharmacol.* **2012**, *83*, 6–15. [CrossRef]

90. Lee, J.; Mitchell, A.E. Pharmacokinetics of quercetin absorption from apples and onions in healthy humans. *J. Agric. Food Chem.* **2012**, *60*, 3874–3881. [CrossRef]

91. Edwards, R.L.; Lyon, T.; Litwin, S.E.; Rabovsky, A.; Symons, J.D.; Jalili, T. Quercetin reduces blood pressure in hypertensive subjects. *J. Nutr.* **2007**, *137*, 2405–2411. [CrossRef] [PubMed]

92. Kressler, J.; Millard-Stafford, M.; Warren, G.L. Quercetin and endurance exercise capacity: A systematic review and meta-analysis. *Med. Sci. Sports Exerc.* **2011**, *43*, 2396–2404. [CrossRef] [PubMed]

93. Jin, F.; Nieman, D.C.; Shanely, R.A.; Knab, A.M.; Austin, M.D.; Sha, W. The variable plasma quercetin response to 12-week quercetin supplementation in humans. *Eur. J. Clin. Nutr.* **2010**, *64*, 692–697. [CrossRef] [PubMed]

94. Moon, J.H.; Nakata, R.; Oshima, S.; Inakuma, T.; Terao, J. Accumulation of quercetin conjugates in blood plasma after the short-term ingestion of onion by women. *Am. J. Physiol. Regul. Integr. Comp. Physiol.* **2000**, *279*, R461–R467. [CrossRef]

95. Hollman, P.C. Absorption, bioavailability, and metabolism of flavonoids. *Pharm. Biol.* **2004**, *42*, 74–83. [CrossRef]

96. Mullen, W.; Edwards, C.A.; Crozier, A. Absorption, excretion and metabolite profiling of methyl-, glucuronyl-, glucosyl- and sulpho-conjugates of quercetin in human plasma and urine after ingestion of onions. *Br. J. Nutr.* **2006**, *96*, 107–116. [CrossRef]

97. Moon, Y.J.; Wang, L.; DiCenzo, R.; Morris, M.E. Quercetin pharmacokinetics in humans. *Biopharm. Drug Dispos.* **2008**, *29*, 205–217. [CrossRef]

98. Chen, X.; Yin, O.Q.P.; Zuo, Z.; Chow, M.S.S. Pharmacokinetics and modeling of quercetin and metabolites. *Pharm. Res.* **2005**, *22*, 892–901. [CrossRef]

99. Patra, M.; Mukherjee, R.; Banik, M.; Dutta, D.; Begum, N.A.; Basu, T. Calcium phosphate-quercetin nanocomposite (CPQN): A multi-functional nanoparticle having pH indicating, highly fluorescent and anti-oxidant properties. *Colloids Surf. B Biointerfaces* **2017**, *154*, 63–73. [CrossRef]

100. Utesch, D.; Feige, K.; Dasenbrock, J.; Harwood, M.; Danielewska-Nikiel, B.; Lines, T.C. Evaluation of the potential in vivo genotoxicity of quercetin. *Mutat. Res.* **2008**, *654*, 38–44. [CrossRef]

101. Okamoto, T. Safety of quercetin for clinical application (Review). *Int. J. Mol. Med.* **2005**, *16*, 275–278. [CrossRef]

102. Pérez-Pastén, R.; Martínez-Galero, E.; Chamorro-Cevallos, G. Quercetin and naringenin reduce abnormal development of mouse embryos produced by hydroxyurea. *J. Pharm. Pharmacol.* **2010**, *62*, 1003–1009. [CrossRef] [PubMed]

103. Vanhees, K.; de Bock, L.; Godschalk, R.W.; van Schooten, F.J.; van Waalwijk van Doorn-Khosrovani, S.B. Prenatal exposure to flavonoids: Implication for cancer risk. *Toxicol. Sci.* **2011**, *120*, 59–67. [CrossRef]

104. Azuma, K.; Ippoushi, K.; Terao, J. Evaluation of tolerable levels of dietary quercetin for exerting its antioxidative effect in high cholesterol-fed rats. *Food Chem. Toxicol.* **2010**, *48*, 1117–1122. [CrossRef] [PubMed]

105. Wang, Y.H.; Chao, P.D.; Hsiu, S.L.; Wen, K.C.; Hou, Y.C. Lethal quercetin-digoxin interaction in pigs. *Life Sci.* **2004**, *74*, 1191–1197. [CrossRef] [PubMed]

106. Mahmoud Hashemi, A.; Solahaye Kahnamouii, S.; Aghajani, H.; Frozannia, K.; Pournasrollah, A.; Sadegh, R.; Esmaeeli, H.; Ghadimi, Y.; Razmpa, E. Quercetin decreases Th17 production by down-regulation of MAPK-TLR4 signaling Pathway on T cells in dental pulpitis. *J. Dent.* **2018**, *19*, 259–264.

107. Choi, J.S.; Li, X. Enhanced diltiazem bioavailability after oral administration of diltiazem with quercetin to rabbits. *Int. J. Pharm.* **2005**, *297*, 1–8. [CrossRef]

108. Beshbishy, A.M.; Batiha, G.E.-S.; Alkazmi, L.; Nadwa, E.; Rashwan, E.; Abdeen, A.; Yokoyama, N.; Igarashi, I. Therapeutic Effects of Atranorin towards the Proliferation of *Babesia* and *Theileria* Parasites. *Pathogen* **2020**, *9*, 127. [CrossRef]

109. Batiha, G.S.; Beshbishy, A.M.; El-Mleeh, A.; Abdel-Daim, M.M.; Devkota, H.P. Traditional uses, bioactive chemical constituents, and pharmacological and toxicological activities of *Glycyrrhiza glabra* L. (Fabaceae). *Biomolecules* **2020**, *10*, 352. [CrossRef]

110. Batiha, G.S.; Beshbishy, A.M.; Guswanto, A.; Nugraha, A.B.; Munkhjargal, T.; Abdel-Daim, M.M.; Mosqueda, J.; Igarashi, I. Phytochemical characterization and chemotherapeutic potential of *Cinnamomum verum* extracts on the multiplication of protozoan parasites in vitro and in vivo. *Molecules* **2020**, *25*, 996. [CrossRef]

MDPI

St. Alban-Anlage 66

4052 Basel

Switzerland

Tel. +41 61 683 77 34

Fax +41 61 302 89 18

www.mdpi.com

Foods Editorial Office

E-mail: foods@mdpi.com

www.mdpi.com/journal/foods